INFORMATION, ENTROPY, AND PROGRESS

PROGRESS

A NEW EVOLUTIONARY PARADIGM

INFORMATION, ENTROPY, AND PROGRESS

A NEW EVOLUTIONARY PARADIGM

Robert U. Ayres

The European Institute of Business Administration
Fontainebleau, France

American Institute of Physics   New York

AIP Press
American Institute of Physics
500 Sunnyside Boulevard
Woodbury, NY 11797-2999

Library of Congress Cataloging-in-Publication Data

Ayres, Robert U.
    Information, Entropy, and Progress/Robert U. Ayres.
        p. cm.
    Includes bibliographical references and index.
    ISBN 0-88318-911-9
    1. Information theory.   2. Evolution.   3. Evolution (Biology)
    4. Economics.   I. Title.
Q360.A95   1992   93-43472
330'.01'154–dc20   CIP

10   9   8   7   6   5   4   3   2   1

Contents

Foreword

Isaiah Berlin divided thinkers into hedgehogs and foxes, those who frame issues in terms of unities and those who see them in terms of their diversities. In creating this book, Robert Ayres is the archetypal hedgehog. Beginning with the most basic, most universal of sciences, thermodynamics, and then building into his world view the concepts of information gathering and self-organization, he constructs a unifying theme of evolution in several arenas. From the viewpoint that all evolutionary phenomena share common characteristics—characteristics that appear naturally through the compound lens of thermodynamics, information and self-organization—he takes his readers through the evolution of the universe, the earth and its chemistry and geology, living organisms, social and economic systems, and finally manufacturing, production and labor. In each area, Ayres is faced with making a network of precise, specialized concepts, all inevitably wrapped in their own jargon, accessible to readers that are each familiar with their own other field and not the field at hand. In doing this, especially in casting the description of each area in terms that fit into Ayres' grand scheme, the author gives the specialists fresh perspectives on their own fields.

The goals of this book are to give the readers a global perspective and a willingness of their own specialities as well as those of others as being part of that unity. The concepts that justify this perspective are, at least in some fields, very precise and therefore most precisely stated in the language of mathematics. Ayres does not hesitate to use mathematical statements where they best express the ideas. Yet the book is a book of words and concepts, not a book of equations and mathematical formalism. It is, in fact, a good read for someone uneasy with equations who nevertheless wants a sense of amateur's mastery over the book's unifying themes.

It is not so important whether Ayres convinces his readers to look at the world just as he does; the interpretive approach that this book presents is a plausible, attractive way to recognize relationships among a lot of things, it is not a thesis based on positing testable causal relations provable to be right or wrong. What is important is that he brings the reader through the process of seeing the world from a point of view different from what they already know. The most effective learning comes from abrasion, from becoming involved enough with ideas to challenge them, to examine them rigorously and to try to assault them. If Ayres is successful, he will stimulate his audience to do just that.

R. Stephen Berry
Chicago, Illinois

Preface

By some strange irony, the last actual step in the preparation of a book such as this seems to be the preface, which precedes all. So, in a sense, this is a sigh of relief and a brief explanation of how the book came to be. It has been a long—very long—process. The idea for the book (at least an early and long since discarded version of it) germinated in the summer of 1978 when I spent a couple of weeks as an unofficial visiting scholar in the economics department of the University of Wyoming, with Prof. Ralph d'Arge. We were trying to write a conference paper together on thermodynamic constraints to the economic growth process. This paper was never written, for various reasons, and I forgot about it for a while in the confusion of moving from Washington D.C. to Pittsburgh PA, and taking up an academic post for the first time.

Life as a Professor at Carnegie-Mellon University was confusing and distracting enough to drive this book project into the background. However it resurfaced after a year or two, when I was asked by Tom von Foerster at AIP to contribute an article for *Physics Today* on "Thermodynamics and Economics" (an article I wrote with Indira Nair, a physicist who works in the Department of Engineering and Public Policy at C-MU). Thus re-inspired, I proposed a joint book project to my old friend and colleague, Allen Kneese at Resources for the Future, Inc. He agreed to collaborate and we struggled through two draft manuscripts and submitted the result to RFF as a possible RFF book. This was in 1985, or thereabouts. Following normal procedures, copies were sent out to a large number of reviewers (about 15, if memory serves), all of whom suggested major revisions. Allen, being busy, sensibly withdrew from the project at that point. (He finally contributed a Foreword in this volume, which my editor decided would be better placed as an Afterword). But for some reason I kept thinking about the project.

In the summer of 1986 I went to IIASA for a year, with the primary purpose of initiating and directing a project to assess the economic and social impacts of computer-integrated-manufacturing technologies. Thinking about the automation of manufacturing problems prompted me to think more deeply about the role of information in manufacturing, and I started to write papers on that topic. (Chapters 8–11 are directly based on that work).

I returned to IIASA each summer after 1986 (and, again, for the academic year 1989-90). During that time I became acquainted with Katalin Martinás, a physicist and thermodynamics specialist, who teaches at Roland Eotvos University in Budapest. She had some very interesting ideas, which led to a productive collaboration. (Our research papers from 1980-90 form the basis for the latter part of Chapter 7. Our collaboration began again in January 1993, after my move to France). I began to rewrite the book, in something like its present form, during my last four summers at IIASA.

The book's emphasis on evolutionary processes in economics emerged, of its own accord, towards the end of my long labors. A conference organized by Ilya Prigogine and others at the University of Texas at Austin, another organized by Luc Soete and Gerry Silverburg at Maastricht, several conversations with Richard Nelson, and two visits to the newly formed Santa Fe Institute, helped clarify my ideas. ("Evolutionary Economics" has meanwhile become a recognized subdiscipline, with its own journal.) However, I must also thank numerous other colleagues whose ideas have been influential over the years. Needless to say, they are not responsible for any remaining errors.

The MS finally went to AIP in summer 1991. It has since undergone yet another revision (mostly in response to a reviewer's constructive comments). I owe a considerable debt to that reviewer, who is no longer anonymous. Thanks Steve! I owe thanks, too, to Helene Pankl who was my secretary at IIASA, Vicky Massimino who was my secretary at C-MU, and Muriel Larvaron, who is my secretary here at INSEAD. Thanks go to Michael Hennelly at AIP. Finally, I owe thanks to my long-suffering wife, Leslie, who has done most of the diagrams and much of the pre-editorial work that word-processors do not eliminate. She has also contributed by her understanding that writing a book like this was not only necessary (to me) but FUN. Thanks again.

Robert U. Ayres
Fontainebleau, France

Introduction: Evolution as Accumulation of Useful Information

Evolution is associated mainly with biology. The concept of evolutionary change can be applied far more broadly, however, and I shall use it in a broad sense in this book, especially in the context of economics and technology.

The terminology used to describe the evolutionary process can also be applied to fields other than biology. In evolution, **ontogenic** development "like the growth of an organism from fertilization to maturity" is essentially predetermined or programmed in considerable detail. **Phylogenic** evolution, on the other hand "like that of a species" is essentially unpredictable, either because of truly random (stochastic) elements or because of complex, non-linear dynamics that lead to deterministic chaos.

In addition, one can distinguish between **genotype** and **phenotype.** The former pertains to potentialities: Genotypic fixation may be likened to fixing the parameters of a model. The latter pertains to actual realizations, as regards both form (morphology) and behavior. Both ontogenic development and phylogenic evolution can be classified as either genotypic or phenotypic.

In theory, this matrix yields four possible types of change (see Table 1.1). Of the four, this book is concerned only with the two that fall under the general heading of **phylogeny**, or inherently unpredictable evolutionary change. The first is primarily genotypic, i.e., the evolution of potentiality. The book begins with the evolutionary determination of physical constants and the abundance of chemical elements in the primordial "Big Bang" (and, later, in stars). It then considers the geochemical and biochemical evolutionary processes that led to the creation of self-replicating polymers, cells, and organisms (genotypes). The second type of change is primarily phenotypic: the evolution of form and/or function. It includes the morphological evolution of galaxies, stars, planetary systems, and the topographical features of Earth.

However, we shall consider the evolution of physical and chemical systems only by way of background. The core of the book is an examination of evolutionary mechanisms and processes in the human sphere: socio-political systems, economic systems, and the progress of science. Again, these systems will be considered through the prism of genotypes and phenotypes, although the distinction is much harder to maintain consistently (and is consequently less useful). Genotypes include populations, laws, and science. Phenotypes include ecosystems, communities, nations, civilizations, socio-economic systems and institutions, and technology. Evidently, actuality depends both on potentiality and on external influences or constraints. It can be taken for granted, too, that genotypes and phenotypes influence the evolution of each other.

The other two theoretical types of evolution fall under the heading of **ontogeny,** or development. Ontogenic changes are inherently predictable. Examples of genotypic ontogeny include the practical applications of scientific knowledge (which are predictable, in general, if the science is sufficiently advanced). Examples of phenotypic ontogeny include the life cycles of planets, stars and galaxies, and living organisms. There are, of course, life-cycle models of such entities as products, technologies, civilizations, and economic systems. This book does not consider ontogenic or developmental models in detail, except insofar as they are contrasted with phylogenic evolutionary models.

The first major hypothesis of this book is that phylogenic evolution—in the physical, biological, and social realms—is inherently unpredictable and largely accidental. It is known that the detailed dynamical behavior of certain classes of non-linear systems is characterized by motions of unpredictable periods and amplitudes. This behavior has been characterized as **deterministic chaos.** The evolving physical, biological, and social systems that interest us are, in fact, strongly non-linear. Thus, all phylogenic evolutionary paths are, very likely, examples of deterministic chaos. The problem of characterizing an evolutionary path, then, is equivalent to the problem of characterizing the region in phase-space (or **strange attractor**) within which the actual path may lie. In general terms, evolutionary paths (attractors) for the first three realms of science (physics, chemistry, and biology) seem to be characterized by (1) increasing **diversity of product,** (2) increasing **complexity,** and (3) increasing **stability** of complex forms.

The second hypothesis of the book is that Darwinian evolutionary "search" processes (in biology, at least) are quite inefficient, in the sense that the search is **myopic.** It tests, compares, and selects only from among near-neighbors in configuration space—thus neglecting most of the theoretical possibilities. Darwinian evolution discovers and freezes-in local optima, rather than global optima. There is no good reason to suppose that the local optima discovered in the course of biological evolution are particularly good from a global perspective.

The myopic search may be likened to the experience of an explorer seeking the highest peak of a mountain range covered by dense cloud. Since he can see only a few meters ahead, the only possible strategy in such a case is to keep walking "up" as long as possible. Eventually the explorer will reach a point from which all paths lead down. This is a peak. But the chances of finding the highest peak by this method are slight. (They are approximately inversely proportional to the number of distinct peaks to be found, that is, to the roughness of the landscape.)

Suppose the cloud lifts somewhat, permitting the explorer to see some distance. A new and much more efficient search strategy (based on **presbyopia**, or far-sightedness) now becomes possible. Even from a starting point deep in a valley, the explorer can see far enough to avoid wasting time on some of the lower and nearer peaks. He will choose the highest peak within his range of visibility and aim for it. As he climbs higher, he can see farther. A higher peak may then become visible. To reach it, the explorer must plot a new route. Using his extended vision, he tries to plot an efficient path to the higher peak—one that does not involve losing more altitude than necessary. Since his vision is imperfect (he cannot see what is behind the nearer hills) he may run into an impassable barrier, such as a gorge or cliff. However, with increasing experience in the mountains, the explorer also develops some skill at guessing where the barriers are more (or less) likely to be found. (In other words, he develops some theoretical understanding of the forces that shape mountains.) He may still run into unpleasant surprises along the way. But eventually he will find a high peak. From its top he may see an even higher one in the distance, and so on.

In the realm of human society, by contrast, evolution has become an increasingly conscious process. It makes increasing use of the more efficient presbyopic search mechanism. Science is one of the mechanisms for extending society's vision and locating promising directions. Science and technology have accelerated the process of social change enormously. Underlying science, of course, are logic and analytical thought—the ability to distinguish between cause and effect. Once consequences are seen to follow from causes, it becomes possible to shape the future—both individually and collectively. (The education of children, saving, investment, and R&D are all examples of presbyopic behavior.) It is this future orientation, widely shared, that is equivalent to the mountaineer's vision. Among other consequences it has led to a noteworthy trend toward consciously *cooperative behavior.*

The third fundamental hypothesis of this book is that biological evolution is a process of accumulating "useful" genetic information. The best measure of evolutionary progress is the ability to store and process information in the brain and central nervous system. Social evolution is, similarly, a process of accumulating "useful" cultural information that is used for social purposes, passed on via social processes, and stored in artifacts (for example, books) as well as in people's memories. In both cases, the term *useful* must be understood as that which assists survival and growth.

In the biological world, this tendency toward a buildup of survival-relevant (SR) and, especially, survival-useful (SU) information is reflected in the increasing number and variability of species, the increasing complexity of organisms and ecosystems, the increasing information-carrying capacity of genes (of dominant species, at least), and the trends toward larger brains, longer lives, and a greater ability (of the dominant species) to modify the natural environment in ways favoring survival and propagation. The relationship between functionality (in the foregoing sense) and information-processing capability is intuitively clear.

In the economic sphere, the tendency toward increasing SU-information is reflected in (1) an increasing structural complexity, (2) the increasing importance of *morphological* tasks, and the increasing importance of creating, storing, processing, and transmitting information, (3) the increasingly conscious, analytical, and *self-referential* nature of political, economic, and industrial management, and (4) the acceleration of the accumulation of knowledge and the shortening of the lifetime of successive generations of technological products.

It is the accumulation of knowledge, including experience of the past, that permits individuals and institutions to look toward and plan for the future. The ability to do this, in turn, is critical to the evolution of cooperative behavior in human society. Again, the relation between these characteristics of economic evolution and information-processing capability seems evident.

The argument proposing similar selection mechanisms for biological and economic evolution draws initially on the obvious analogies between biological organisms, communities, and ecosystems, and human social (and economic) systems.[1] In human society, generally, evolution seems to favor increasing organizational complexity and functional diversity. It also clearly favors the creation, accumulation, and transmission of knowledge by extra-genetic mechanisms. Such mechanisms have sharply increased both the rate and focus of evolutionary change.

The fourth and final major hypothesis of the book is that boundaries of the evolutionary path, or attractor, for human social and economic systems are determined by an extended (presbyopic) version of Herbert Simon's notion of **bounded rationality** for economic transactions. The basic principle of bounded rationality is as follows: Two parties will engage in a transaction only if each party is better off after the transaction. In the extended version, parties are presumed to be able to remember the results of previous transactions and to develop theories about the relationship between causes and consequences. Thus, "better off" may be interpreted in the sense of future expectations, rather than immediate profit.

The evolutionary attractor corresponding to this principle is a tendency to increase (1) embodied structural information and (2) memory. In the social realm, the tendency is to maximize the embodiment of self-referential information and the capacity to process (acquire, reduce, store, and transmit) information. More simply, the direction of evolution is characterized by an increasing ability of organisms and systems to *sense* the state of the environment, to *assess* its risks and opportunities, to learn and remember appropriate responses, and to transmit this useful information to other individuals in the community and to succeeding generations. Yet no single one of these sub-objectives can be given absolute priority due to their intricate linkages. The "objective function" of evolution cannot be characterized precisely, but the process itself is governed by a logic analogous to bounded rationality (or "satisficing," rather than "maximizing").

Obviously the functional activities noted above (sensing, learning, remembering, and transmitting) are accomplished in a variety of ways by different organisms and systems. One of the most significant evolutionary innovations of humans was the *extra-somatic* storage and processing of information. The first step in this direction, of course, was the invention of pictographs and written language, and the creation of books and libraries. This was followed only in the very recent past—historically speaking—by the introduction of technological devices to enhance the human senses (i.e., to enhance the rate of information acquisition) and to enhance the human brain (i.e., to process and reduce the acquired information more efficiently). The conscious awareness of the nature of these processes and the ability to understand and modify them deliberately to improve the functionality of social systems is perhaps the most significant development of all.

Social systems in general, and the economic system in particular, differ from biological organisms in several crucial respects: First, they have no natural lifespan (though they can die), so there is no analog of inheritance or generations. They have nothing like the genetic system and sexual reproduction for transmission of information between generations. (Evolution in social systems is Lamarckian, rather than Darwinian.) Social systems also lack natural physical boundaries between themselves and their environment—they don't have skins—although they are usually localized and bounded geographically by other social systems. In the case of the economic system the nearest analog of a boundary is the scope of the market: Goods and services with well-defined market prices are internal to the system, and goods and services that are not exchanged or for which no prices can be determined are external.

Nor do social systems possess a "hard-wired" central nervous system or brain (as animals do), though all social systems do have more or less specialized functions for sensing and communication. Biological organisms developed sensory organs and a central nervous system primarily to sense the external environment, although higher animals have also developed internal (proprioceptive) channels for regulating their metabolism. Most sensing and learning in social systems, by contrast, are done at the "cellular" level (by humans, of course) and only the organizational aspects thereof, together with remembering and transmittal, are truly social functions. It is interesting to note that the economic system is effectively defined by the scope of its internal communications system: the price system. Market prices are the signals (the "invisible hand") by which the market regulates itself. By the same token, the economic system lacks a well-developed mechanism for sensing the condition of the environment in which it is embedded. This is a fundamental weakness that also threatens the long-run survival of human civilization, unless it can be rectified by the creation of new "sensory organs" via conscious socio-political processes.

The book concludes with a discussion of economic growth from the perspectives of evolution and information. The discussion emphasizes the interaction between economic growth, on the one hand, and the availability of external resources and the accumulation of technological knowledge as an (internalized, hence embodied) substitute for external resources, on the other. It also stresses the difference between short-term (myopic) and long-term (presbyopic) investment and regulatory policies. It points out that the selection mechanisms that have hitherto been functioning in the human economic and social spheres are essentially short-term, whereas the evolutionary path implied by such policies may be incompatible with long-term survival, due to neglect of environmental damage. Finally, it speculates briefly on possible adjustments that could be made in the economic system as it now functions to introduce longer-term considerations into the selection mechanism.

ENDNOTES

1. Many less-obvious, but no less-salient, analogies that seem to support the hypothesis have been pointed out by "general systems" theorists [e.g., Miller, 1978].

Chapter 1

Concepts, Definitions, and Analogies

In assembling the material for this book gradually over a number of years, I have noticed that some of the conceptual building blocks for the four major theses I hope to expound are commonly used rather casually without definition. (I am often guilty of doing this myself.) Even a very basic concept such as "equilibrium" or "information" is sometimes introduced without explanation, presumably because it is assumed to be common knowledge among specialists or sufficiently clear from context. Given the scope of my enterprise, however, it is important to begin by introducing and defining some basic ideas with reasonable care, if not mathematical rigor. I think it is desirable, also, to provide background on the development of the ideas in question. I will address the concept of information in considerable detail in the following chapter because of its central role in the book.

1.1. IRREVERSIBILITY AND EQUILIBRIUM IN MECHANICS AND THERMODYNAMICS

A good place to start is with the terms **reversibility** and **irreversibility**, as applied to physical or economic processes. As might be expected, reversibility can be characterized in several ways that may or may not be completely equivalent. The first of them is probabilistic: A process that can go in two directions is reversible if two (or more) directions are equally probable for all starting points, given a random perturbation or "nudge." The process will go the other way with an equal and opposite nudge. The motion of an incompressible billiard ball on a frictionless surface is reversible in this sense, for instance. Underlying this way of thinking about reversibility is the notion of starting from a steady (or stationary) state. These terms will be discussed later.

A more sophisticated definition of reversibility introduces another fundamental concept, **time**, which I will not attempt to define other than to say that time is what is measured by clocks. A dynamic process is reversible if its history and its future are statistically indistinguishable. Suppose, for instance, a process is recorded in moving pictures, and the film is played both forward and back-

1

ward in time. The criterion for reversibility is that an observer would not be able to determine from internal evidence which direction is which. In mathematical terms, a reversible dynamic process is one in which reversing the direction of time (i.e., replacing the variable t by its negative $-t$ in the equations of motion) leaves those equations unchanged. In other words, the equations of motion for a reversible process are invariant with respect to time reversal.

As a matter of more than incidental interest, Newton's laws of motion for masses moving in a gravitational force field are reversible in the sense of being invariant to the direction of time. In fact, as the mathematicians Laplace, Lagrange, and Hamilton showed, this is also true for a general class of dynamical systems in which the force field is independent of time itself and independent of velocity. Such systems are called **conservative** because the sum of kinetic energy plus potential energy is unchanging, i.e., conserved. All the basic force fields known to physics are conservative in this sense, as they must be to satisfy the first law of thermodynamics, namely, the law of conservation of energy. Actually, the mathematical proofs of the existence of a conserved quantity in this class of dynamical systems preceded (and helped inspire) the full conceptual development of the fundamental quantity we now call **energy**.[1] It is interesting and perhaps significant that Galileo, Huygens, Leibnitz, and Newton—who basically formulated the laws of mechanics—had no such general concept of energy conservation.

In view of the connection between reversibility and conservation in mechanics, it is important to emphasize that other so-called forces are not conservative but, rather, *dissipative*. Examples include friction (sliding or rolling), electrical and thermal resistance, and viscosity in fluids. "Friction forces" can be incorporated as a velocity-dependent dissipation or drag term in a Newtonian equation of motion, but there are two consequences of doing so. First, the sum of kinetic plus potential energy is no longer conserved, but declines monotonically. To maintain the principle that total energy is conserved in a dissipative system, a new kind of energy must be introduced into the system, namely internal energy.[2] The total amount of internal energy in the system increases over time to balance the decline in kinetic plus potential energy. The experimental verification of this equivalence by Julius Robert Mayer (c. 1841), later confirmed by James Joule, William Thompson (Lord Kelvin) and others, led to the formal enunciation of the modern law of energy conservation.[3]

The second consequence of introducing dissipative-force terms—like friction—into the equations of motion is, of course, that the equations cease to be invariant with respect to time reversal. In effect, because of dissipative forces, high-quality energy is transformed over time into low-quality energy. Such a dynamical system is non-reversible [Mosekilde *et al.*, 1980]. In fact, non-reversibility is a characteristic of dissipative systems. Some economists and others [e.g., Kelsey, 1988] have questioned the applicability of the notion of irreversibility in economics on the grounds that there is no conserved quantity (analogous to energy) in economics that can be "dissipated." The category "dissipative systems" can be mathematically generalized, however, as argued by Radzicki [1988]. Specifically, a dynamical system with a negative divergence of flow can be defined as dissipative.[4] While the legitimacy of this argument for extending the notion of dissipative from physical systems to social (or other) systems is still being debated, I accept the argument for generalization.

Having said this, it is possible to define **thermodynamics** as the general theory of macroscopic irreversible (i.e., dissipative) physical systems whose behavior can be defined in terms of a relatively small number of extensive and intensive variables.[5] In fact, the term *macroscopic* implies that the system can be described with a small number of variables. An extensive variable has the property that the value for the system as a whole is the sum of the values for all component subsystems—the number of molecules in a gas, or its volume, for example. The population of a nation or its GNP are also examples of extensive variables. An intensive variable, by contrast, has a value for the system as a whole, when it is in a state of **equilibrium**, that is equal to the values for all the component subsystems. Examples of intensive variables for a gas would be temperature and pressure. For a nation, examples might be average family size or average income.

The concept of equilibrium as used above deserves clarification, since the term is also used in quite different ways, for example, in mathematics, mechanics, thermodynamics, and economics. (I will return to it below.) In mathematics, equilibrium is the end state of a global maximization (or minimization) problem, i.e., a balance of forces. Such a balance can be **static** or **dynamic**. An example of the former is a marble sitting at the bottom of a bowl. Examples of the latter would be a pendulum, a spinning gyroscope, or the moon orbiting the earth. In each case, the force of gravity is counteracted by changes in angular momentum, often called centrifugal forces. In mechanics, of course, it is well known that the equations of motion can always be derived from a variational principle, namely, by the maximization or minimization of a mathematical construct.[6] There are two generic types of objective functions in mechanics, known as the Lagrangian and the Hamiltonian functions, respectively. Either can be used, depending on convenience and physical interpretation of physical data.[7]

In thermodynamics, the true equilibrium state is strictly static: It is a state where "nothing happens or can happen."[8] In equilibrium, kinetic energy is allocated uniformly among all degrees of freedom. In effect, no processes take place in an equilibrium state. All physical processes are, necessarily, associated with disequilibrium and the approach to equilibrium, rather than with equilibrium itself. In thermodynamics, the equilibrium is essentially uninteresting. However, it is of interest that the equilibrium state in thermodynamics—like mechanics—is the end state of a maximization or minimization problem.[9]

In the case of thermodynamics, the extremum problem involves minimizing a generalized potential energy function,[10] or maximizing a generalized dissipation function that has come to be called **entropy**. Curiously, while thermodynamics evolved from the study of energy, it was the maximization version of the extremum principle that was discovered first, by the German physicist Rudolph Clausius in 1847. Clausius defined entropy S in terms of a path integral over its differential, such that $dS = dQ/T$, where dQ is a differential of heat flow and T is the absolute temperature. He showed that $dS > 0$ for all irreversible thermodynamic processes. It was proved later (1909) by the mathematician Constantin Carathéodory that irreversibility defined in a purely mathematical sense is a sufficient condition to guarantee the existence of a non-decreasing integrating function. Entropy is merely the name of that function. (Its physical interpretation is still being argued.) The Austrian physicist Ludwig Boltzmann offered his famous statistical definition of entropy ($S = k\log Z$, where k is a constant and Z is the number of possible states of the system) in 1853.[11] As will be discussed below, thermodynamic equilibrium is sometimes defined as that state corresponding to maximum entropy.

The concept of reversibility (and irreversibility) is also applicable to computers. Some years ago, computer scientist Rolf Landauer pointed out that all real computers lose information as they compute. Hence they are irreversible, in the sense that computational operations cannot be run backward. One cannot proceed from the answer to the starting point. Information is lost at the micro-level because conventional logic (**and-gates** and **or-gates**) has fewer output channels than input channels. It necessarily generates the same output signals for different input signals. When such logic is used, one cannot trace the source of an intermediate state of the computer uniquely to its original state. Around 1970 a question arose among some physicists and computer scientists as to whether a reversible computer was theoretically possible. The answer turned out to be affirmative. But such a computer must be designed to use only reversible logical elements, i.e., logic elements with a unique one-to-one relationship between input signals and output signals.[12]

This discussion on computers may appear to be irrelevant to the main theme of this book, but it is not. The key point is that a reversible computer has no practical application. Real computers are deliberately designed to lose intermediate information, i.e., to be irreversible. The practical benefit of any data-processing operation in practice is *data-reduction*. In other words, the computer converts a large quantity of low-grade input information ("dirty data") into a smaller quantity of higher-grade intermediate information ("clean data"), and thence into still-higher-grade output information ("the answers"). What this process boils down to is upgrading information. I will argue later that the evolutionary process can be characterized in much the same way.

1.2. ENTROPY AND THE SECOND LAW OF THERMODYNAMICS

In 1865 the existence of a non-decreasing function (previously defined as **entropy**) for irreversible processes was proposed by Clausius as a general law of physics, now known as the **second law of thermodynamics**. The law states that an isolated system not in equilibrium will always tend to approach equilibrium; therefore, the entropy function always increases in any isolated system.[13] Clausius's proposed general law, based on irreversibility, gave rise to one of the most persistent (and, at times, bitterest) disputes in the history of science. The fundamental difficulty was—and remained until very recently—that no such general law of continuous dissipation and global irreversibility (or "time's arrow," in Arthur Eddington's phrase) can be derived from Newton's equations of motion (which, as noted above, are *reversible*).

Ludwig Boltzmann made a bold attempt to resolve the difficulty by combining Newtonian mechanics with statistics. His treatment depended on the assumption that the probability distribution function describing possible macrostates of a gas—positions and velocities of the entire ensemble of molecules—is a simple product of the probability distribution functions of each molecule independently of all the others. This approach ultimately became known as **statistical mechanics**. One of his key results, based on the independence condition, was the so-called H-theorem (*c.* 1872), which seems to confirm that entropy (using the statistical definition) will, with overwhelmingly high statistical probability, tend to increase.

The validity of the H-theorem was challenged immediately. In the first place, Boltzmann's theory depended on the existence of atoms and molecules—a point not universally accepted by physicists in the late 19th century. Second, the statistical independence assumption seemed unjustified to many. Boltzmann defended this assumption on the basis of the extremely large number of individual molecules in a gas. The real trouble, however, arose from the reversibility of Newton's equations. In 1876 the Austrian physicist Joseph Loschmidt suggested a *Gedanken* ("thought") experiment that clearly underlined the difficulty. Suppose a gas is released at 12:00 (noon) in the corner of an evacuated room, with its molecules moving independently according to Boltzmann's statistical assumption, but also obeying Newton's laws of motion. Now suppose at 12:05 "the hand of God" reverses the direction of all the molecules. It follows from reversibility that at exactly 12:10 all the molecules must return to the corner of the room. Surely its entropy must also have the same value as at the beginning. Yet Boltzmann's H-theorem suggests that entropy has increased continuously. Loschmidt noted that there seems to be a contradiction.

Boltzmann responded to this by asserting that the H-theorem is a statistical statement, which must be true in the overwhelming majority of scenarios because the number of microstates associated with equilibrium is overwhelmingly greater than the number associated with any non-equilibrium state. In effect, he conceded that occasional exceptions might be possible. Yet Loschmidt's argument clearly implies that, for every microstate with increasing entropy, there must be another in which the directions of all particles are reversed, with decreasing entropy. Boltzmann's explanation certainly didn't satisfy the mathematicians, who felt that a real theorem should be true all the time, not just most of the time.

Henri Poincaré came up with a mathematical proof of his own in 1894 that added fuel to the fire. He proved that any group of three or more particles moving according to Newton's equations in a confined space displays "nearly" periodic motion and will eventually return to a state "arbitrarily close" to its initial state. Moreover, this "return from the diaspora" is **recurrent** — it happens again and again. Furthermore, as time goes on, the probability of a recurrence approximating the original state (within any given degree of closeness) continues to increase. In other words:

> "Whereas **reversibility** means that, by going back in time (or, what amounts to the same thing, by reversing directions), particles return exactly to where they started from, **recurrence** states that, by moving forward in time, particles nearly return to where they started from, and do so repeatedly" [Shinbrot, 1987, p. 36; emphasis added].

In 1896 the German mathematician Ernst Zermelo used Poincaré's proof directly to attack Boltzmann's rebuttal of Loschmidt's argument. In effect, Poincaré had proved that a gas released in the corner of an evacuated chamber—as Loschmidt had supposed—must return again and again, unendingly, to that corner. Thus, entropy must also be a recurrent quantity. It seems to follow that the H-theorem cannot describe the behavior of a "real" gas (i.e., a gas that also satisfies Newton's equations of motion). Boltzmann argued once again that, while Poincaré's proof was valid, the times of recurrence for a real gas, due to the extremely large number of particles involved, must be so long ("far in excess of the age of the universe") that such recurrences are not a realistic possibility. Though this argument was widely rejected as feeble in 1900 (and Boltzmann committed suicide from depression in 1906), it is now regarded by most physicists as the correct view. In fact, final mathematical proof of its validity—in the sense of (statistical) consistency with micro-reversibility—seems likely to be produced within the next few years.[14]

1.3. IRREVERSIBILITY IN ECONOMICS

Irreversibility as a concept has been used in several different ways (mostly loosely) in economics. At first sight, it is natural to associate irreversibility with dissipation, as in physics. As noted above, however, the concept can be defined independently, and rigorously, in mathematics. Thus, physical irreversibility has its economic implications. These need not be a consequence of "dissipation," even in the generalized sense discussed in footnote 4. Yet the relevance of the notion has remained unclear enough for Paul Samuelson [1983, p. 36] to say that "In theoretical economics there is no irreversibility concept."

In fact, theoretical economics does consider a kind of macro-irreversibility that is a visible consequence of such dissipative (i.e., thermodynamic) processes as aging, "wearing out," and resource exhaustion. This has been particularly emphasized by Nicholas Georgescu-Roegen, who wrote: "The popular maxim 'you can't get something for nothing' should be replaced by 'you cannot get anything but at a far greater cost in low entropy'" [Georgescu-Roegen, 1971, p. 279; also 1976; 1979].

The inability of the economic system to operate without continuous inputs of free energy ("available work") was clearly recognized and stated as long ago as 1922 by the Nobel laureate chemist Frederick Soddy [1922, 1933]. It was also certainly implicit in the axiom formulated by Tjalling Koopmans [1951] as the impossibility of producing an output without an input, which he described whimsically as "the impossibility of the Land of Cockaigne." Koopmans's axiom, however, can also be interpreted as an implication of the conservation of mass (first law of thermodynamics), so it is not an unambiguous statement about irreversibility. Gerard Debreu [1959] later introduced an explicit irreversibility axiom, related to that of Koopmans, in the context of a general equilibrium model. The topic has also been discussed by Wittmann [1968, pp. 39-42], Bryant [1982], Faber and Proops [1985], and Brody et al. [1985].

All the discussions of irreversibility in economics cited above have been directly or indirectly derived from the second law of thermodynamics and "time's arrow." Faber and Proops [1986] and Bausor [1986] have, in addition, attempted to introduce a more precise notion of the uni-directionality of time in economic processes.

In 1951, however, a very different—non-physical—concept of economic irreversibility, known as Ville's axiom, was formulated:

"No [price] path exists which moves always in the preferred direction but ends at its starting point."

This axiom was set forth as a necessary condition for the existence of a differentiable **total utility function** that depended only on the quantity of goods exchanged. A substantially equivalent but more transparent statement is the following:

"There is no spontaneous transaction between two economic decision-makers resulting in negative (or zero) surplus values for either party" [Ayres and Martinàs, 1990].

This definition is fundamental to the derivation of a non-decreasing "progress function" in economics, discussed in Chapter 7.

Transactional irreversibility arises from the rationality of exchanges (trades) in the marketplace. If individual (or firm) A is willing to exchange his apples for oranges from B, it is because he has more than enough apples and too few oranges. He is presumably not willing to exchange any of his oranges to get even more apples. (It is not the activity of trading per se that motivates the exchange, but the relative preferences of the two parties.) It follows that exchanges occur only when they leave both parties better off in some immediate sense. It also follows that voluntary exchanges do *not* occur if they leave either party worse off in terms of his/her own preferences.

It is important to note that *rationality* in decision-making behavior can be defined in several ways. For instance, *perfect rationality* is an idealized case in which each party achieves the optimum result, taking into account all possibilities of exchanges and fully reflecting the preferences (utility functions) of all actors in the market. The notion of perfect rationality is important to the core theorem of neo-classical economics, the existence of **general equilibrium**. (This theorem is discussed later in this chapter and in Chapter 6, section 3.)

On the other hand, the simplest and most general form of rational behavior is called *bounded rationality*. Formulated by Herbert Simon [1955, 1959, 1982], it merely presupposes that each party to the transaction has "common sense" and does not voluntarily engage in transactions that leave it worse off (i.e., impair its economic state). A variety of possible decision-making criteria would satisfy bounded rationality. This concept is also important later in the book (Chapter 7).

1.4. THERMODYNAMIC CONSTRAINTS IN ECONOMICS

The fact that economic activity involves the transformation of physical materials and fuels into new forms (that in turn ultimately produce services) is enough to ensure that economic "laws of motion" must be consistent with, and subject to, physical constraints. We have established that economic activities are inherently dissipative and governed by the second law of thermodynamics (increasing entropy). This has both deep theoretical implications and important practical ones. Theoretically, dissipation implies that homothetic economic growth (i.e., growth without technological and structural change), is impossible. This point is discussed in Chapter 6.

The number of economicsts and physicists attempting to integrate thermodynamic constraints into theoretical economic models in other ways continues to grow [see, for example, Ayres, 1978; Burness *et al.*, 1980; Mirowski, 1984; Ayres and Nair, 1984; Faber and Proops, 1985; Faber *et al.*, 1990; Perrings, 1987; England, 1990; Khalil, 1990].

The practical implications of economic activity being governed by the laws of thermodynamics relate to the extent to which economic activity is "sustainable" in terms of its impact on the environment. As noted, there can be no output without inputs of materials and energy. Thus, economic activity at present clearly depends to some degree on the extraction and use of fossil forms of energy and high-quality mineral resources left from primeval geophysical and chemical processes. These resources will be exhausted within centuries, if not decades.

On the other hand, the present economic system also utilizes the environment as a "sink" for waste materials, on an enormous (and unsustainable) scale. That waste emissions are an automatic consequence of resource extraction can be inferred directly from the laws of physics, notably the conservation of mass [Ayres and Kneese, 1969, 1989; Kneese *et al.*, 1970]. In the current absence of adequate policies or incentives for reducing waste emissions, or recycling them, most waste materials are being dispersed into the environment as pollutants. Some of them are toxic to humans or other species, with the consequence that the biological environment is being threatened. Moreover, the physical environment itself—especially the atmosphere—is now showing many signs of adverse changes attributable to human activities .

In simple terms, the burning question for policy-makers is whether the global environment can be protected without an end to economic growth [WCED, 1987]. Or, putting the question in even starker form, Will the fulfillment of the economic-growth imperatives of the poorest two-thirds of the human race inevitably destroy the environment that sustains life on earth? Is it possible to decouple economic growth from natural-resource use? These theoretical questions are addressed explicitly in Chapters 6, 8 and 12.

1.5. EQUILIBRIUM, STEADY STATE, AND CHAOS

An important distinction must be made here. A state of equilibrium is not the same as a **steady state** or a **stationary state**. A steady state is a temporary condition of constancy of some subsystem variable(s) maintained by currents or flows (i.e., motion) within the system or between the subsystem and the environment. Steady states can be associated with **local** (as opposed to global) minima in some generalized potential function, or maxima of some dissipation function (i.e., entropy). Local extrema can be far away from the global extremum, just as the nearest hill need not be the highest.

A steady state that is not an equilibrium state can exist in non-linear dynamic systems. (A steady state in a linear system must be an equilibrium state.) The equations of motion for non-linear systems can have multiple solutions. In non-linear feedback **points, limit cycles, quasi-periodic motion**[15] and **deterministic chaos**. Limit cycles are simple or complex multi-variable oscillations with definite periodicity [see for example, Smale, 1973; and Casti, 1984]. Examples include metronomes and heartbeats. An example familiar to non-mathematicians is the Lotka-Volterra system of equations that describe—in ultra-simplified terms, of course—predator-prey relationships in a simple ecosystem [Lotka, 1956; Volterra, 1937; Peschel and Mende, 1986]. Deterministic chaos is characterized by non-repetitive oscillations with variable and unpredictable periods and amplitudes [Crutchfield *et al.,* 1986].

An important feature of chaotic systems is that the solution space (**phase space**) is naturally divided into allowed and forbidden regions. The allowed regions—where non-linear dynamical systems spend almost all their time—have been given the evocative name **strange attractors** [Ott, 1981]. This expresses the idea that such a system is attracted away from its initial conditions, in a way analogous to the attraction of a particle by a potential field. Attractors are generalizations of orbits. Thus, the stable equilibrium is a point attractor, and the periodic oscillatory orbit is a limit cycle, as noted above. An example of chaotic motion—irregular and unpredictable, but nevertheless deterministic and constrained—might be the flight path of a moth around a light.[16]

The old distinction between deterministic behavior and random behavior has been significantly blurred in recent years by the realization that **chaos** is virtually indistinguishable from true randomness [see, for example, Jensen, 1987]. The essential characteristic of non-linear dynamic systems (for certain ranges of parameters) is an extreme sensitivity of results to starting conditions. In more precise mathematical language, the solutions for starting conditions that differ only slightly diverge at an exponential rate known as the average Liaponov exponent, also known as the Kolmogorov-Sinai entropy. This phenomenon has been demonstrated in quite simple mathematical models [e.g., Lorenz, 1963; May, 1975; Rössler, 1976]. The extraordinary sensitivity of some non-linear dynamical systems to infinitesimal differences in starting conditions has been called the "butterfly effect," based on a joking reference to the notion that the flutters of a butterfly's wings in the Amazon might be enough to shift the climate over the northern hemisphere from one trajectory to another [Gleick, 1987, attributed to Lorenz].

It now appears likely that—because of this "butterfly effect"—quite a few non-linear dynamical systems of great importance to our everyday lives are essentially unpredictable, except in the very short run, although their trajectories remain confined within some region of phase space defined as a strange attractor. The weather/climate system is one example. Other likely candidates include insect populations, fish populations, and the stock market. Under these circumstances the normal meaning of prediction must be reconsidered. In the long run the only possibility, for such systems, is to identify the boundaries of the strange attractor for the system in question. Unpredictability can thus be attributed either to truly random or stochastic processes, or to deterministic chaos.

Another key characteristic of non-linear dynamical systems is the phenomenon of **self-organization**, discussed later. Whereas the "butterfly effect" refers to the exponential divergence of trajectories in the neighborhood of a single strange attractor, *hyperselection* refers to the transient stage of evolution when a system is able to "choose" among disjoint attractors. Still another name for the same phenomenon is "lock-in/lockout." Thus, the selection of one evolutionary path may be largely accidental, and yet that path may be dominant for a long time in evolutionary terms. This happens because feedback controls or—in the economic case—"returns to adoption" favor the established choice over most competitors [Arthur *et al.*, 1987; Arthur, 1988].

Modern open-system thermodynamics has focused attention on the self-organizing properties of certain non-linear chemical and bio-chemical systems. Such systems may be located in local minima of the generalized thermodynamic potential function, which corresponds to a local maximum of the entropy function. This explains why such self-organizing systems are often dissipative (as it was noted above, all irreversible phenomena are necessarily dissipative). *Self-organization,* in this context, connotes spatial order, stability (or, more precisely, homeostasis: the tendency to resist disruption or disordering due to external disturbance), and persistence. Dissipative systems are not always self-organizing or far away from thermodynamic equilibrium. Living organisms, however, are clearly examples of self-organizing dissipative systems.[17]

In economics, the term **general equilibrium** is widely used. It refers to a state of local utility maximum, in the special sense (defined by the Italian economist Vilfredo Pareto, and called by his name) that no individual transactor can improve his utility except at greater utility-loss to another or others. Having reached this state, rational utility-maximizers could assume they would obtain no further benefit from exchanges, and trading would therefore come to a halt. In this sense, the general-equilibrium condition posited by economists is static. It resembles the thermodynamic equilibrium in which "nothing happens."

The general-equilibrium state is also defined as a state in which all markets *clear*, meaning that a set of scalar prices exists such that all goods offered for sale find buyers and all would-be customers find goods for sale at the market price. The existence of such an equilibrium, first conjectured by the French economist Leon Walras [1877], was finally proven by Abraham Wald in the 1930s for several models. More general proofs were given later by Kenneth Arrow and Gerard Debreu [1954] and by Lionel McKenzie [1954]. All proofs of the existence of general equilibrium are subject to a number of very restrictive and unrealistic assumptions.

Subsequent work has removed some of restrictions on the generality of the proof, although other difficulties appear to be fundamental and unavoidable. One of the most unrealistic of the assumptions underlying the original proof of the existence of general equilibrium is that of perfect rationality and foresight. The modified concept of bounded rationality, mentioned previously, appears far more in accord with observed human behavior. Unfortunately, if one adopts the notion of bounded rationality, one must give up the handy analytic device of profit (or utility) maximization.

A more fundamental difference between the economic and thermodynamic concepts of equilibrium is the fact that economists have no concept, or measure, of distance from equilibrium. This springs from the assumption (in mainstream economics, at least) that all observed data represent equilibrium conditions. Thus, thermodynamics is essentially the study of approach to equilibrium. By contrast, neo-classical economics—as currently practiced—purports to be the study of equilibrium itself. The former is essentially dynamic, the latter essentially static. This topic is discussed in greater depth in Chapter 6, section 3.

1.6. ORDER, STRUCTURE, AND ORGANIZATION

Another set of related concepts that deserves discussion is **order, structure,** and **organization**. Of the first two, the second is narrower, since structure always implies order, but order does not always imply structure. The antonym of order is **disorder** or **randomness**. Because the state of (global) thermodynamic equilibrium implies maximum entropy, disorder and/or randomness, it has been tempting to assume the correspondence is general, i.e., that maximum order corresponds to minimum entropy. This notion may have started with Boltzmann.

Thermodynamicists have been at pains in recent years to clarify the situation. Entropy is a quantitative and rigorously defined notion; order (or disorder) is not. For this reason, it is always clear (in principle) whether entropy increases or decreases in a given process, but not always clear whether order does. In some cases (perhaps most), increasing disorderliness seems to correspond roughly to increasing entropy. But in several well-known cases, the correspondence does not hold [see McGlashan, 1966; and Wright, 1970]. Orderliness is a measure of non-randomness, regularity, and hence of predictability. In physics, the notions of short-range order and long-range order—in contrast to disorder—are quite well established, albeit not rigorously defined.

A periodic crystalline solid exemplifies long-range order in the following sense: If the locations of a few molecules are known, the lattice parameters corresponding to their intermolecular force constants can be determined with reasonable precision. Based on this information, the locations of all other lattice sites are predictable, even at great distances. A liquid exemplifies short-range order: The location of a molecule and the parameters of the intermolecular force constants suffice to determine the probable locations of other nearby molecules. But there is no correlation beyond a few molecular diameters. A gas is disordered in the sense that there is no correlation between the location of any one molecule and any other.

Order also implies (or may imply) regularity and short- or long-range correlations between events in time. The tick-tock of a clock or the orbiting of the moon around the earth or the earth around the sun illustrate temporal orderliness of a simple periodic kind. On deeper reflection a set of related chemical reactions in a stationary state also reflects order, as does a pattern of words, lines, numbers, colors, or shapes. *Pattern* is another word for order of a more complex nature. In fact, the business of science is to discern regularities—patterns—in nature and to discover simple rules or algorithms that can reproduce them, hence explaining them in some sense.

Structure, used loosely, implies order (or pattern) embodied in an inhomogeneous medium. This usage also applies perfectly well to the buildings constructed by humans, which are commonly called "structures" by virtue of having been constructed. Of course, the word is often used as a synonym for *order*. Primarily, however, it is used to reflect a stable combination of spatial differentiation of form (inhomogeneity, anisotropy), as in astronomy. It may also imply specialization of function, as in a biological organism.

Fortunately, the term **stability** can be given a straightforward quantitative meaning in some cases, at least. It is related to the notion of equilibrium, in the sense that an equilibrium can be more or less stable. In classical mechanics, the degree of stability of an equilibrium is related to the amount of effort (energy) required to destroy it. Alternatively, it is (inversely) related to the length of time the equilibrium would likely survive in the presence of random disturbances. Thus, the inverse of mean average lifetime is an excellent measure of stability in all cases where it is observable. However, this wouldn't be helpful in other cases, such as social systems or ecosystems. Again, a general measure is lacking but probably not essential.

Stability may be an inherent property of some structures in some circumstances. A three-legged stool, for instance, is said to be inherently stable (although it can be deliberately tipped over). An airfoil may be designed to be inherently stable against perturbations, more or less independent of the pilot. Stabilization can also be accomplished by active control as in the case of a bicycle. Modern high-speed aircraft, for instance, are not inherently stable, but must be actively controlled by the pilot, or some surrogate, on a continuous basis. "Active control" implies a control system, consisting of "feedback" (sensors), a computer or brain to interpret the sensory inputs, and some means of "feedforward" to make appropriate adjustments.

Organization is evidently another close relative of structure. An organization has structure; so does an organism. A structure is, in some sense, organized. Again, stability—or durability—is also implied. What, then, is **self-organization**? For our purposes, it can be thought of as a structure or organization that is self-controlled, self-stabilizing, or homeostatic. (The study of self-regulating control systems in organisms and machines was first formalized by the mathematician Norbert Wiener [1948], who coined the term **cybernetics**.) Self-organization is, therefore, an aspect of cybernetics, at least from one perspective. The connection between self-organization and path-dependence (or hyperselection) was noted earlier. The connection between self-organization, information, and intelligence has been explored by W. Ross Ashby [1960], Francisco Varela [1975], and Heinz von Foerster [von Foerster and Zopf, 1962], among others.

1.7. DIVERSITY AND COMPLEXITY

Diversity is virtually a synonym for variety. It denotes the existence of a large number of possibilities or variations. As applied, in particular, to an ecosystem or an economic system, it connotes the existence of many species, many niches, a wide variety of jobs, business opportunities, available products and services, product models, landscapes, and so forth. A simple count of such possibilities might be feasible in some situations, but it is difficult to set forth a simple rule specifying in all cases what would be counted.

Consider the problem of defining diversity in the automobile market: Should we count manufacturers (like Ford and Toyota), "nameplates" (Chevrolet, Buick, Cadillac), models (Monte Carlo, Seville, Corolla), or allowed combinations of options?

Intuitively, diversity appears to be an aspect of complexity. Greater diversity seems to imply greater complexity. What, then, is **complexity**?

A typical dictionary defines *complex* as "consisting of a number of connected parts or elements; complicated; involved; intricate." The only clue this definition offers as to how different *degrees* of complexity might be measured is in terms of the number of component parts. This approach, however, must be used with care when comparing, say, large animals (elephants, for example) with smaller ones (mice). The large organism probably has more cells (and more molecules), but this alone does not make it more "complex" as we understand the word intuitively. A similar problem arises in comparing the computers of 1960 with those of today. Modern computers are enormously more powerful than the earlier versions, but also contain far fewer "parts," if a chip is considered a part. This is because many circuit elements are now routinely compressed into a single chip. On the other hand, one cannot simply measure complexity in terms of circuit elements, since many mechanical parts (e.g., in keyboards and disk drives) have no electronic function.

On reflection, it is clear that we need both a more general and a more precise way of measuring complexity. One notion that has been suggested is that complexity is the *quantity of information* needed to describe (or the quantity "embodied in") an object, system, or organism. One might, perhaps, use Shannon's formal definition of information [Shannon, 1948], which will be discussed in Chapter 2. This definition does not distinguish randomness from complexity, however. The most random configuration of a gas has the highest Shannonian information content, but the lowest complexity in terms of behavior or properties.

Another appealing idea is using algorithmic information content: the size of the shortest algorithm or computer program needed to describe a complex object (say, a geometric shape). This approach is especially appropriate for measuring "morphological information" (discussed in Chapter 10). However, algorithmic information content would also be very large for a completely random number with a great many digits—one that cannot be expressed, for example, as a ratio of natural numbers or some other simple rule. This also does not accord with our intuitive understanding of complexity.

To get around such difficulties, Charles Bennett has suggested a measure called "logical depth," which involves three things: (1) a computer, (2) an optimum computer program for producing the given result (or its numerical representation), and (3) a way to count the minimum number of machine cycles needed to derive the result on the computer via the program. [See Bennett and Landauer, 1985.] Unfortunately, the scheme cannot be implemented. There is no set of rules (i.e., no program) for constructing optimal programs and, for very deep reasons, one can never know that a given program is optimal.

Another scheme, called (by analogy) logical "breadth," has been suggested by Heinz Pagels and Seth Lloyd. Rather than focusing on the "best" program to generate a digital representation of a complex object, it seeks to describe the steps by which the object itself was originally produced. For instance, to define the complexity of a product of biological evolution, Pagels and Lloyd would seek to describe the evolutionary events themselves in terms of the amount of (Shannonian) information processed. Unfortunately, this scheme, also, suffers from problems of practical implementation, especially in dealing with objects whose origins are unknown or indescribable, such as art objects or works of literature.

In fact, *complexity* (along with related terms like *order* and *structure*) appears to be essentially indefinable in any way that would permit objective measurement. One is reduced, in practice, to saying something like "I don't know what it is, but know it when I see it." Perhaps the notion is indefinable except in anthropic terms. This indefinability raises several interesting points that will be touched on later.

1.8. BIOLOGICAL CHANGE: DEVELOPMENT VERSUS EVOLUTION

An **organism** can be defined, for our purposes, as a basic unit of life that is capable of self-reproduction. (Looser usages are common enough but cannot be commended.) The simplest organisms are **cells**. In addition to **reproduction**— either sexual or asexual—all cells carry on other activities, including **metabolism** (ingestion of food, digestion, synthesis of proteins and other biochemicals, excretion of wastes) and **repair** of damage. The term *metabolism* could be interpreted broadly enough to include reproduction and repair functions, but it is conceptually cleaner to distinguish them [see Casti, 1987]. Some cells are also able to **sense** the environment and to respond by moving along chemical, thermal, or optical gradients toward or away from certain stimuli.

A number of other biological and evolutionary concepts need to be introduced. A *species* is defined as a population of organisms similar enough to interbreed successfully, in the sense of producing viable self-reproducing offspring. The term *similarity,* here, refers to genetic compatibility or **genotype**, as contrasted with manifest physical structure and behavior, or **phenotype**. The genotype refers essentially to genetic potential, while the phenotype refers to the actual realization of that potential (see Thesis). It is of some importance to note that genotype is a direct co-determinant of phenotype, whereas the converse does not hold true, at least in the direct sense. An externally induced change in the phenotype of an organism is not passed on to subsequent generations; i.e., it is not heritable. It affects heritability, however, to the extent that the phenotype (influenced by environmental conditions) influences the selectivity of characteristics.

It is also important to note that a population of organisms consists of a distribution of genotypes and phenotypes. These distributions can be termed **macro-genotype** and **macro-phenotype**, respectively.

Every multi-cellular organism undergoes a process of **development** as it grows and interacts with its environment. It begins as a single cell (created by the merger of egg and sperm) and passes through various stages from embryo to adult. Biologists generally identify two broad categories of developmental processes: **constructive** and **limiting**. The former are subcategorized into **growth**,

morphogenetic movement, and **differentiation.** Growth, in this taxonomy, means increasing biomass. Morphogenetic movement, by contrast, refers to changes in form—as illustrated, for example, by the metamorphosis of tadpoles into frogs and caterpillars into butterflies. Differentiation, in this case, refers to the increasing specialization (and distinguishability) of the parts of an organism as it develops [Bonner, 1952].

The major theme of the book is **evolution.** The term in ordinary usage carries all sorts of incidental implications (e.g., "evolutionism" vs. "creationism"), depending on the context. Indeed, it is the nature of the evolutionary path that attracts so much interest from both scientists and laypersons. We humans arrived very late on the evolutionary scene, yet in a short time we have achieved absolute dominance over all other species. The temptation is powerful to interpret evolution as a kind of ladder-climbing process that began with very simple organisms and culminated (if not ended) with ourselves.[18]

Many scientists have thought that evolution—like ladder-climbing—must be **irreversible** in some sense, although it has never been quite clear how the notions of reversibility/irreversibility apply in this case. (Evidently dissipation is not involved.) One school avoids this problem by defining evolution as "the changing of something into something else over time" [Faber and Proops, 1989, p. 38]. This definition is clearly too broad, however, since it does not even exclude cyclic or periodic processes. A better definition is probably the following:

"Evolution can be defined physically as a historical irreversible process consisting of an unlimited series of self-organization steps. Each step is triggered by a critical fluctuation ... leading to an instability of the original semi-stable state" [Ebeling and Feistel, 1984].

This definition might be faulted on the niggling grounds that it seems to confuse definition with explanation, but it is helpful for all that. It includes the elements of irreversibility and instability, for instance. To be sure, the Ebeling-Feistel definition does not specifically include the elements of variation and selection, which many scientists think to be central to the idea of evolution. Indeed, the existence of a characteristic selection, or "search," algorithm seems to distinguish evolutionary processes from either deterministic change or optimization processes. Modern biologists and computer scientists have, in fact, sought to simulate evolutionary processes by leaving the "objective" of evolution unspecified and focusing solely on the selection algorithm.[19]

Another aspect that Ebeling and Feistel leave ambiguous is the inherent predictability of the path. To be sure, their emphasis on instabilities and fluctuations certainly implies a degree of unpredictablity. But the distinction between evolution and programmed change or growth (development) should be stronger. In biology, and in this book, evolution is explicitly distinguished from programmed growth or development. The evolutionary history of a biological species is called **phylogeny,** in contrast to the developmental history ("life cycle") of an individual, or **ontogeny.** Note that phylogeny comprises both

TABLE 1.1. *Typology of Change.*

	Phylogeny	Ontogeny
Genotype	Darwinian evolution of a species	?
Phenotype	Lamarckian evolution of a social or economic system	Development of a fetus "Life cycle" of a star "Life cycle" of an ecosystem

phenotypic and genotypic evolution, whereas ontogeny is primarily phenotypic development (Table 1.1).

The critical distinction between ontogeny and phylogeny is that ontogeny (development) is essentially predictable, in contrast to phylogeny, or genotypic evolution. This is precisely because it consists of the unfolding of a pre-programmed pattern contained in the genotype. (Hence the textbook phrase "Ontogeny recapitulates phylogeny.") Developmental history is programmed by the genes, although there is a stochastic element arising from environmental interactions. On the other hand, phylogeny consists of an interplay between genotypic evolution and phenotypic development. Hence it is clearly not predetermined or predictable in any detail. (The question remains whether the future path of phylogeny can be characterized as some kind of "order from chaos," or whether it is purely a "random walk" [see, e.g., Montroll and Shuler, 1979]).

1.9. ECOSYSTEM, BIOSPHERE, AND GAIA

An **ecosystem** is a self-organizing and persistent collection of interacting populations occupying the same territory. It is not necessarily stable or homeostatic, since it may exhibit a developmental character ("life cycle"). An important concept in ecology, which I will make some use of, is that of **ascendancy**. Recent work by Ulanowicz [1986] and others to quantify this concept is worthy of note. The antonym of ascendancy is **retrogression**. These terms, along with **climax**, are basically equivalent to the concepts of **adolescence, maturity,** and **senescence** from the old life-cycle model of ontological development, which has been successively adopted by many of the branches of social science, including anthropology.

An ecosystem can evolve both ontogenically and phylogenetically, on different time scales, of course. An ecosystem that is disturbed (by a forest fire, for example) tends to retrace a definite macro-phenotypic sequence, beginning with fast-growing, short-lived species and shifting toward a preponderance of slower-growing but long-lived species. The latter (climax) stage may be very stable (e.g., a tropical rain forest), or quite unstable (like some northern conifer forests), depending on circumstances. Of course, as species genotypes evolve, the interspecies relationships may change and the macro-phenotype may change in parallel. This would correspond to phylogenic (as opposed to phenotypic) evolution at the ecosystem level.

Another important concept in ecology is that of **trophic level**. A trophic level in an ecosystem is defined by the feeding habits of organisms, organisms feeding on the same things being ipso facto on the same level. The lowest level consists of photosynthesizers (e.g., phytoplankton). The next level is that of **herbivores** or animals exclusively living on plants (e.g., zooplankton), and above that are various **carnivores** (predators) and parasites, all of which meet their metabolic energy needs by consuming other animals. These groupings are arranged in a hierarchical manner often called a food chain. The food-chain hierarchy bears some resemblance to the structure of a social or business hierarchy, with ordinary workers or primary producers at the bottom of the pyramid, various levels of processors and/or managers in the middle levels, and top management at the apex. Such pyramidal structures can be used to model and compare energy flow, materials flow, and information flow in organizations.

The concept of **metabolism** deserves a few more words. It comprises the sum total of all chemical- and energy-transformation processes in a living system. The concept applies to all living systems, from cells to ecosystems to economies. There are, in fact, two basic categories, namely **anabolic**, or constructive, processes and **catabolic**, or destructive, processes. Photosynthesis and protein synthesis are examples of the first, while fermentation, **respiration**, and **digestion** are examples of the second. All living organisms utilize both types, as do ecosystems. In general, photosynthesizers play the role of anabolites in an ecosystem, while grazing animals, predators, parasites, and decay organisms play the role of catabolites.

The economic analogy of biological metabolism ("industrial metabolism") is the set of materials- and energy-transformation processes that convert raw materials into finished materials and final forms of energy (such as electric light). Again, one can identify both anabolic and catabolic processes in industry. The separation and reduction of metals from ores or of chemical intermediates from petroleum, and the creation of shapes in solids by cutting or grinding, illustrate the catabolic type, while the synthesis of alloys or plastics, and the welding and assembly operations in manufacturing, illustrate the anabolic type.

A new term coming into use is **anthroposphere**. It denotes the "built environment" and the material goods, both in use and discarded, associated with industrial metabolism. It is, in some sense, the industrial analog of **biomass**.

The word **biosphere**, connoting the entire earth as a single ecosystem, was coined by the Austrian geologist Edward Suess in his famous book *The Face of The Earth* [Suess, 1904–1924]. The concept has been defined as a "particular envelope of the terrestrial crust, a layer permeated by life" [Vernadsky, 1924, cited by Grinevald, 1987], and reinterpreted to mean "the actual layer of vitalized substance enveloping the earth" [Teilhard de Chardin, 1977, cited by Grinevald, 1987]. Vladimir Vernadsky emphasized, particularly, the fact that man exists within the biosphere and cannot exist without it. He also emphasized that the biosphere itself—all terrestrial plants and animals—are interconnected with one another through competition, nutrition, predation, parasitism, and decay, and with the non-biotic solar, geological, hydrological, and atmospheric environment through photosynthesis, respiration, and transpiration.

The term **noösphere** was apparently introduced first by the French geologist and eco-philosopher, Pierre Teilhard de Chardin in a number of essays written in the early 1920s, though not all published at the time. It was first popularized in two books of Edward LeRoy [1927, 1928], and later adopted by the great Russian geo-scientist Vladimir Vernadsky [1945 and other works]. The term has been used by Teilhard de Chardin and Vernadsky to distinguish the effects of human activity on the **biosphere** from the biosphere itself. The noösphere includes, but is more encompassing than, the notion of industrial metabolism mentioned above.

One final term, popularized within the past two decades, is the **Gaia hypothesis**, which originally connoted the somewhat mystical idea that the earth itself— all the spheres from the atmosphere and the biosphere to the geosphere—together constitute a living system or super-organism that actively maintains the environment in a way that optimizes the conditions for life. (This idea was originally expressed by the British geologist-physician James Hutton in 1785.) The modern exponent of this idea is the British atmospheric scientist-inventor James Lovelock [1972, 1979].

The original version of the hypothesis was attacked on two grounds. The first was that it implies "altruism on a global scale"—an idea that smacks of teleology and is particularly unappetizing to molecular biologists steeped in Darwinian theories of natural selection [Doolittle, 1981; Dawkins, 1982]. The second objection was that geological evolution had previously seemed explicable in terms of chemical and physical forces alone, so that living organisms merely **adapted** to existing conditions [Holland, H. D. 1984]. Others—emphasizing co-evolution—questioned the possibility of homeostatic climate regulation by exclusively biological means [Schneider and Londer, 1984]. Lovelock believes he has answered the latter objection by proposing a simplified but plausible model ("daisy world") exhibiting precisely the postulated homeostatic characteristics [Lovelock and Watson, 1983].

The "new" Gaia hypothesis still holds that the biosphere is the essential homeostatic regulating mechanism for the earth's atmosphere, without which it would not continue to be able to support life. Lovelock attaches special emphasis to the thermodynamic disequilibrium of the chemical composition of the earth's atmosphere. (In fact, the atmosphere in equilibrium would contain neither free oxygen nor free nitrogen; it would consist almost entirely of carbon dioxide, like the atmospheres of Venus and Mars.) Lovelock also stresses that, in the "daisy world," the earth's temperature must be maintained near its present level and the atmospheric oxygen concentration must remain between 15 percent and 25 percent. Dissenters still hold that other (purely geochemical) mechanisms could accomplish the same result. This argument is not likely to be settled for many years, although the Gaia proponents appear to be gaining ground.

1.10. MYOPIA VS. PRESBYOPIA; SELFISHNESS VS. ALTRUISM

Myopia is medical jargon for near-sightedness. The metaphor for economic behavior is obvious, and the term has been freely applied to distinguish behavior based on short-term considerations from behavior based on longer-term factors. The antonym for myopia is **presbyopia,** or far-sightedness. A question of considerable interest, which will be explored from several points of view in later chapters, is whether one can operationally equate myopia with selfishness, and presbyopia with altruism. One of the major puzzles of evolutionary biology has been the evolution of "social" behavior and "altruism." The so-called kinship, or consanguinity, theory seems to provide an explanation of apparently altruistic behavior among social insects having the same mother (ants, termites, bees, etc.) in terms of myopic or "selfish" selection criteria at the genetic level. The kinship theory does not explain altruism in less closely related groups, however.

The behavior of insects is programmed and hence involuntary, which seems to distinguish it from the "voluntary" altruism of higher animals. Yet, the latter may also be programmed socially if not genetically. (The demonstrated willingness of young Shiite Muslim fundamentalists to sacrifice their lives in battle would seem to be an illustration of the point.) Yet, undeniably, there are numerous examples of truly voluntary, conscious, and unprogrammed self-sacrifice, not only on behalf of children or family members, but quite frequently on behalf of abstract ideologies, especially religions or nations.

Clearly, some kinds of self-sacrifice, possibly including the development of formal codes of ethics and morality (like "the golden rule") can be explained as far-sighted, at least under some assumptions. This has been called "reciprocal altruism." It is a generalization of the normal concept of bargaining (*quid pro quo*), with the *quid* de-linked somewhat from the *quo*, both in time and in quantity. It is evident, of course, that many kinds of economic behavior (e.g., saving and investing) also exemplify far-sightedness. The question still remains: How can we explain the evolution of such behavior? This question will be considered in Chapter 6.

In economic systems the terms borrowed from biology are not necessarily directly applicable, although one can see a fairly clear analogy between a firm and an organism. A collection of firms in the same business also resembles a species in some respects, but not in others. One key difference is that there is no economic analogy to genes or bi-sexual reproduction. Moreover, multi-cellular organisms in biology have a natural life span, whereas firms do not (though they can die). Firms do not reproduce in the biological sense. They can spawn new firms, but not necessarily of the same kind. Nor is there any analogy in biology to the conglomerate merger.

It follows that neither genotype nor phenotype has any clear analogy in economics. Thus there is no obvious basis for distinguishing between economic development and economic evolution, except in terms of ontogeny and phylogeny. To be sure, the "product life cycle" does have an ontogenic aspect, as do theoretical Marxism and some of the more recent "stage" theories of economic development [see, e.g., Rostow, 1960]. The more-deterministic theories of economic change are clearly unsuccessful, however, and the evolution of technology itself far more closely resembles phylogeny than ontogeny. These points are discussed at much greater length in Chapter 6.

ENDNOTES

1. The concept of mechanical energy in roughly its modern form was introduced by the German physicist Wilhelm Rankine around 1845. It was extended to include all forms of energy by Hermann Helmholtz in 1847 [see Higatsberger, 1986].

2. Indeed, thermodynamics as a discipline began with the studies of the French engineer N. L. Sadi Carnot on the fundamental nature of the process by which heat energy is converted to kinetic energy by a steam engine. His great treatise "On the Motive Power of Fire," published in 1824, achieved much more than that, however. It established the basic conceptual language for the study of irreversible non-equilibrium systems—about which more later.

3. It is also an interesting footnote of history that Mayer was repeatedly rebuffed in his attempts to publish his results and to this day has received little credit for his monumental discovery.

4. Let $x(t)$, $y(t)$, $z(t)$, ... be a systems state (or stock) variables and let x, y, z,... be the corresponding flow variables. Suppose
   (1) $x = P[x(t),y(t),z(t),...]$
   (2) $y = Q[x(t),y(t),z(t),...]$
   (3) $z = R[x(t),y(t),z(t),...]$.
   Then a system is "dissipative" if its divergence of flow is negative, viz.,
   (4) $div\ [x(t),y(t),z(t),...] = P/x + Q/y + R/z + ... < 0$.

5. It is worth noting that extensive and intensive variables in generalized systems are sometimes called *primal* and *dual* variables, respectively. Duality is a property of linear systems and has been exploited in the development of practical methods of analyzing the behavior of such systems since the time of Lagrange and Hamilton. Wassily Leontief, Lev Kantorovich, John von Neumann, Tjalling Koopmans, and George Dantzig have all contributed to 20th century developments in this area, particularly in mathematical economics and linear programming.

6. Indeed, the use of extremum principles in physics has been enormously sucessful. So much so, that it has become a (nearly) standard methodology and has been widely adopted in other fields, including control engineering, ecology, and economics. In fact, the standard neo-classical assumption of "profit maximization" is an example. Another example is the "energy throughput maximization" principle suggested by Alfred Lotka as a law of biological evolution. Nevertheless, despite the power of the concept, it is becoming clearer that it has inherent limits in applicability to very

complex systems. This point will be made more strongly elsewhere in the book.

7. Both these constructs have also been used for a long time in economics, despite the lack of any conserved quantity analogous to energy. See note 4 above.

8. Thus, as will be pointed out later, the early radiation-dominated universe during the first million years after the Big Bang, could not have been in thermodynamic equilibrium, although it is sometimes assumed to have been.

9. Garrett Birkhoff [1944] pointed out long ago that a variational principle can be found that is appropriate to almost any physical or mathematical theory.

10. In the case of thermodynamics, in particular, the most generalized potential energy function—that function whose minimum corresponds to thermodynamic equilibrium—was not formally identified as such until 1969 (although Helmholtz's and Gibbs's free energy were clearly special cases). This function is usually called *available work* or just *availability*, although it has also been named *exergy* by some European workers and *essergy* by some Americans.

11. This formula is engraved on Boltzmann's tomb in Vienna.

12. Conceptually reversible logic elements ("Fredkin gates") have been worked out [Fredkin and Toffoli 1982], and conceptually reversible computer architectures based on reversible elements have also been demonstrated (as simulations) by Edward Fredkin and Norman Margolus, and independently by Charles Mitchell

13. An isolated system is one exchanging neither energy nor matter with any other system. Humans can create localized approximations of isolated systems for experimental (or industrial) purposes, and in such local conditions a sort of thermodynamiac equilibrium is possible. However, the only natural example of an isolated system is the universe itself. The presumed increase of entropy in the universe must correspond to—and can perhaps define—the forward direction of time. Nernst called the final equilibrium state of maximum entropy *Wärmetod*, or "heat-death."

14. Despite the unsatisfactory status of the underlying theory, since the turn of the century Boltzmann's statistical approach has achieved wider and wider acceptance, especially as it appears compatible with quantum mechanics. By the 1950s even the mathematicians "no longer saw the problem as how to prove Boltzmann wrong but how to prove him right" [Shinbrot, 1987]. In 1974 the American mathematician Oscar Lanford proved an important result: While all finite systems of Newtonian particles exhibit recurrence, Boltzmann's equation does correctly describe the behavior of a Newtonian system in the limiting case that the particles are infinitesimally small and infinitely many (so that the recursion time is also infinite). In this case, the H-theorem becomes a true theorem. Boltzmann's intuitive arguments have, to a large extent, been confirmed by Lanford's theorem. The Lanford results have been further extended by several mathematicians (Shinbrot, Illner, Pulvirenti, Hurd) to show that Boltzmann's equation also correctly describes the **approach** to equilibrium in more realistic cases. The resolution of the Boltzmann-Loschmidt-Zermelo debate seems to be as follows: While a confined gas does not relax into a true thermodynamic equilibrium and will indeed exhibit recurrent behavior, the time periods for recurrence are much longer than the life of the universe. On the other hand, the state of the gas becomes effectively indistinguishable from that of a true

equilibrium quite soon after it is released. Most of the time it is in such a quasi-equilibrium state—just as Boltzmann said.

15. Poincaré's recurrences would be an example of this.

16. It is of incidental interest, perhaps, that strange attractors can be subclassified into four geometric forms. These are **points, closed curve, torus,** and **chaos.** Point attractors exist for either linear or non-linear systems; in the linear case they correspond to true equilibrium solutions. In the non-linear case the point attractor corresponds to a steady state. Closed-curve attractors correspond to oscillations with definite periodicity. Tori and chaotic attractors correspond to solutions that oscillate, but in a non-repetitive manner. Thus they have infinite periodicity. The difference between the last two has to do with their "information-preserving" or "information-generating" characteristics. See Chapter 2.

17. A point worth emphasizing here, though it will be repeated later, is that living systems evidently exist in (and could not exist in the absence of) just such localized thermodynamic extrema, corresponding to very high (local) rates of dissipation of available free energy. This in turn implies a local maximum in the rate of entropy creation [Prigogine *et al.* 1972].

18. One attribute we see in ourselves and possibly a few other "higher" species is "intelligence," or the ability to learn and communicate. This we presume to be related to brain structure, since intelligence, as we normally use the word, is only roughly correlated to brain size. It seems to be the cerebral cortex, in particular, that gives humans their superior ability to learn and communicate. This feature of the brain is relatively recent, and most animals like mollusks, crustacea, fish, insects, and arachnids have no cerebral cortex, and simple organisms like worms, zooplankton, and plants lack even a central nervous system.

19. See, for example, Holland. J. H. [1975, 1988], Kauffman [1988], Kauffman and Levin [1987], and Schwefel [1987, 1989].

REFERENCES

Arrow, Kenneth J. and Gerard Debreu, "Existence of an Equilibrium for a Competitive Economy," *Econometrica* 22(3), 1954.

Arthur, W. Brian, "Self-Reinforcing Mechanisms in Economics," in: Anderson, Arrow and Pines (eds.), *The Economy As an Evolving Complex System* :9-32 [Series: Santa Fe Institute Studies in the Sciences of Complexity], Addison-Wesley, Redwood City, Calif., 1988.

Arthur, W. Brian *et al.*, "Path-dependent Processes and the Emergence of Macro-structure," *European Journal of Operations Research* 30, 1987 :294-303.

Ashby, W. Ross, *Design for a Brain*, John Wiley and Sons, New York, 1960. 2nd edition.

Ayres, Robert U., Resources, Environment and Economics, N.Y.: John Wiley and Sons, 1978.

Ayres, Robert U. and Allen V. Kneese, "Production, Consumption and Externalities," *American Economic Review*, June 1969. [Reprinted in *Benchmark Papers in Electrical Engineering and Computer Science*, Daltz and Pentell (eds.), Dowden, Hutchison and Ross, Stroudsuerg, 1974 and Bobbs-Merrill Reprint Series, New York 1].

Ayres, Robert U. and Allen V. Kneese, "Externalities: Economics and Thermodynamics," in: Archibugi and Nijkamp (eds.), *Economy and Ecology: Towards Sustainable Development*, Kluwer Academic Publishers, Netherlands, 1989.

Ayres, Robert U. and Katalin Martinas, *A Non-Equilibrium Evolutionary Economic Theory*,

Working Paper (WP-90-18), International Institute for Applied Systems Analysis, Laxenburg, Austria, 1990.

Ayres, Robert U. and Indira Nair, "Thermodynamics and Economics," *Physics Today* **37**, November 1984 :62-71.

Bausor, R., "Time and Equilibrium," in: Mirowski (ed.), *The Reconstruction of Economic Theory* :93, Kluwer-Nijhoff Publishing Company, Boston, 1986.

Bennett, Charles H. and Rolf Landauer, "The Fundamental Physical Limits of Computation," *Scientific American* **253**(1), 1985.

Birkhoff, G. D., "One Can Manage to Obtain a Variational Principle Appropriate to Almost Any Physical or Mathematical Theory," *Scientific Monthly* **58**, 1944 :54.

Bonner, J. T., *Morphogenesis: An Essay on Development*, Princeton University Press, Princeton, N.J., 1952.

Brody, Andras, Katalin Martinas and Konstantin Sajo, "Essay on Macro-economics," *Acta Oec.* **36**, 1985 :305.

Brundtland, G. H. (ed.), *Our Common Future*, Oxford University Press, New York, 1987. [Report of the WCED]

Bryant, J., "A Thermodynamic Approach to Economics," *Energy Economics* **4**, 1982 :36-50.

Burness, Stuart *et al.*, "Thermodynamics and Economic Concepts as Related to Resource-Use Policies," *Land Economics* **56**(1), February 1980 :1-9.

Casti, John L., "Simple Models, Catastrophes and Cycles," *Kybernetes* **13**, 1984 :213-229.

Casti, John L., *Newton, Aristotle and the Modelling of Living Systems*, Abisko Workshop on System Modelling, Abisko, Sweden, May 1987. [Written version of presentation]

Crutchfield, J. *et al.*, "Chaos," *Scientific American* **255**, December 1986 :46-57.

Dawkins, Richard, *The Extended Phenotype*, W. H. Freeman and Co., New York, 1982.

Debreu, Gerard, *Theory of Value*, John Wiley and Sons, New York, 1959.

de Chardin, Teilhard, *The Future of Man*, Publ., City, State, 1977. [Translated by Norman Denny, London, Collins, Fount paperbacks]

Doolittle, W. Ford, "Is Nature Really Motherly?", *Co-evolution Quarterly*, Spring 1981. [Also in Barlow, 1992, p. 235].

Ebeling, Werner and Rainer Feistel, "Physical Models of Evolution Processes," in: Krinsky (ed.), *Self-Organization: Autowaves and Structures Far from Equilibrium*, Chapter 6 :233-239, Springer-Verlag, New York, 1984.

England, Richard W., *Mechanical, Thermodynamic and Dissipative Phenomena: On the Relationship Between Economics and Physics*, Meeting on Ecological Economics, the Balaton Group, Csopak, Hungary, Aug. 30–Sept. 4, 1990.

Faber, Malte, Horst Niemes and Gunter Stephan, *Entropy, Environment and Resources*, Springer-Verlag, Berlin, 1987.

Faber, Malte and John L. R. Proops, "Interdisciplinary Research Between Economists and Physical Scientists: Retrospect and Prospect," *Kyklos* **38**(4), 1985 :599-616.

Faber, Malte and John L. R. Proops, "Time Irreversibilities in Economics," in: Faber (ed.), *Studies in Austrian Capital Theory, Investment and Time; Lecture Notes in Economics and Mathematical Systems*, Springer-Verlag, Berlin, 1986.

Faber, Malte and John L. R. Proops, *Evolution in Biology, Physics and Economics: A Conceptual Analysis*, Discussion Paper (131), University of Heidelberg Department of Economics, Heidelberg, Germany, January 1989. [Mimeo]

Fredkin, Edward and Tomasso Toffoli, "Conservative Logic," *International Journal of Theoretical Physics* **21**(3/4), 1982.

Georgescu-Roegen, Nicholas, *The Entropy Law and the Economic Process*, Harvard University Press, Cambridge, Mass., 1971.

Georgescu-Roegen, Nicholas, "The Economics of Production," in: *Energy and Economic Myths: Institutional and Analytic Economic Essays*, Pergamon Press, New York, 1976.

Georgescu-Roegen, Nicholas, "Myths About Energy and Matter," *Growth and Change* **10**(1), 1979.

Gleick, J., *Chaos: Making a New Science*, Viking Press, New York, 1987.

Higatsberger, Michael J., *The Genesis of the Concept of Entropy*, The Fifteenth International Conference on the Unity of the Sciences, Washington, D.C., November 1986.

Holland, John H. *Adaptation in Natural and Artificial Systems*, University of Michigan Press, Ann Arbor, Mich., 1975.

Holland, H. D., *The Chemical Evolution of the Atmosphere and the Oceans*, Princeton University Press, Princeton, N.J., 1984.

Holland, John H., "The Global Economy as an Adaptive Process," in: Anderson, Arrow and Pines (eds.), *The Economy As an Evolving Complex System* :117-124 [Series: Santa Fe Institute Studies in the Science of Complexity 5], Addison-Wesley, Redwood City, Calif., 1988.

Jensen, Roderick V., "Classical Chaos," *American Scientist* **75**, March/April 1987 :168-181.

Kauffman, Stuart A., "The Evolution of Economic Webs," in: Anderson, Arrow and Pines (eds.), *The Economy as an Evolving Complex System*, Chapter 2 :125-146 [Series: Santa Fe Institute Studies in the Science of Complexity 5], Addison-Wesley, Redwood City, Calif., 1988.

Kauffman, Stuart A. and S. Levin, "Towards a General Theory of Adaptive Walks on Rugged Landscapes," *Journal of Theoretical Biology* **128**, 1987 :11-45.

Kelsey, David, "The Economics of Chaos or the Chaos of Economics," *Oxford Economic Papers* **40**, 1988 :1-31.

Khalil, Elias L., "Entropy Law and Exhaustion of Natural Resources: Is Nicholas Georgescu-Roegen Defensible?" *Ecological Economics* **2**, 1990 :163-178.

Kneese, Allen V., Robert U. Ayres and Ralph d'Arge, *Aspects of Environmental Economics: A Materials Balance Approach*, Johns Hopkins University Press, Baltimore, 1970.

Koopmans, Tjalling C. (ed.), *Activity Analysis of Production and Allocation* [Series: Cowles Commission Monograph] (13), John Wiley and Sons, New York, 1951.

Le Roy, Edouard, *L'exigence idealiste et le fait de l'evolution*, Boivin, Paris, 1927.

Le Roy, Edouard, *Les origines humaines et l'evolution de l'intelligence*, Boivin, Paris, 1928.

Lorenz, E. N., "Deterministic Non-periodic Flow," *Journal of the Atmospheric Science* **20**, March 1963 :130-41.

Lotka, Alfred J., *Elements of Mathematical Biology*, Dover Publications, New York, 1956. Reprint edition. [Original title: *Elements of Physical Biology*, 1924]

Lovelock, James E., "Gaia As Seen Through the Atmosphere," *Atmospheric Environment* **6**, 1972 :579-580.

Lovelock, James E., *Gaia: A New Look at Life on Earth*, Oxford University Press, London, 1979.

Lovelock, James E. and Andrew J. Watson, "Biological Homeostasis of the Global Environment: The Parable of Daisy World," *Tellus* **358**, 1983 :264-289.

May, R. M., "Simple Mathematical Models with Very Complicated Dynamics," *Nature* **26**, June 1975 :459-67.

McGlashan, M. L., "The Use and Misuse of the Laws of Thermodynamics," *Journal of Chemical Education* **43**(5), 1966 :226-232.

McKenzie, Lionel W., "On Equilibrium in Graham's Model of World Trade and Other Competitive Systems," *Econometrica* **22**, 1954 :147-161.

Mirowski, Philip, *Physics and the Marginalist Revolution*, 1984.

Montroll, Elliott W. and K. E. Shuler, "Dynamics of Technological Evolution: Random Walk Model for the Research Enterprise," *Proceedings of the National Academy of Sciences* **76**, 1979 :6030-6034.

Ott, Edward, "Strange Attractors and Chaotic Motions of Dynamical Systems," *Review of Modern Physics* **53**, 1981 :655.

Perrings, Charles, *Economy and Environment: A Theoretical Essay on the Interdependence of Economic and Environmental Systems*, Cambridge University Press, Cambridge, England, 1987.

Peschel, Manfred and Werner Mende, *The Predator-Prey Model: Do We Live in a Volterra World?* Akademie-Verlag, Berlin, 1986.

Prigogine, Ilya, Gregoire Nicolis and A. Babloyantz, "Thermodynamics of Evolution," *Physics Today* **23** (11/12), November/December 1972 :23-28(Nov.) and 38-44(Dec.).

Radzicki, Michael J., "A Note on Kelsey's The Economics of Chaos or the Chaos of Economics," *Oxford Economic Papers* **40**, 1988 :692-693.

Roessler, O. E., "An Equation for Continuous Chaos," *Physics Letters* **57A** July 1976 :397-398.

Rostow, W. W., *The Stages of Economic Growth*, Cambridge University Press, Cambridge, England, 1960.

Samuelson, Paul A., "Rigorous Observational Positivism: Klein's Envelope Aggregation; Thermodynamics and Economic Isomorphisms," in: Adams and Hickman (eds.), *Global Econometrics: Essays in Honor of Lawrence R. Klein* :1-38, MIT Press, Cambridge, Mass., 1983.

Schneider, Stephen H. and Randi Londer, *The Coevolution of Climate and Life*, Sierra Club Books, San Francisco, 1984.

Schwefel, Hans-Paul, *Collective Intelligence in Evolving Systems*, Discussion Paper, Informatics Center, University of Dortmund, Dortmund, Germany, 1989*a*.

Schwefel, Hans-Paul, *Collective Learning in Evolutionary Processes*, International Symposium on

Evolutionary Dynamics and Non-Linear Economics, University of Texas, Austin, Tex., April 1989*b*.

Shannon, Claude E., "A Mathematical Theory of Communication," *Bell Systems Technical Journal* **27**, 1948.

Shinbrot, Marvin, "Things Fall Apart," *The Sciences*, N.Y. Academy of Sciences, May/June 1987 :32-38.

Simon, Herbert A., "A Behavioral Model of Rational Choice," *Quarterly Journal of Economics* **69**, 1955 :99-118. [Reprinted in *Models of Man*, John Wiley and Sons, New York, 1957]

Simon, Herbert A., "Theories of Decision-Making in Economics," *American Economic Review* **49**, 1959 :253-283.

Simon, Herbert A., *Models of Bounded Rationality*, MIT Press, Cambridge, Mass., 1982.

Smale, Stephen, "Stability and Isotropy in Discrete Dynamical Systems," in: Peixoto (ed.), *Dynamical Systems*, Academic Press, New York, 1973.

Soddy, Frederick, *Cartesian Economics*, Hendersons, London, 1922.

Soddy, Frederick, "Wealth, Virtual Wealth and Debt," in: *Masterworks of Economics: Digests of 10 Classics*, Dutton, New York, 1933.

Suess, Edouard, *The Face of the Earth*, Clarendon Press, Oxford, England, 1900-24. [5 volumes. Translated from the German by Hertha B. C. Sollas]

Ulanowicz, Robert E., *Growth and Development: Ecosystems Phenomenology*, Springer-Verlag, New York, 1986.

Varela, Francisco G. J., "A Calculus of Self-Reference, *International Journal of General Systems* **2**, 1975 :5-24.

Vernadsky, Vladimir I., *La geochimie*, Felix Alcan, Paris, 1924.

Vernadsky, Vladimir I., "The Biosphere and the Noosphere," *American Scientist* **33**, 1945 :1-12.

Ville, Jean, "The Existence Conditions of a Total Utility Function," *Review of Economic Studies* **19**, 1951 :123-128.

Volterra, V., "Principes de biologie mathématique," *Acta Biotheoretica* **3**(1), 1937 :18-30.

von Foerster, Heinz and G. W. Zopf Jr. (eds.), *Principles of Self-Organization*, Pergamon Press, New York, 1962.

Walras, Leon, *Elements d'economie politique pure*, Corbaz, Lausanne, 1874.

Wiener, Norbert, *Cybernetics: Control and Communications in the Animal and the Machine*, John Wiley and Sons, New York, 1948.

Wittmann, W., *Produktionstheorie*, Springer-Verlag, Heidelberg, Germany, 1968.

Wright, Paul K., "Entropy and Disorder," *Contemporary Physics* **11**(6), 1970 :581-588.

Chapter 2

Information and Thermodynamics

2.1. THEORETICAL FRAMEWORK: WHAT IS INFORMATION?[1]

Information is commonly used in three ways: (1) in the semantic sense (as data), (2) in the pragmatic sense of "knowledge," and (3) in a formal technical sense as the resolution of doubt or uncertainty.[2] In the following sections I will introduce a more complete typology of the third type of information with two major categories: **D-information** (basically Shannonian),[3] and a subset relevant to evolutionary selection or survival, called **SR-information**, to be defined in section 5. I will attempt to explain how the more traditional usages can be fitted into the new typology. Knowledge and data, for instance, are both relevant to selection in the human sphere.

Information in the Shannonian sense is a function of the *a priori* probability of a given state or outcome among the universe of physically possible states. (This type of information is a measure of **distinguishability** or **doubt**, hence, **D-information**.) The higher the number of physically possible alternative states, the more a given selection or set of equivalent selections will reduce uncertainty.

As a matter of historical interest, the first formal definition of D-information H was proposed by an engineer at Bell Labs [Hartley, 1928] in the context of telegraphic communications:

$$H = -K\log_2 f, \tag{2.1}$$

where f is the frequency with which a given code element (for example, a letter of the alphabet expressed in Morse code) appears in a given message. Here the frequency f is defined naturally as $f = Z_1/Z_0$, where Z_1 is the number of times the code element (e.g., letter) occurs in the message and Z_0 is the total number of elements in the message. However, the Hartley definition is not easily generalizable to other situations and has been resurrected only in discussions of Shannon's more complete theory.

To proceed logically, it is helpful to step back from the original preoccupation of communications engineers with channel capacity, codes, and messages and simply consider an experiment with W possible different outcomes. (If the outcomes are equally likely, then $f = 1/W$). We seek to define a measure of uncertainty $S = S(W) = S(1/f)$ with three properties:

$$(i) \quad S \geqslant 0;$$

$$(ii) \quad S(1) = 0;$$

$$(iii) \quad S(AW) = S(A) + S(W).$$

The first (non-negativity) condition is included only for convenience. The second is the requirement that when there is only one possible outcome ($W = 1$), there is no uncertainty ($S = 0$) The third condition is the requirement of additivity of uncertainty. If two independent experiments can generate A outcomes and W outcomes, respectively, their combined uncertainty must be the sum of the individual uncertainties. It can be proved rigorously that the only mathematical function meeting these conditions is the logarithmic form $S \simeq \log W$. This is equivalent to (2.1) [Yaglom and Yaglom, 1983]. Thus, Hartley's definition of **information** is, essentially, the complement of a natural definition of **uncertainty**.

Shannon defined information formally as the reduction of uncertainty or, more precisely, as the difference between uncertainties in two situations X, X'

$$H = S\langle Q|X \rangle - S\langle Q|X' \rangle, \tag{2.2}$$

where $S(Q|X)$ is the uncertainty with regard to question Q in situation X. The question Q can be anything. In the special case of W equally probable outcomes, the uncertainty function S has already been identified above as $S = K \log W$, which is formally identical to Boltzmann's definition of entropy, except for the coefficient K.

The constant K in equation (2.1) is a scale factor that is fixed as soon as one selects a unit of measurement of information. We can permanently fix $K = 1$ by defining a *bit* of information as the amount provided by the elimination of doubt between two equally probable outcomes. The number of bits of information required to solve a problem (e.g., to decode a message) corresponds exactly to the minimum number of distinct binary decisions necessary to make the correct final choice. This is a function only of the number of possible binary choices and the *a priori* probability of each outcome.

Imagine the possibilities are *cells* (in a dungeon, for example), and that one seeks a particular prisoner known to be in one of them. Suppose the searcher has no other information, whence all cells can be assumed to be equally probable *a priori*. (This is a good illustration of the pragmatic use of the word *information* as knowledge.) Assume W cells in the dungeon. There are many alternative search strategies. One of the simplest is sequential interrogation of the jailer: *Is the prisoner in cell 1? Is the prisoner in cell 2?...* The searcher might be lucky and hit on the correct choice immediately. Or, conversely, he might be unlucky and not find the prisoner until the next-to-last ($W - $ 1st) question. These outcomes are equally probable. If this sequential strategy were adopted, the average (or expected) number of questions required would be $W/2$.

On reflection, a much more efficient strategy for uncertainty reduction presents itself. It is the well-known "tree search." The searcher could ask questions that reduce the number of remaining possibilities by a factor of 2 each time. For instance, if the cells are numbered $1,...,W$, one could ask: *Is the prisoner located in the subset of cells from 1 to $W/2$ inclusive?* Whether the answer is yes or no, the uncertainty is reduced by a factor of 2. If the answer is yes, the prisoner is known to be in the first group of addresses $(1,2,...,W/2)$; if the answer is no, the prisoner must be in the second half $(W/2 + 1,...,W)$ In either case, the same procedure is repeated until two possibilities remain. The information gained by each of these questions is exactly 1 bit if the number of possibilities remaining at that stage is divisible by 2 and slightly less than 1 bit if not. The smallest number of such questions (or decisions) is $\log_2 W$. (Obviously, if W is not an even number, $W/2$ might not be an integer; but for large values of W, this complication can be ignored.)

In the real world, one cannot always proceed quite so efficiently, because possible states are seldom, in fact, equally probable. For instance, consider the game Twenty Questions. Suppose you are thinking of a famous author, and I have to identify that author. Possible questions might include

Is the author male?
Does his/her name begin with letters A through M (the first half of the alphabet)?
Is the author living?
Does (did) the author live in England?

In all cases, the *a priori* probability of a *yes* answer is likely to be different from the probability of a *no*. For instance, considerably more than half of all famous authors are male, and well over half have names beginning with letters in the first half of the alphabet. On the other hand, probably fewer than half of all famous authors are living. Thus, the values of *yes* versus *no* answers are unequal in terms of reducing uncertainty, and the information elicited by each question is normally either more than 1 bit or less than 1 bit, depending on the case.

Let P_i be the *a priori* probability of the ith state (or event). The information provided by an actual realization of that event (i.e., the answer to the question is yes) therefore reduces uncertainty (adds information) by an amount

$$H_i = \log_2\left(\frac{1}{P_i}\right) = -\log_2 P_i. \tag{2.3}$$

For example, if 3/4 of all famous writers are male, then a "yes" is worth less than 1 bit, because the number of possibilities was cut by a factor of 1/4 rather than by a factor of 1/2. In fact, the value of a *yes* in this case is $\log_2(4/3)$

$= 2 - \log_2 3 = 0.4147$ bits. On the other hand, the value of a *no* is $\log_2 4 = 2$ bits! The information provided by a series of affirmative answers is the simple sum of their individual information values, provided the probabilities are truly independent of one another. For non-independent states i, j, one must adjust by adding or subtracting a term representing the information value corresponding to the deviation of joint probability from a simple product of probabilities. While the assumption of independence is generally adequate, one must always be alert for the exceptional cases.

Returning for a moment to the Twenty Questions example, it was pointed out that the values of a *yes* and a *no* are generally unequal. It is therefore natural to ask: What is the *probable* or *expected* amount of information elicited by each question? Evidently, the information value of each outcome must be weighted by its probability, P_i. It is helpful to note that probabilities add up to unity:

$$\sum_i P_i = 1, \quad i = 1,...,N. \tag{2.4}$$

It follows that

$$\langle H \rangle \cong - \langle \log_2 P_i \rangle \cong \frac{\Sigma_i P_i \log_2 P_i}{\Sigma_i P_i} \cong - \sum_i P_i \log_2 P_i. \tag{2.5}$$

This expression for expected information was first introduced by Claude Shannon [1948] of Bell Telephone Laboratories, also in the context of communications.

As an illustration of this formula, consider the question *Is the famous author (you are thinking of) male?* If, in fact, 3/4 of all famous authors are male, then the answer *yes* on your part adds 0.4147 bits of information to my stock, whereas the answer *no* is worth 2 bits. But *yes* has a probability of 3/4, while *no* has a probability of 1/4. Thus, the expected amount of information to be gained from this question is

$$\frac{3}{4} *0.4147 + \frac{1}{4} *2 = 0.811 \text{ bits}$$

or somewhat less than 1 bit.

This way of looking at information opens a link to physics via statistical mechanics. One may consider the case of a homogenous physical system (of atoms or molecules) that is changing or evolving over time. At an given moment it is in a definite state, but a moment later it will be in another state. Let P_i be a set of state probabilities ($P_i \geqslant 0$), satisfying (2.4). When a definite observation is made, confirming that event i (among all possible events) has occurred (i.e., the system is in the ith state), the information gained by that observation is given by (2.3). But the original probability of that event (state) was only P_i. Another measurement might find the system in quite a different state. The expected amount of information provided by an observation of the system is the average value of the expression (2.3) over many observations. This average is exactly (2.5).

In reality, we have no means of observing the microstates of large systems (e.g, gases) so precisely that the exact state of each constituent molecule can be determined at a moment in time. One can say, however, how much information would be provided by such an observation, provided all microstates are equally probable and that the total number of such microstates is W. In this case, the information value of the (hypothetical) observation would be (from equation 2.3),

$$H \cong - \log_2\left(\frac{1}{W}\right) = \log_2 W. \qquad (2.6)$$

This is the maximum value of uncertainty. In the more realistic case of unequal probabilities, Shannon's formula (2.5) must be used. It can be evaluated numerically if an expression for the probability distribution of states is available.[4]

If an *a priori* probability distribution $\hat{p}_i$ exists (where $\hat{p}_i$ is the estimated or assumed probability of occurrence of the ith outcome), then actual measurement of the true probabilities p_i or frequencies will reduce the uncertainty by an amount

$$H \cong - [\log\hat{p}_i - \log p_i] \cong \log\left(\frac{p_i}{\hat{p}_i}\right). \qquad (2.7)$$

This expression, first introduced by Kullbach [1951] has been called *surprisal* (a term introduced by R. Levine) in information theory. It is invariably non-negative [Aczel and Daroczy, 1975].

2.2. JOINT AND CONDITIONAL PROBABILITIES

For reasons that will be clear shortly, it is particularly interesting to consider **networks** from an informational perspective. To be concrete, consider a network consisting of N Christmas tree lights that can be on or off. The network (system of lights) can be in different macrostates at different times. Consider two macrostates, b and a. (For concreteness, a occurs before b). Now define

$p(a_j)$ = probability that the jth light is lit in macrostate a.
$p(b_i)$ = probability that the ith light is lit in macrostate b (later).
$p(a_j b_i)$ = combined or joint probability that the jth light is lit in macrostate a **and** the ith light is lit in macrostate b.
$p(b_i|a_j)$ = conditional probability that the ith light in macro state b is lit *if* the jth light was lit in macrostate a.

Bayes's theorem states that the conditional probability is given by

$$P(b_i|a_j) = \frac{P(a_j,b_i)}{p(a_i)} \tag{2.8}$$

which is equivalent to

$$P(a_j,b_i) = P(a_j)p(b_i|a_j). \tag{2.9}$$

The *a priori* uncertainty with regard to the condition of the ith light in state b (from 2.3) is $-\log p(b_i)$. This can be reduced by resolving the uncertainty with regard to the condition of the jth light in (prior) state a, assuming we know the conditional probabilities relating the two, viz.,

$$H^{ij} = -\log p(b_i) - [-\log(b_i|a_j)]$$

$$= \log(b_i|a_j) - \log p(b_i)$$

$$= \log\left(\frac{b_i|a_j}{p(b_i)}\right)$$

$$= \log\left(\frac{p(a_j,b_i)}{p(a_j)p(b_i)}\right). \tag{2.10}$$

Averaging over all possibilities

$$\langle H(b,a)\rangle = \sum_{i,j} p(a_j,b_i)\log\left(\frac{p(a_i,b_i)}{p(a_j)p(b_i)}\right) \tag{2.11}$$

which is always non-negative.

In Chapter 8 (section 2) it is argued that manufacturing can be considered a kind of information-transformation system. The shaping, forming, finishing, joining, and assembly of parts are all concerned with the imprinting, or embodiment, of morphological information into or onto materials. A natural extension of this idea is to consider how the manufacturing process might be described in the language of communications. This conceptualization opens the way to using some of the methodological tools that have been developed in information theory.

For example, a machine tool can be thought of as a communications channel. It "operates on a variable x to produce a specified value of that variable ... if a certain value, say $x=L$, is chosen ... the machine will produce values randomly but in such a way that their mean (e.g., L) and variance are known.... The question ... is: How much information can such a channel process?" [Wilson, 1980, p. 205]. This question is now in a form that can be addressed using standard results from probability theory, using the known characteristics of the probability distribution of outputs. The average information transferred by the machine is defined by a continuous-variable version of equation (2.11),

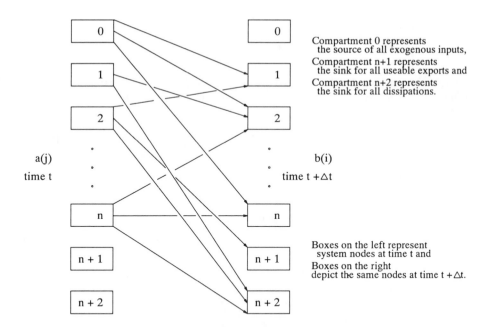

FIGURE 2.1. Temporal Representations of Flows Among the Compartments of a System. Source: Reproduced with Permission from Hirata and Ulanowicz, 1984.

converting summations to integrals. The channel capacity is the maximum value of this number.

At this point it is useful to introduce another interpretation of the network and the macrostates *a, b*. The network can be generalized from Christmas tree lights to a processing system consisting of a source, generalized *nodes* where processing takes place, and two *sinks*, one for *useful outputs* and one for *waste outputs* (Figure 2.1). One can also envision a conserved *medium of exchange* that flows through the entire system, beginning with the source and ending at one of the sinks. (The medium of exchange could be energy, water, electricity, "mass," or money.)

Instead of a light being "on" or "off" in the *i*th or *j*th node, suppose a quantum of the medium is present or not present. The probability definitions given above are easily modified by substituting quantities or flow rates for abstract probabilities.

Let X_{ij} define the flow increment (quantity in time increment dt) from the *i*th node to the *j*th node. The sum of all instantaneous outputs from the *j*th node is

$$X_j = \sum_{i=0}^{N+2} X_{ji} \qquad (2.12)$$

while the sum of all inputs to the *i*th node is

$$X_i' = \sum_{j=0}^{N+2} X_{ji} \tag{2.13}$$

(where the prime indicates that a very short period of time dt has elapsed). The total throughput of the system remains constant, by assumption, so

$$T = \sum_{j=0}^{N+2} X_j = \sum_{i=0}^{N+2} X_i'. \tag{2.14}$$

In effect, one can now equate flows with probabilities:

$$p(a_j) \to \frac{X_j}{T},$$

$$p(b_j) \to \frac{X_j}{T},$$

$$p(a_j b_i) \to \frac{X_{ji}}{T},$$

$$p(b_i | a_j) = \frac{p(a_j b_i)}{p(a_j)} \to \frac{X_{ji}}{X_j}.$$

Substituting into (2.11) yields

$$\begin{aligned}
\langle H(b,a) \rangle &= \sum_{i,j}^{N+2} \left(\frac{X_{ji}}{T}\right) \log\left(\frac{X_{ji}T}{X_j X_i}\right) \\
&= \sum_{i,j}^{N+2} \left(\frac{X_{ji}}{T}\right) \log\left(\frac{X_{ji}}{T}\right) - \sum_{j}^{N+2} \left(\frac{X_j}{T}\right) \log\left(\frac{X_j}{T}\right) - \sum_{i}^{N+2} \left(\frac{X_i}{T}\right) \log\left(\frac{X_i}{T}\right),
\end{aligned} \tag{2.15}$$

utilizing (2.12) and (2.13).

It can be proved that the first term is always non-positive, the second and third terms are both non-negative, and the expression as a whole is also non-negative [Aczel and Daroczy, 1975].

The first application of the network-information concept to the problem of ecosystem stability (resilience) was MacArthur's [1955], based on an earlier suggestion by Odum [1953]. The network-information perspective has been revived and elaborated by Ulanowicz [1986], who defines a new variable, which he identifies with the ecological concept of **ascendancy** A of an ecosystem, to be the product

$$A = \langle H \rangle T, \tag{2.16}$$

where $\langle H \rangle$ is the *network information* and T is the *energy throughput*. Ulanowicz suggests that (2.16) is a generalization of Lotka's principle (see Chapter 3, section 8). In the form (2.15), in particular, the first term (non-positive) is interpreted by Ulanowicz to be a measure of the *resiliency* of the system, while the second and third terms (non-negative) can be interpreted as measures of the *magnitude* of the system. The combined expression reflects an apparent tradeoff between resiliency and efficiency, and suggests that an appropriate maximization principle for biology should combine the two factors. It must be emphasized that (2.16) is a hypothetical relationship. Its value in ecosystem analysis remains to be tested empirically.

In this section I have presented the network-information formulation in a perfectly general form. It can be applied to a wide variety of other systems, including economic systems. Thus, it would be interesting, for instance, to see whether an expression similar to (2.16) could be given a reasonable economic interpretation. This sort of questions will be considered later.

2.3. INFORMATION AND ENTROPY

In Chapter 1 it was pointed out that entropy was introduced originally to thermodynamics as a generalized measure of energy dissipation. Clausius, who needed to state the second law of thermodynamics rigorously, defined entropy S in differential form,

$$dS = \frac{dQ}{T}, \tag{2.17}$$

where dS is the change in entropy in a closed system due to a reversible physical process in which a differential quantity of heat dQ flows along a gradient from a higher temperature to a lower temperature. Entropy is a maximum when all temperature gradients are eliminated and heat ceases to flow. This is the state of equilibrium.

In 1872 Boltzmann stated his famous H-theorem, which defined entropy in terms of the number of possible microstates ($S = k \log W$) and proved that entropy will tend to increase as the distribution of microstates becomes more predictable (i.e., during approach to equilibrium). Meanwhile, James Clerk Maxwell in 1871 raised the question whether "intelligence" offered a possible way to avoid the implications of the second law of thermodynamics.

Maxwell's idea was to suppose an intelligent being "whose faculties are so sharpened that he can follow every molecule in its course." He also supposed a container divided into two parts, A and B, connected by a small hole. Imagine that this being (Maxwell's Demon), which is capable of seeing and determining the velocity of each molecule, is positioned where "he opens and closes this hole so as to allow only the swifter molecules to pass from A to B, and only the slower ones to pass from B to A. He will thus, without expenditure of work, raise the temperature of B and lower that of A in contradiction to the second law of thermodynamics" [Maxwell, 1871].

A resolution of this apparent paradox was offered by Leo Szilard [1929] and refined by Leon Brillouin [1951]. The basic argument they made was that the "intelligent being" postulated by Maxwell cannot distinguish between swift molecules and less swift ones without sensory information. Sensory information requires that the molecules be "illuminated" by electromagnetic radiation. This, in turn, requires energy. Thus, the Demon must necessarily *expend* useful energy to *gain* information. This was perhaps the first really clear link between information and entropy. It led Brillouin to identify information as *negative entropy* or *negentropy*.

A link between Bayes's theorem in probability reasoning from cause to effect and the second law of thermodynamics was postulated by J. P. Van der Waals in 1911 [see Tribus and McIrvine, 1971]. The link between entropy and information was also emphasized in 1930 by G. N. Lewis, who asserted flatly that "gain in entropy means loss in information—nothing more" [*ibid.*]. The modern probabilistic restatement of Boltzmann's theorem is due to Slater [1938], who used the definition

$$S = -k \sum_i f_i \ln f_i, \qquad (2.18)$$

where f_i is the fraction in microstate i. (Obviously, the fraction f_i in the microstate is normally equal to the probability of being in that state.)

In 1948 Shannon identified his expression for expected uncertainty[5] (2.5) as entropy. Consistency between the two definitions of entropy was demonstrated by Brillouin [1951, 1953, 1962], but he did not attempt to derive either one from the other. This last step was taken by E. T. Jaynes [1957a,b], who showed that Shannon's measure of uncertainty can be taken as a primitive and used to derive state-probabilities.

Jaynes introduced a formal entropy-maximization principle, which Myron Tribus [1961a,b] subsequently used to show that all the laws of classical thermodynamics can also be derived from Shannon's uncertainty measures. In summary, it is now established to the satisfaction of virtually all physicists that information is the reduction of uncertainty and that uncertainty and entropy are essentially identical (not mere analogs). Nevertheless, the controversy has not been completely settled. One of the problems is that information (in its normal sense, at least) is qualitative — an *intensive* variable, in thermodynamic terms— whereas entropy is an *extensive* variable [Spreng, 1990]. I think the problem can be resolved by introducing the notion of *useful information*, or SR-information, as is done later in this book.

2.4. THERMODYNAMIC INFORMATION

Assume a thermodynamic subsystem with internal energy U, volume V, pressure P, temperature T, and molar composition n_i, embedded in an environment (which is itself in thermal equilibrium) with pressure P_0, temperature T_0, and chemical potential v_{i0}. Then S is the entropy (uncertainty) of the subsystem as long as it is distinguishable, and S_0 is the entropy it would have if it reached equilibrium with its environment $(P \to P_0, T \to T_0)$. Thermodynamic information H can now be defined for an initially distinguishable material subsystem as follows:

$$H = S_0 - S = (S_0 - S_f) + (S_f - S). \qquad (2.19)$$

Here S_f is the entropy of the same material system after it has reached thermal equilibrium with the environment, but without any material exchange with the environment (i.e., no diffusion); S_0 is the entropy of the final diffused state. Evidently, H is a measure of the initial *distinguishability* of the subsystem from its environment (i.e., before the process starts).

From classical thermodynamics it can be proved that the entropy of the final equilibrium state when the subsystem is indistinguishable from its environment can be written as

$$S_0 = \frac{U + P_0 V - \Sigma_i v_{i0} n_i}{T_0}. \qquad (2.20)$$

Now, subtracting S from both sides and using (2.19), one obtains

$$H = S_0 - S = \frac{U + P_0 V - T_0 S - \Sigma_i v_{i0} n_i}{T_0}. \qquad (2.21)$$

Multiplying both sides of (2.21) by T_0 yields a new variable that we denote B:

$$B = H T_0 = U + P_0 V - T_0 S - \sum_i v_{i0} n_i. \qquad (2.22)$$

The quantity B, first discovered by Josiah Willard Gibbs in 1878 [see Gibbs, 1928], was rediscovered in 1965 by Robert Evans. Evans later proved that this expression is, in fact, the most general form of thermodynamic potential and the most general measure of **available useful work**, sometimes called **essergy** [Evans, 1969]. (Note, by the way, that in the equilibrium case, when $B = 0$ equation (2.22) simply reduces to (2.20). All other thermodynamic potentials are also special cases.) B was defined, initially, as a measure of information. As such, B is also the most general measure of disequilibrium (or distinguishability) for any system describable in terms of thermodynamic variables.

It is convenient, however, to split the right-hand side of (2.21) into two parts, so that thermodynamic information H has two components,

$$H_{\text{therm}} = H_{\text{chem}} + H_{\text{diff}} = H_{\text{chem}} - H_{\text{conc}} \tag{2.23}$$

where

$$H_{\text{chem}} = \frac{U + P_0 V - T_0 S}{T_0} = \frac{Q_c}{T_0} - S, \tag{2.24}$$

where Q_c is the heat of combustion (see Table 2.2) and

$$H_{\text{conc}} = \frac{1}{T_0} \sum n_i \nu_{i0} = -H_{\text{diff}}. \tag{2.25}$$

The first component H_{chem}, defined by (2.24), can be regarded as the information lost (or gained) as a result of thermal and/or chemical processes alone. Similarly, H_{conc} is the information gained by concentration or "unmixing" processes (or lost by diffusion or mixing processes). Note, for future reference, that a physical condensation involving phase separation is also, in entropic terms, an "unmixing" process.

As it happens, the concentration-diffusion information (2.25) is the easiest to compute, at least for ideal gases. It is also the *only* term of interest in a few special cases. Gold mining and diamond mining, for instance, are examples of pure physical concentration activities with no chemical transformations involved. This is also true of most metal ore beneficiation processes (e.g., by flotation), although beneficiation is generally followed immediately by a chemical transformation (reduction).

The information term $H_{\text{conc}} = -H_{\text{diff}}$ associated with increased concentration (or unmixing) can be computed by making use of Boltzmann's statistical definition of entropy of diffusion. For an ideal gas, this has been shown to be

$$H_{\text{conc}} = R \sum_i n_i \ln\left(\frac{X_{ci}}{X_{0i}}\right), \tag{2.26}$$

where R is the ideal constant, n_i is the number of moles of the ith molecular species being diffused, X_{ci} is the mole fraction of the ith species in the concentrated state before diffusion, and X_{0i} is the mole fraction of the ith species in the diffused state.[6] Evidently (2.26) represents the amount of information lost when a concentrated material, otherwise in equilibrium, is diffused into an environmental sink, such as the oceans or atmosphere. Obviously, the same expression can be interpreted as the (thermodynamic) information initially *embodied in the concentrated material*. See Table 2.1 for examples of information embodied by physical concentration and potentially lost by diffusion.

Let us now consider the other component of distinguishability information, as shown in equation (2.24). The most important chemical process in the industrial sphere is combustion. Electric power generation, process heat, and space heat all involve combustion of hydrocarbon fuels in air. So-called carbothermic reduc-

TABLE 2.1. *Information Added by Physical Separation (or Lost by Diffusion), Using the Ideal Gas Approximation.*

Element or Compound	Concentration or Diffusion Process	X_c Concentrated Mole Fraction	X_0 Dispersed Mole Fraction	X_c/X_0	$H_{cal}/°K$ Per Mole
Oxygen	Separation from air	1	0.21	4.75	+3.09
Nitrogen	Separation from air	1	0.79	1.266	+0.47
Helium	Separation from air	1	$5*10^{-6}$	$2*10^{-5}$	+24.24
Helium	Separation from natural gas (0.3% by vol.)	1	$3*10^{-3}$	$3.3*10^{-2}$	+6.95
CO_2	Separation from coke combustion products	1	0.21	4.75	+3.09
CO_2	Diffusion to air from coke combustion products	0.21	$3*10^{-4}$	700	−13.00
NaCl	Separation from seawater	1	10^{-2}	10^2	+9.13
MgCl	Separation from seawater	1	10^{-3}	10^3	+13.70
Gold	Separation from seawater (0.11 ppb by wt.)	1	10^{-12}	10^{12}	+54.79
Gold	Separation from average ore (7 gm/ton)	1	10^{-6}	10^6	+27.39
Gold	Dispersion to earth's crust	1	$10*10^{-10}$	$2*10^9$	−53.59

Source: Calculated by author based on data from the U.S. Geological Survey and the U.S. Bureau of Mines [Ayres, 1987].

tion of metal oxides can also be considered, for purposes of this analysis, as a combustion process, where the oxidant is an oxide ore rather than pure oxygen.

The **heat of combustion**, introduced in equation (2.24), is normally denoted as

TABLE 2.2. *Heat of Combustion Q_c & Available Work B, Assuming Water Product Condensed to Liquid Form (High Heating Value). Energies are in kcal/mole fuel.*

Fuel	Formula	Heat of Combustion Q_c	Available Work* B	% Change $100(Q_c-B)/Q_c$
Hydrogen	H_2	68.2	55.9	−18.0%
Ethane	C_2H_6	373	349	−6.4%
Propane	C_3H_8	531	502	−5.5%
Ethylene	C_2H_4	337	317	−5.9%
Liquid Octane	C_8H_{18}	1307	1261	−3.5%

*This is the "basic" available work, not including a diffusion contribution.

ΔH_c—but to avoid confusion with H, our measure of information, we use the symbol Q_c.[7] It is tabulated in standard references, such as the *Chemical Engineers Handbook* [Perry and Chilton, 1973] and is defined as the sum

$$Q_c = \Delta U + P_0 \Delta V, \qquad (2.27)$$

which corresponds to the first two terms on the right-hand side of equations (2.22) and (2.24). Some representative values for fuels are given in Table 2.2. Heat of combustion is a special case of **heat of formation**, which is applicable to processes other than oxidation/reduction.

Most combustion and reduction processes (except in internal combustion engines) occur at constant pressure the pressure of the atmosphere so P_0 is fixed. In cases when fuel and oxidant are both initially in vapor form, there is no change in volume ($\Delta V = 0$), since n moles of inputs are converted into exactly n moles of outputs (assuming H_2O in the combustion products to be in vapor form). However, for liquid and solid fuels the volume of reaction products exceeds the volume of inputs. This means some work has to be done on the atmosphere to make room for the added volume of the reaction products. This work is not recoverable or "useful." If the volume of reaction products (at standard temperature and pressure) is smaller than the volume of inputs (fuel plus oxidant), the atmosphere actually does work on the system. This work can be "bootlegged," in principle, and becomes part of the available work or essergy.

In any case, the first two terms on the right-hand side of (2.24) can be computed easily and are tabulated in standard references. It remains to compute the third term, the entropy change of the system ΔS due to an irreversible heating process. This can be broken down into two components, as follows:

$$\Delta S = \Delta S_1 + \Delta S_2, \qquad (2.28)$$

where ΔS_1 results from the change in composition from reactants to products (mixing and virtual dispersion of reactants, plus virtual concentration of reaction products), while ΔS_2 results from heating of the reaction products in the flame. The first term ΔS_1 is, in general, positive but quite small (0.5% assuming H_2O as vapor), arising mainly from the initial mixing of fuel and air. The basic formula for entropy of mixing has been given in (2.26) and only the numerical values remain to be determined.

The second term in (2.28) is quantitatively dominant in almost all cases. To calculate the entropy of heating (ΔS_2), one may recall the standard textbook definition of entropy change in terms of heat flow:

$$\Delta S_2 = \int_{T_0}^{T_c} \frac{dQ}{T}, \qquad (2.29)$$

where the integration is carried out from T_0 to the combustion temperature T_c. Now, subject to very general conditions, one can express heat flow rate dQ in terms of dT and dP:

$$dQ = \left(\frac{\delta Q}{\delta T}\right)_p dT + \left(\frac{\delta Q}{\delta p}\right)_T dP = C_p dT + \Lambda_p dP. \qquad (2.30)$$

Here, C_p is defined as the heat capacity at constant pressure (a function of temperature only) and Λ_p is the so-called latent heat of change of pressure.

There are two important cases to consider. The first is **isobaric** heating, i.e., heating by combustion at constant (atmospheric) pressure. This applies to cooking, baking, space heating, and water heating. It also applies, strictly speaking, to the combustion process in industrial boilers and steam power plants, although most thermodynamic analyses of steam (or other external combustion) engines neglect this aspect. The second case, to be considered later, is that of non-isobaric heating. It occurs mainly in internal combustion engines (Otto cycle, Diesel cycle, and Brayton cycle).

For isobaric (constant pressure) heating, $dP = 0$ and the second term of (2.30) can be dropped. Also, as a reasonable first approximation, one can usually assume a constant average value for C_p over the temperature range in question. Thus, in this case

$$\Delta S_2 = nC_p \int_{T_0}^{T_c} \frac{dT}{T} = nC_p \ln \frac{T_c}{T_0}, \qquad (2.31)$$

where T_c is the combustion temperature, n is the number of moles of combustion products, and

$$C_p = aR, \qquad (2.32)$$

where R is the ideal-gas constant.

One can derive an approximate expression for the combustion temperature T_c:

$$T_c = T_0\left(1 + \frac{h}{na}\right), \qquad (2.33)$$

where h and a are measured constants. As a reasonable approximation for hydrocarbon fuels: $h = 19.5M$, $a = 4.8$, $n = 0.6M$, and $h/na \approx 6.77$, where M is the molecular weight of the hydrocarbon.

A useful general approximation for the loss of available useful work by combustion at constant pressure, not including any contribution from the diffusion of combustion products into the atmosphere, follows directly from (2.31), (2.32), and (2.33):

$$B_c = B\left[1 - \frac{na}{h} \ln\left(1 + \frac{h}{na}\right)\right] \approx 0.7B. \qquad (2.34)$$

In words, even if we make no allowance for the diffusion of combustion products into the atmosphere, about 30 percent of the initially available chemical energy B becomes unavailable in any combustion process simply as a result of the entropic loss due to isobaric heating of combustion products.

From the structure of (2.30) it can be seen, incidentally, that the isobaric case is the most favorable, because any change in pressure $(dP > 0)$ will increase the magnitude of dQ and hence of ΔS_2 (2.29).

2.5. SURVIVAL-RELEVANT INFORMATION

Uncertainty-reducing information, or distinguishability information, which I have denoted **D-information**, is a quantity that exists independently of any reference system or observer. It is simply a measure of the non-randomness (or inherent distinguishability) of subsystems, given any specified boundaries. D-information can be regarded as a fundamental variable for describing the natural world. In bounded systems where thermodynamic variables are definable, D-information is proportional to negative entropy. It therefore has the same dimensions as entropy and is an extensive variable, like entropy.

Whether D-information is a conserved quantity is still an open question. If information can be equated to negative entropy (as most physicists believe), then the global increase in entropy predicted by the second law of thermodynamics implies a global decrease in D-information. On the other hand, a few physicists argue that D-information is as much a conserved quantity as energy. This implies that when D-information appears to be "lost" in a dissipative process, it is merely dispersed and unavailable rather than actually destroyed.

By contrast, it is helpful to introduce another category: **Survival Relevant Information**, namely, information relevant to evolutionary selection processes. SR-information can be either useful (U) or harmful (H) to the chances of being selected. The subcategory **SU-information** is what most people mean when they speak of "information" without further qualification. By similar logic, **SH-information** can be roughly equated to erroneous information (misinformation) as well as intentional deception (disinformation). Clearly, usefulness or harmfulness as applied to SR-information is definable only in terms of a specific local reference system.[8] It is important to note that, whereas thermodynamic information, discussed previously, is an extensive variable with dimensions of entropy, SR-information is an intensive variable.

In the biological case, the useful information content of genes is defined in reference to the organism itself: It is the *architectural* information needed to specify the physical and chemical structure of a functional cell, as well as the *procedural know-how* to build it. Obviously the harmful subcategory comprises those (non-neutral) genetic variants that result in impaired functions.

In physical science, the usefulness of information (a theory or a model) is determined in reference to how well the theory or model fits reality.[9] In technology and economics, the usefulness of information is ultimately determined by its marginal contribution to functional performance or productivity. In science generally, information is useless if it is redundant or superfluous and adds nothing to the predictive or explanatory value of a theory. Harmful information corresponds to erroneous data and/or defective theories or models. In technology and economics, harmful information may take the form of defective product designs, defective materials, defective parts, defective assemblies, erroneous communications, or erroneous data anything that results in avoidable costs for inspection, error detection and correction, checking, auditing, redesign, rework, repair and maintenance, or recovery from accidental loss and destruction.

SR-information is definitely *not* conserved, nor are its subcategories SU-information and SH-information. The fact that an item of *useful* knowledge can be copied and transmitted many times at almost zero cost has been a source of conceptual difficulty to economists trying to fit it into the usual framework of commodity exchange.[10]

It has been said that "information" (meaning SU-information) is unlike other commodities in that it can be sold (or given away) without being lost to the seller. Thus, with every such transaction, the total quantity of SU-information in circulation increases, but its quality or (survival) value does not change as a result of such a transaction.[11] This property is derived from SU-information being an intensive, not an extensive, variable. In economic terms the quality-value of information and its quantity-value must be treated quite differently. An increase in quality-value (perhaps the product of research), leaving quantity unchanged, can benefit society, irrespective of *ownership*. On the other hand, the aggregate quantity-value of extra copies of the same SU-information (leaving quality-value unchanged but increasing availability or utilization), may or may not be independent of the extent of circulation. It is convenient to label these cases a and b.

In case a the added quantity-value to those individuals or firms obtaining extra "copies" of a piece of SU-information (disregarding acquisition, retrieval, or decoding costs) is more or less balanced by a loss to the original (monopolist) discoverer. It has a fixed value to society, independent of how the ownership of that information is divided. An example of this kind of fixed total value would be information about the location of a petroleum deposit. In case b, however which can be categorized as *catalytic* in some sense the aggregate socio-economic value of a piece of SU-information is an increasing function of the number of copies that exist (i.e., the number of people who have it). Examples might include knowledge of how to avoid unwanted pregnancies, how to avoid infection by the AIDS virus, or how to make farmland more productive. This is also true of SU-information that could increase the effectiveness of research (e.g., better measurement techniques) or the effectiveness of technology transfer.

Of course, a complete economic theory of SR-information would have to establish the fundamental criteria for differentiating between economic usefulness and harmfulness and for evaluating these attributes quantitatively. It would also have to provide criteria for determining economic value on the basis of the extent of diffusion (case *a* versus case *b*). It would incidentally have to treat the problem of ownership rights, exchange mechanisms, markets, and prices. This, in turn, raises questions of public policy with respect to providing efficient incentives for the creation of SU-information (R&D) and its transfer and/or protection, especially where the public interest in diffusion is critical. The present book does not attempt to deal with all these issues, though it attempts to identify some elements of a framework for such analysis.

2.6. SUMMARY

Information is an everyday word that most people understand perfectly well in context. The word really has two common usages in addition to its technical meaning, however. It is used in the semantic sense both as *data* and as *knowledge*. This book also uses it in the more technical sense as a measure of the resolution of doubt or uncertainty. This usage, introduced in 1948 by Claude Shannon [Shannon and Weaver, 1949; Cherry, 1978] is linked with the normal semantic usage insofar as the availability of data can be regarded as reducing uncertainty. Quite clearly, however, in many cases the technical usage is inconsistent with the normal one.

More recently, it has been helpful to reinterpret Shannonian information as a fundamental quantitative measure of *distinguishability*. Specifically, the information content of (or embodied in) a subsystem is proportional to the degree to which it can be distinguished from its environment. The notion of distinguishability clearly implies the existence of an observer whose role is to define the boundaries between the subsystem and the larger system in which it is embedded. This implicit observer is often neglected in formal mathematical treatments of information theory, resulting in unnecessary confusion and ambiguity.

In physics, where the theory is most complete, Shannonian information can be defined in terms of entropy, which can be calculated by the methods of statistical mechanics. In turn, statistical entropy can be calculated by the methods of thermodynamics. It follows that information in this domain can be directly related to thermodynamic potential energy. The basic relationship is

$$H = \frac{B}{T_0},$$ (2.35)

where H is the Shannonian information content of a subsystem, B is the most general thermodynamic potential of that subsystem with respect to its environment (known as *available useful work*) and T_0 is the temperature of the environment.[12] Actually, the definition of *environment* is often at the disposal of the observer. In general, the information content of any arbitrary subsystem can be defined with respect to any convenient reference system.

TABLE 2.3. *Information: Definitions and Reference States.*

Domain	Definition	Reference State
Probabilistic systems	$H_{prob} = -\sum_i P_i \log_2 P_i$ $P_i = $ *a priori* probability of *i*th state	Random distribution or "white noise"
Physical composition	$H_{conc} = \dfrac{B_{conc}}{T_0} = \dfrac{-\sum_i n_j v_{i0}}{T_0}$ $T_0 = $ Temperature of reference state $n_i = $ Occupancy number of *i*th species $v_{i0} = $ Chemical potential	Earth's crust at temperature T_0
Chemical state	$H_{chem} = \dfrac{B_{chem}}{T_0} = \dfrac{U + P_0 V - T_0 S}{T_0} = \dfrac{Q_c}{T_0} - S$	Standard temperature and pressure of Earth's atmosphere, T_0, P_0
Morphology	See Chapter 10	Application-dependent
Cybernetics and ergonomics	See Chapter 11	Random distribution

In the economic domain, some activities (notably energy conversion, chemistry, and metallurgy) are readily described in terms of thermodynamic language. Others, ranging from metalworking, assembly, and construction to services (including "information processing") are not easily reduced to thermodynamics or statistical mechanics. In areas of manufacturing and construction where chemical changes are not involved notably, materials-shaping and assembly it is easier to bypass thermodynamics and work directly with Shannonian information [Ayres, 1987]. Statistical quality control (SQC) in manufacturing is also amenable to this formalism. In the case of services, it is easier to relate Shannonian information with information in the semantic sense (data and knowledge). Fortunately, in recent decades, due especially to the work of Fritz Machlup [1962, 1979, 1980], it has become fashionable to think of many economic services in informational terms. It also helps that the analysis of human labor in functional terms ("human factors" and ergonomics) has increasingly used the language of Shannonian-information theory [see, for example, Miller, 1978]. It need hardly be mentioned that the operation of computers, now central to most sectors of the economy, is particularly well suited to information-theoretic description.

In the various non-thermodynamic applications of information theory noted above, the choice of reference state (environment) is crucial for consistent quantitative treatment. To compute the information-content of anything, one must decide at the outset what is being compared to what. In Shannon's communications theory, the information content of a message is a measure of its non-randomness which means the reference state, in this case, is random "noise."

For compositional information, the reference state is normally taken to be the composition of the earth's crust or in some cases, the ocean or the atmosphere. For thermodynamic information, the reference state is taken to be "standard temperature and pressure" (STP) of the earth's atmosphere. Table 2.3 summarizes these three basic cases.

The choice of reference state is less obvious (and more arbitrary) in other cases. To calculate the information content of geometric shapes, for instance, should form be compared to "formlessness" (the analog of randomness)? Or would it make more sense to describe shaped surfaces in terms of distortions of a perfect hypothetical flat plane? The former might be more appropriate for the molding of a piece of clay, whereas the latter might be more appropriate for an industrial stamping or forming process that begins with sheet metal (or paper).

Or, as suggested by Kullbach [1951], should one take into account the functional aspects (purposes) of shapes, thus implicitly comparing each shape with the idealized shape needed to achieve the function? For instance, one could postulate the existence of an ideal lifting body part (wing) or an ideal turbine blade. Of course, one can rarely characterize the ideal shape, even for a very well defined function. This approach is therefore not feasible for quantitative calculations, in general.

For purposes of this book, it will be convenient to characterize evolution in the language of information theory. In brief, it seems likely that all physical processes and transformations (including phylogenic evolution), can be described in terms of two fundamental information quantities. These are, respectively, pure uncertainty-reducing or distinguishability information (**D-information**) and evolutionary survival-relevant information (**SR-information**). Evolutionary survival-relevant information may be useful, neutral, or harmful. It is convenient to identify the subcategories survival-useful **SU-information**, and survival-harmful **SH-information**, meaning information making a positive or negative contribution to evolutionary survival.

In the biological domain, SU-information corresponds to selectively beneficial alleles (genetic variants), while neutral and harmful types correspond to neutral or harmful alleles. In the domain of human economic activities, SU-information includes both (1) thermodynamic and morphological information embodied in material goods and structures and (2) human **knowledge** contributing to the stability of society and the production of goods and services (including the production of new knowledge itself). Neutral SU-information corresponds to non-productive materials (i.e., information embodied in such materials) or useless knowledge. SH-information corresponds to defective materials or designs, erroneous data, or counterproductive knowledge. In the domain of computers, SU-information is the (good) data and the program that enables the computer to obtain a result meaningful to the programmer, while SH-information is "garbage" data or programming errors.

Later chapters will suggest that the evolutionary mechanisms that account for natural selection in both biological and socio-economic systems do so by increasing both the selective strength and quantity of SU-information, while tending to eliminate SH-information. (A considerable amount of survival-neutral information may be carried along with the SU-information, of course.)

ENDNOTES

1. Sections 2.1 and 2.2 are taken from IIASA RR-88-1 [Ayres 1988].

2. A classic reference is Shannon and Weaver [1949]. For a good recent dicussion, see Cherry [1978] or Young [1987].

3. Claude Shannon [1948] of Bell Telephone Laboratories and MIT, was the first to formulate a mathematical theory of information based on concepts of doubt or uncertainty.

4. Conversely, Jaynes [1957a,b] showed that the best (least biased) probabilty distribution of microstates can itself be estimated (absent other information) by maximizing expected information (2.5) using a Lagrangian formalism.

5. Shannon identified *entropy* as uncertainty; information is its complement.

6. Note that $R = kA$, where $A = 6.02*10^{23}$ (Avogadro's number) and $k = 3.298*10^{-24}$ cal/°K (Boltzmann's constant). Hence, $R = 1.985$ cal/°K per mole or 8.31 joules/°K.

7. In practice, the heat of combustion for a fuel is approximately given by $-Q_c = anR(T_c - T_0)$, where a is a constant, T_c is the combustion temperature, and n is the number of moles. Any oxidation process resulting in an oxide that is solid or liquid at standard temperature and pressure (S.T.P.) would fall into this category in principle. However, in the case of water (H_2O), the oxidation product is initially in vapor form and condensation occurs later and separately, if ever.

8. The analogy between useful available work and total energy is relevant here. Total energy is an absolute quantity that depends on one arbitrary specification (absolute zero of temperature) but is independent of local reference systems. On the other hand "available" work is a function of the local reference system. Clearly the heat energy stored in the ocean is not available to a potential user on the surface of the earth, because there is no thermal gradient. On the other hand, if the same reservoir were surrounded by empty space (or located on Mars), a great deal of energy could be extracted from it and used to operate a heat engine.

9. For computers, the usefulness of the information being processed is defined in terms of the purpose of the programmer, i.e., the problem being solved. Anything that uses up memory or takes time is harmful, although outright errors in programming or data transcription are more harmful than (say) performing calculations with inefficient algorithms or unnecessary precision.

10. The problem is partly conceptual: U-information is really a vector with two distinct

dimensions. One dimension can be defined as strength and the other as quantity. As applied to technology, the first dimension is a measure of the "state of the art," while the second is a measure of availability or extent of penetration. Scientific research contributes to the first. Publication and repetition merely add to the second.

11. However, each time an item of U-information is reproduced for transmission, there is some probability of a copying error, resulting in a gradual process of contamination and degradation. Also, the value of information to a user clearly depends on the cost of retrieval, decoding, and other processing.

12. The application of information theory to statistical mechanics was worked out by Jaynes [1957a,b]; the application to thermodynamics was mainly due to Tribus [1961a,b; also Tribus and McIrvine, 1971]. The proof that B is the most general thermodynamic potential was due to Evans [1969].

REFERENCES

Aczel, J. and Z. Daroczy, *On Measures of Information and Their Characteristics*, Academic Press, New York, 1975.

Ayres, Robert U., *Manufacturing and Human Labor As Information Processes*, Research Report (RR-87-19), International Institute for Applied Systems Analysis, Laxenburg, Austria, July 1987.

Ayres, Robert U., "Complexity, Reliability and Design: Manufacturing Implications," *Manufacturing Review* 1(1), March 1988 :26-35.

Brillouin, Leon, "Physical Entropy and Information," *Journal of Applied Physics* 22(3), 1951 :338-43.

Brillouin, Leon, "Negentropy Principle of Information," *Journal of Applied Physics* 24(9), 1953 :1152-1163.

Brillouin, Leon, *Science and Information Theory*, Academic Press, New York, 1962. 2nd edition.

Cherry, Colin, *On Human Communications*, MIT Press, Cambridge, Mass., 1978. 3rd edition.

Evans, Robert B., *A Proof That Essergy Is the Only Consistent Measure of Potential Work for Chemical Systems* Ph.D. Thesis [Dartmouth College, Hanover, N.H.], 1969.

Gibbs, Josiah Willard, "Thermodynamics," in: *Collected Works*, 1, Longmans and Yale University Press, New Haven, Conn., 1928 [Collection]

Hartley, R. V. L., "Transmission of Information," *Bell Systems Technical Journal* 7, 1928.

Hirata, H. and R. E. Ulanowicz, "Information Theoretical Analysis of Ecological Networks," *International Journal of Systems Science* 15, 1984 :261-270.

Jaynes, Edwin T., "Information Theory and Statistical Mechanics, I," *Physical Review* 106, 1957a :620.

Jaynes, Edwin T., "Information Theory and Statistical Mechanics, II," *Physical Review* 108, 1957b :171.

Kullbach, S., *Information Theory and Statistics*, John Wiley and Sons, New York, 1951.

MacArthur, R., "Fluctuations of Animal Populations and a Measure of Community Stability," *Ecology* 36, 1955 :533-536.

Machlup, Fritz, *The Production and Distribution of Knowledge in the U.S.*, Princeton University Press, Princeton, N.J., 1962.

Machlup, Fritz, "Uses, Value and Benefits of Knowledge," *Knowledge* 1(1), September 1979.

Machlup, Fritz, *Knowledge and Knowledge Production* 1, Princeton University Press, Princeton, N.J., 1980.

Maxwell, James Clark, *Theory of Heat*, Publ., City, 1871.

Miller, James G., *Living Systems* 5, McGraw-Hill, New York, 1978.

Odum, Eugene P., *Fundamentals of Ecology*, Sanders, Philadelphia, 1953. 2nd edition. SE11}

Perry, Robert H. and Cecil H. Chilton, *Chemical Engineer's Handbook*, McGraw-Hill, New York, 1973. 5th edition.

Shannon, Claude E., "A Mathematical Theory of Communication," *Bell Systems Technical Journal* 27, 1948.

Shannon, Claude E. and Warren Weaver, *The Mathematical Theory of Communication*, University of Illinois Press, Urbana, Ill., 1949.

Spreng, Daniel, *Substitutional Relationships Between Energy, Time and Information*, Mimeo 1990.

Szilard, Leo, "Über die Entropieverminderung in einem thermodynamischen System bei Eingriffen intelligenter Wesen," *Zeitschrift für Physik* **53**, 1929 :840-856.

Tribus, Myron, *Thermostatics and Thermodynamics*, Van Nostrand, New York, 1961a.

Tribus, Myron, "Information Theory as the Basis for Thermostatics and Thermodynamics," *Journal of Applied Mechanics* **28**, March 1961a :1-8.

Tribus, Myron and Edward C. McIrvine, "Energy and Information," *Scientific American* **225**(3), September 1971 :179-188.

Ulanowicz, Robert E., *Growth and Development: Ecosystems Phenomenology*, Springer-Verlag, New York, 1986.

Wilson, David R., *An Exploratory Study of Complexity in Axiomatic Design*, Ph.D. Thesis [Massachusetts Institute of Technology, Cambridge, Mass.], August 1980.

Yaglom, A. M. and I. M. Yaglom, *Probability and Information*, D. Reidel, Boston, 1983. 3rd edition. [English translation from Russian]

Young, Paul, *The Nature of Information*, Praeger Publishers, New York, 1987.

APPENDIX:
CONNECTION AMONG ESSERGY, AVAILABILITY, EXERGY AND FREE ENERGY

Name	Function	Comments
Essergy	$U + P_o V - T_0 S - \sum_i V_0 N_i$	This function was formulated for the special case of an existing medium in 1878 (by Gibbs) and in general in 1962. Its name was changed from "available energy" to "exergy" in 1963 and from "exergy" to "essergy" (i.e., "essence of energy" in 1968.
Avail-ability	$U + P_0 V - T_0 S - (E_0 + P_0 V_0 - T_0 S_0)$	Formulated by Keenan in 1941, this function is a special case of essergy (*Eq.* 2.22.).
Exergy	$U + P_0 V - T_0 S - (E_0 + P_0 V_0 - T_0 S_0)$	Introduced by Darrieus (1930) and Keenan (1932), this function (which Keenan has called the "availability in steady flow") was given the name "exergy" by Rant in 1956. This function is a special case of essergy (*Eq.* 2.22.).

Free   Helmholtz: U-TS

Energy   Gibbs: U + PV-TS

The functions E-TS and E + PV-TS were introduced by von Helmholtz and Gibbs (1973). The two functions are Legendre transforms of energy which were shown by Gibbs to yield useful alternate criteria of equilibrium. As measures of the potential work of systems, these two functions are special cases of the essergy function (*Eq.* 2.22.).

Source: [*Evans* 69]

Chapter 3

Physical Evolution: From the Universe to the Earth

3.1. THE EVOLUTION OF MATTER

Biological and socio-economic systems are, of course, not the only ones to evolve. In this chapter we shall see that something like phylogenic evolution (of the genotypic kind) has occurred in the physical world of matter and energy. The distribution of radiation and matter in the universe, by atomic species and by physical form (stars, galaxies, galactic clusters, etc.) constitute the present traces of those early evolutionary processes. In this sense, they correspond to the fossil remains and geological evidence that has enabled scientists to retrace the bio-geological history of the earth. As physicists now see it, the evolution of matter can be characterized as a series of "phase changes," analogous to the condensation of a vapor to a liquid or solid.

In this century, and especially the last few decades, most physicists have become convinced that the basic laws of physics can be expressed as either **symmetry principles** or **conservation laws** if stated with sufficient generality. This statement requires some explanation, since *symmetry* implies that some quantity is invariant (i.e., conserved) under a mathematical transformation of some sort. For instance, the law of conservation of momentum is a consequence of translational invariance or homogeneity of Euclidean space; the law of conservation of angular momentum is, similarly, a consequence of the rotational invariance (isotropy) of Euclidean space. It turns out that energy conservation is a consequence of translational invariance in time (considered the fourth dimension in Einsteinian space). So far as we know, none of these basic symmetries or their corresponding conservation principles have been violated anywhere in the universe.

Taking the relationship between symmetry and conservation laws as a starting point, however, physicists began looking for more-exotic examples of conserved quantities and/or symmetries. Thus, we talk about symmetry with respect to electric-charge reversal (conservation of charge), symmetry with respect to matter-antimatter reversal (conservation of baryon number), symmetry with

51

respect to "mirror image" (parity conservation), and so on. Modern particle physics has added a number of completely new "dimensions" for such symmetries. They include "strangeness," "color," and "flavor." Without trying to summarize these difficult concepts into a few paragraphs—surely a doomed endeavor—I would add only that the cases of asymmetry (non-conservation) can apparently be explained within a more general framework of symmetrical laws. In other words, many of the most important phenomena in the physical world, including the masses of the elementary particles themselves, actually result from *broken* symmetries.

How does a fundamental symmetry or a conservation law get broken? We can only answer by invoking analogies with more-familiar situations, which tend to make even some physicists a trifle uncomfortable. The key seems to be spontaneous (quantum) fluctuations such as the spontaneous creation and destruction of particle-antiparticle pairs in the underlying matrix of everything, called the "vacuum."

To see how fluctuations might work to lock in asymmetries in the classical realm, consider a fluid. Every region of the space occupied by the fluid looks like every other part: It is homogeneous and isotropic. But now suppose the temperature drops below the freezing point. The resulting crystalline solid is no longer homogeneous or isotropic. It has preferred directions, determined by short-range forces that seldom come into play at higher temperatures. The translational and rotational symmetry of the homogeneous fluid is now broken. The laws governing the behavior of the solid are different at least, phenomenologically from the laws governing the fluid. It is as though new physical laws came into being when the fluid froze.

From the cosmological perspective, there are several conspicuous examples of asymmetry:

1. Entropy is *increasing* in the universe as a whole.
2. Time flows *forward* (not backward).
3. The universe is *expanding* (not contracting).
4. Matter is not equally balanced with antimatter
5. Parity (mirror symmetry) is not conserved in all interactions, although it is conserved in most of them.

The observed lack of symmetry on the macro-scale requires explanation. The basic equations (laws) of general relativity and quantum mechanics, which we currently accept, show no preference for one direction in time over the other. Nor do they explain why matter should be preferred to antimatter, or why rules that are obeyed in the real world are not obeyed in the "mirror-image" world. Yet the universe does exhibit such preferences.

Seeking explanations for these apparent preferences has proved to be very fruitful.[1] In a fundamental sense, the evolution of matter and structure in the cosmos is the evolution of asymmetry, or **symmetry-breaking**. Physicists now think that some of the characteristics of the present universe, including its apparent laws, are actually results of a kind of "big freeze"—which will be identified later as the GUT transition—that occurred a tiny fraction of a nano-second after the universe began in the Big Bang. (This was the first of several such "freezes.")

It is not yet clear whether the first three asymmetries on the foregoing list are independent of one another. (Most physicists think not.) Thus, Arthur Eddington essentially equated (1) and (2) by calling entropy "time's arrow." George Sussman [1985] has suggested that the increase in cosmic volume could be a "strong cause" of entropy production. I will return to this controversy later.

In the 19th century, and well into the present one, the universe was assumed to be static. This notion had to be abandoned in the 1920s in the face of contrary astronomical evidence. The universe is now known to be expanding at a rate first measured by the American astronomer Edwin Hubble in 1929. Hubble observed that distant galaxies in every direction were receding from our own, and that the farther the galaxy, the faster it was moving away. On the basis of these obser-vations (and other assumptions—for example, about the structure of stars) he calculated the age of the universe to be about 2 billion years. This estimate has been repeatedly revised upward as astrophysics has progressed. The estimated age of the universe now stands between 10 and 20 billion years, most likely around 16 billion.

The expansion of the universe is thought to have originated with a single explosion of inconceivable violence, known as the Big Bang. Curiously, the Big Bang was first postulated in 1922 before Hubble's observational verification of the expansion of the universe by Russian mathematician Alexander Friedmann, who worked it out as an implication of Einstein's general theory of relativity.[2] In the light of recent astronomical observations, a number of cosmologists have expressed doubt that the Big Bang theory is still viable.[3] The primary difficulty lies in reconciling some extremely large cosmological inhomogeneities—notably, the recently discovered "superclusters" [Tully and Fisher, 1987]—with the observed homogeneity of the background radiation that supposedly originated in that primordial event.[4]

Another problem with the Big Bang theory, to which I shall return later, is that it seems to require a large amount of invisible and undetectable "dark" matter. The density of observable matter is insufficient in itself to account for the gravitational mass needed by the Hubble expansion [de Vaucouleur, 1970].[5] This inconsistency has led to the "dark matter" hypothesis and has given credibility to the notion that massive "black holes" are likely to be found in the centers of galaxies. I think, however, that the Big Bang theory will survive and this problem will be traced to Einstein's theory of gravitation, on which many of the calculations, including the prediction of "black holes," are based.

In principle, the equations of motion of all the mass in the universe, hence the rate of expansion as a function of time, are determined by two sets of equations. The fundamental equations of general relativity, relating gravitation to mass-energy, constitute one set.[6] The second set—coupled to the first—describes the quantum mechanical processes governing the conversion of energy from its primordial form (mainly very hot photons) to the other forms, especially neutrinos, electrons, neutrons, and protons. Taken together, these equations describe extremely non-linear systems, since the mass-energy conversion *rates* depend on the photon and mass-energy *density* and *temperature*. But these quantities, in turn, depend on the volume of the universe, which itself is a function of the distribution of mass-energy.

This point needs to be emphasized because of the well-known tendency of non-linear dynamical systems to have multiple steady-state solutions and to exhibit bifurcations or "catastrophes" (in René Thom's definition). Phase transitions such as the freezing or "boiling" of water are examples of such sudden shifts from one steady state to another. As will be discussed in the next section, one phase change, the "freezing of the vacuum," apparently initiated the expansion of the universe.

As expansion continued and the radiation (and matter) cooled, other phase changes occurred. More and more of the total energy of the system now belongs to the rest-mass of the matter and its associated gravitational field. Whether the universe will continue to expand forever, or reach a maximum volume, depends on the actual amount of gravitational mass (and residual radiation) in the universe, as compared to the kinetic energy of the expansion itself. Significant uncertainties still exist with regard to these numbers especially the gravitational mass and radius of the universe today. Most cosmologists, however, seem to think the expansion will eventually cease and the universe will begin to contract once again, a few billion years from now.

3.2. FROM THE FIRST PICOSECOND TO THE NEXT THREE MINUTES

Current theories of elementary particle physics allow a reasonably good understanding of the phenomena that can be observed at energies of up to about 100 GeV (roughly the energies reachable by large accelerators at CERN and Fermilab).[7] These energies correspond to the temperature of the universe about 10 picoseconds (10^{-11} seconds) after the Big Bang. Thus, the events that occurred earlier (at still higher temperatures) cannot yet be explained by any theory that is subject to experimental confirmation. The theories are still quite speculative and not altogether satisfactory [see, for instance, Fritzsch, 1984, 1985; Marx, 1984; Motz, 1986; and Hawking, 1988].

In particular, the predominance of matter over antimatter, the fourth major macro-asymmetry listed above, definitely requires some kind of theory to explain what happened before the universe was 10^{-11} seconds old. The prevailing view is called somewhat optimistically the **Grand Unified Theory,** or GUT. It holds that, prior to 10^{-35} seconds, particles, antiparticles, and photons were equally abundant. The very hot, tiny universe was in thermal equilibrium and (it is assumed) all forces except gravity were equal in range and strength and all masses were zero. There were no observable asymmetries at that time (even if there had been any observers!)

Then (according to current Big Bang theory) something strange happened. As I have mentioned, it can only be described as a kind of phase transition similar to the freezing of a supercooled liquid, except that it was empty space "the vacuum" itself—that froze. When a liquid freezes to form a crystalline solid, asymmetries (preferred directions) are created in the crystal structure. GUT postulates that, in the same way, the fundamental forces were distinguished from each other by the freezing in of preferred rest masses. This, in turn, resulted in those forces having different ranges. The force responsible for all the asymmetries that resulted from the freezing of the vacuum is called the Higgs field. It is unobservable by any means known to us and can only be deduced from its effects.

A fairly radical variant on the original theory has been suggested by physicist Alan Guth, essentially in response to the "flatness problem" noted by Dicke and Peebles. Guth exploited the notion that the early cooling and expansion of the universe could be understood in terms of phase transitions to increasingly condensed states of matter. He modified the standard Big Bang model to allow for the possibility of a phenomenon analogous to supercooling. In effect, Guth postulates that the expansion of space and the cooling rate can be temporarily decoupled in certain regions. It turns out that very short transition lags in small "supercooled" regions can have enormously magnified consequences. To be precise, suppose a phase transition that would took place when the universe was only 10^{-35} seconds of age was delayed in a "supercooled" domain by a factor of 60 (so that it takes place at about 10^{-33} seconds). In this case the domain could have expanded, meanwhile, by a factor of 10^{25}! Without going into details, suffice it to say that this theory—called "inflation"—may provide a resolution of the "flatness" problem.

Guth's version of GUT postulates that the freezing of the vacuum released latent heat that triggered an extremely rapid inflation of the vacuum (i.e., of space itself). The rapid expansion of space inflation caused a departure from thermal equilibrium. This, in turn, led to the next major discontinuity in the evolution of the universe, the so-called quark transition. As cooling and expansion continued, photons were constantly interconverted into quarks, and vice-versa, by interacting with very massive X particles: viz.,

$$\text{photons} \leftarrow \rightarrow X \leftarrow \rightarrow \text{quarks.}$$

At higher temperatures quark annihilation occurred more frequently than the creation process. But cooling changed the picture. As the quarks cooled, they no longer had enough energy to create the heavy X. They were therefore unable to annihilate each other via the X route (except by quantum "tunneling"). Instead, they formed bound states, which are observable as **baryons** (protons, neutrons) or **mesons**. Baryons have odd spins and consist of three quarks of different "colors," while mesons have even spins and consist of a quark and an antiquark.

It is noteworthy that the number of baryons is a conserved quantity in all known reactions (i.e., the reactions we can artificially induce and see).

But how did it happen (as most physicists believe, with the notable exception of Alfvén and his followers) that the number of baryons left over was not exactly matched by the number of antibaryons? Admittedly the mismatch was small: not more than 1 part in 10^9 or 10^{10}. Still, had the number of baryons been exactly equal to the number of antibaryons, there would be no solid matter in the universe, only radiation. (The photon/proton ratio is roughly 10^9 to 10^{10}. Had this ratio been much greater, gravitation would have caused the universe to recollapse long ago. If it had been much smaller, there would be no stars or planets.) This asymmetry, too, was a consequence of the freezing of the vacuum. The baryon condensation was the second major phase transition in the history of the universe. It began when the universe was a microsecond (10^{-6} sec) old and was complete in a millisecond (10^{-3} seconds).

The cooling and condensation continued (Table 3.1), but the rate of change was slowing down (so to speak) rapidly. By the time the universe was only 10^{-2} seconds old, only electrons, positrons, neutrinos, antineutrinos, and photons were left, with a (relatively) small number of neutrons and protons. Between the first hundredth of a second and the first 3 minutes the temperature dropped from 100 billion °Kelvin (°K) to 1 billion °K. Most of the electrons and all the positrons also disappeared (by annihilation), leaving mainly the neutrinos, antineutrinos, and photons [Weinberg, 1978]. Meanwhile, the few residual protons and neutrons interacted fairly intensively, yielding a balance of 14 percent neutrons and 86 percent protons at the end of the first 3 minutes.

Figure 3.1 shows the history of the universe in terms of temperature and radius beginning (more or less) with the GUT transition and continuing to the present. Note that the entire history of our galaxy is compressed into a tiny region at the extreme right of the chart.

3.3. FROM 3 MINUTES TO 1 MILLION YEARS

The next step in the evolution of matter was again a kind of condensation: the formation of stable deuterons (hydrogen nuclei). For 700,000 years or so, this process accelerated, even as the rate of cooling was decelerating. Eventually helium nuclei (more stable than hydrogen nuclei) also began to accumulate. At this point in time (700,000 years, more or less), the temperature had fallen to about 3,000 K and the hydrogen and helium nuclei combined to form electrically neutral atoms. Matter, as we know it, appeared for the first time.

Without free electrons in the particle-photon soup, the universe became transparent to radiation for the first time. In effect, the radiation left from the early universe decoupled from the matter. This can be regarded as the third major phase transition in the history of the universe.

At the time of the third transition the universe was less than 1 million years old and a thousand times smaller than it is now. Another landmark event

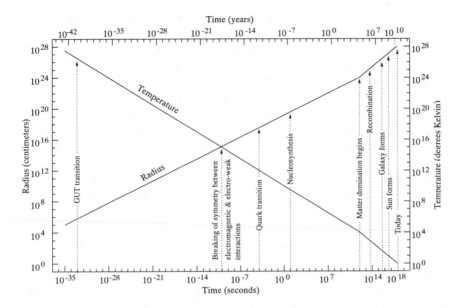

FIGURE 3.1. History of the Radius and Temperature of the Universe. Source: Adapted from Krauss, 1989; Figures B.1 and B.2.

occurred about the same time (perhaps a little earlier, perhaps a little later): the transition from a regime in which most of the energy was in the form of radiation to a regime in which most of the energy was embodied in matter (recalling Einstein's equation $E = mc^2$). In effect, the earlier period can be called "radiation-dominated," as contrasted to the "matter-dominated" regime that exists now.

Perhaps the most conclusive evidence for the Big Bang, and the age of the universe, is the **black-body** background radiation. This is directionless thermal

TABLE 3.1. *Stages in the Evolution of the Cosmos.*

Age (in Seconds)	Dominant Phenomena
$< 10^{-43}$	Quantum gravity effects
$10^{-43} - 10^{-33(35)}$	Phase transition ("freezing of vacuum"); inflation; "symmetry-breaking"
$10^{-33} - 10^{-6}$	Quarks; electro-weak forces
$10^{-6} - 10^{-3}$	Baryons (nucleons, mesons)
$10^{-3} - 10^{+2}$	Leptons (electrons, neutrinos)
$10^2 - 10^3$	Atomic nuclei
$10^3 - 3*10^{13}$	Atoms, molecules
10^6 (secs)$-2*10^{10}$ (years)	Galaxies, life

radiation that dates from the early period when matter and radiation were still in thermodynamic equilibrium. This is consistent with the observed (Planck) spectral distribution. The radiation was originally emitted at short wavelengths, corresponding to the comparatively high temperature (3,000 °K) at which the final transition took place. It has, of course, shifted toward longer and longer wavelengths into the microwave region as the universe has expanded (the "red shift"). The temperature of the black-body radiation today (2.9 °K) is exactly inversely proportional to the size of the universe, so a thousandfold increase in volume corresponds exactly to a thousandfold drop in radiation temperature.

The physicist George Gamow was the first to suggest that all the chemical elements might have been created in the primordial "fireball" itself. (The analogy between the Big Bang and the explosion of an atomic bomb must have been in the minds of many physicists in the 1950s.) Gamow's ideas were eventually confirmed with regard to the relative abundances of the lightest elements (hydrogen, helium, and lithium), which were approximately stabilized within the first 700,000 years after the Big Bang.[8] Current theory holds that the light elements, including the common metals, were formed early. With regard to the heavier elements, however, it is now clear that other (non-equilibrium) mechanisms were mainly responsible.

With two colleagues (Ralph Alpher and Robert Herman), Gamow also predicted, in 1948, that the temperature of the residual background radiation should be 5 °K. The search for a directionless residual background radiation was first suggested by the physicist Robert Dicke in 1965. Such radiation was actually observed (at a lower-than-expected temperature) soon afterward, in 1967, by Arno Penzias and Robert Wilson, in the course of early experiments with satellite communication.

By the end of the first million years, matter had condensed into two stable and electrically neutral forms, consisting of hydrogen or helium atoms (and some other light elements), plus neutrinos and antineutrinos. The tendency of the two forms of matter to interact with each other or with electromagnetic radiation had decreased by orders of magnitude. In effect, the two types of matter and radiation have pursued independent evolutionary paths since that early time. So far as we know, the only change in electromagnetic radiation has been a continued cooling and frequency shifting toward lower frequencies. Neutrinos presumably have also lost energy as the universe has expanded. Only the heavier forms of matter (the chemical elements) have continued to evolve.

3.4. SINCE THE FIRST MILLION YEARS

Since the first million years, the evolution of the universe has slowed significantly, and most of the action has centered around the creation of stars, galaxies, and larger structures by aggregation and condensation of hydrogen and helium from the primordial soup. Since matter became electrically neutral and transparent to radiation-hence invisible-the main driver of these changes has probably been gravity. However, a major mystery remains to be unraveled, as mentioned in the first section of this chapter.

Cosmologists have noted that the visible universe now has a very complex structure on a wide range of scales from the planetary to the supra-galactic [Beckenstein, 1985; Tully and Fisher, 1987; Horgan, 1990]. Stars were once thought to be distributed uniformly. Later, they were found to cluster instead. Some tight clusters (galaxies) contain as many as a million suns. Galaxies take a variety of forms, from flat disks and spiral nebulae to well-formed spheroids and ellipsoids. Galaxies, in turn, cluster in groups, chains, and even galactic clusters with thousands of galaxies in them. Finally, there are known to be superclusters consisting of chains of clusters of galaxies. One is the "great wall," 500 million years in extent, interspersed by great voids. The largest such void spans 300 million light years. On still larger scales, however, the universe is thought to be relatively homogeneous.

Gravitation is almost certainly the main organizing force. Theoretical models (beginning with the work of the British astrophysicist James Jeans) have shown that, if "seeds" of higher densities are present in a cloud of uniformly dispersed matter, the cloud will tend to collapse into stars and on a larger scale into galaxies. The shapes of galaxies can also be explained, to a degree. The spiral nebulae, for example, can be understood as a gravitational wave slowly traveling around a flat disk-shaped galaxy.

Yet, if gravity is indeed the organizing force for large-scale structures, there remains a difficulty. The known rate of expansion of the universe and the observed density of mass are inconsistent, at least if Einstein's theory is used to make the calculations. Observed gravity-related phenomena on the cosmic scale require much more mass than can be accounted for by stars. Between 90 percent and 99 percent of the mass of the universe is "missing." Various suggestions have been made, ranging from massive "black holes" (a popular idea, but see note 3) to attributing more mass than previously thought to "hot dark matter" e.g., neutrinos [Krauss, 1990] or, simply dust. This is one of the deepest questions facing cosmologists today.

In addition to gravitational organization, the cooling and expanding universe evolved via continuing creation (by nuclear fusion) of the heavier elements. This is still going on in the centers of stars. Once a star is formed by gravitational collapse of a gas cloud, it follows a relatively predictable (i.e., ontological) life cycle. The life cycle of stars is often portrayed in terms of the so-called Hertz-sprung-Russell diagram (see Figure 3.2). This represents all the starts in a particular globular cluster (known as 47 TUC) classified in terms of two variables: color (a measure of temperature), along the horizontal axis, and brightness or luminosity along the vertical axis. The diagonal band is known as the **main sequence.** Stars are distributed along it according to mass. They spend 90 percent of their lifetime at a more or less stationary point, albeit with a slight drift toward the left (greater brightness) as they age. A young star first appears

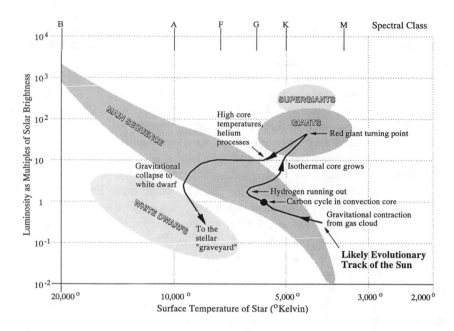

FIGURE 3.2. H-R (Hertzsprung-Russell) Diagram of Stars. Source: Adapted from Hoyle and Wickramasinghe, 1978; Figures 11 and 12.

at the lower right on the main sequence; as it ages it gets brighter and hotter. Eventually it undergoes a critical transition (the "turn-off") and begins its second career as a red or blue giant, before becoming a white dwarf or (if it is very large) exploding as a supernova.

Our sun is fairly typical of a star in the prime of life. Current theory holds that the sun was born 4.7 billion years ago, starting as a detached cloud of gas. The original gas cloud formed when the universe was 1 million years old, apparently consisted of 90 percent hydrogen and 9 percent helium. Gravitational condensation took about 100 million years. The sun began a self-sustaining hydrogen-fusion reaction when the central density became high enough. Energy is produced by a series of proton-proton chain reactions that also produce heavier elements as a by-product. (Or, perhaps the heavy elements are the main product and energy is the by-product!)

The details of these nuclear processes were first worked out by Hans Bethe in the late 1930s. During its first 100 million years or so, the sun was a young red giant (in contrast to an old red giant at the end of its life). It gradually heated up as it shrank in diameter and increased in density. As the temperature rose, the fusion-reaction rate increased till the power output reached a level where the outward pressure of radiation balanced the inward force exerted by gravity. Finally, about 4.5 billion years ago, the shrinkage ceased and the sun reached maturity at its present size and became a main-sequence star. The sun is now middle-aged; its probable life span is 10 billion years. The rate at which a star burns its nuclear fuel is proportional to the cube of its mass. Thus, a star with twice the mass of the sun would last only one-eighth as long, while a star with 10 times the solar mass would have a life expectancy of only 100 million years or so.

At present, the composition of the core of the sun has reached 30 percent helium. Its core density is 180 times the density of water. About 40 percent of its energy output is now derived from hydrogen-fusion reactions and 56 percent from helium-beryllium-lithium fusion reactions. As the sun ages, its hydrogen fuel will be depleted more and more, the core will become denser, and more of its energy will be derived from helium fusion. In the final stages, it will derive its energy from the fusion of heavier elements.

Since the sun became a mature main-sequence star and began consuming its hydrogen, it has gradually increased its temperature and heat output by about 30 percent altogether. The mechanism responsible is similar to the greenhouse effect, in which a buildup of carbon dioxide (and other gases) in the earth's atmosphere, captures and re-radiates infrared radiation from the earth's surface, thus acting like a blanket. In the case of the sun, the mechanism is that hydrogen is more transparent to radiation than helium. Thus, the more hydrogen the sun converts to helium, the hotter its interior becomes. This, in turn, increases the rate of hydrogen "burning" (or fusion). The gradual warming of the sun means that the intensity of solar illumination of the earth has increased significantly since the earth was first formed. This is a point of some importance in the context of the geological evolution of the earth and the so-called Gaia hypothesis, as will be noted later.

If it does not explode, the core of the aging star continues to produce energy (by a chain of fusion reactions known as the carbon cycle) until the helium supply is also exhausted. This last stage takes only a few tens of millions of years. Finally, the core cools down, becoming a white dwarf, and finally a neutron star, or black dwarf. White dwarf stars are found below and to the left of the main sequence in Figure 3.2.

A few explanatory words on sources of nuclear energy may be appropriate here. Figure 3.3 plots the nuclear-binding energy against the atomic number of the elements. The graph shows an available-energy minimum (actually a binding-energy *maximum*), which can be approached from either the left or the right. In the region on the left, self-sustaining (i.e., *exothermic)* fusion is possible, given high enough temperatures and pressures to start the process. On the right-hand side, especially the far right, self-sustaining fission is possible, again given suitable conditions.

If all possible nuclear reactions were allowed to proceed to equilibrium—the state of maximum nuclear-binding energy—in a sufficiently hot nuclear "furnace," light nuclei would tend to fuse, while (in the presence of thermal neutrons) heavy nuclei would tend to split apart by fission into lighter fragments. The ferrous metals, notably chromium, iron, and nickel (atomic weights

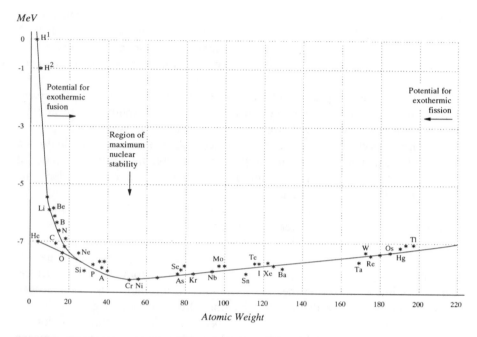

FIGURE 3.3. Nuclear-binding Energy (Potential) of Elements per Nucleon (MeV). Sources: Aston, 1935, 1936: Bainbridge, 1932, 1933.

52-59), actually have the most stable nuclei, i.e., the greatest nuclear-binding energy. It can hardly be a coincidence that these elements predominate in the earth's molten core.

Note that heavy elements (i.e., heavier than the ferrous metals) would not exist in a universe where all reactions had proceeded to thermal equilibrium, since they can be built up step by step, only by **endothermic** processes, which require external energy sources and are not inherently self-sustaining. Such reactions can occur, of course, in any region of sufficiently intense nuclear reaction among lighter elements already present, much as the trans-uranic elements such as plutonium are created by the bombardment of uranium isotopes by neutrons within a nuclear reactor.[10] But, except in the centers of stars, the *rate* of nuclear fission and fusion processes among stable isotopes is effectively zero. It follows, therefore, that heavy elements could not have been created in significant quantities to begin with, or survived long, in a steady-state (near equilibrium) stellar interior. Thus, the heavy elements in the crust of the earth must have been created in a non-equilibrium environment such as a supernova explosion. As one would expect, the cosmic abundance of elements declines rapidly with atomic weight (Figure 3.4).

With some stars the expansion that occurs late in the life cycle can be quasi-explosive, resulting in a very rapid consumption of the remaining fuel. This is

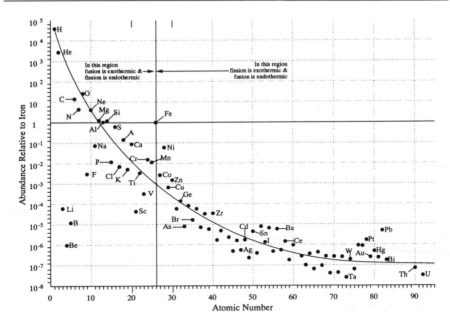

FIGURE 3.4. Cosmic Abundance of Elements Relative to Iron. Source: Hoyle and Wickramasinghe, 1978; Appendix 1.

called a **nova.** Or, more rarely, the expansion may be followed by a gravitational collapse that in turn results in an extremely violent event, a **supernova.** Exploding supernovae probably account for most of the dispersed heavy-element matter in the universe, including that found in the earth's crust. In other words, the matter in the earth, and probably the other inner planets, did not originate in the same gas cloud from which the sun was formed, but are almost certainly leftovers from an earlier supernova.[11] It is likely that this heavy mass, already cool, was somehow captured by the gravitational field of our nascent sun, perhaps as dust that later agglomerated into planetary masses.

In this context, it is interesting to note that interstellar clouds contain "dust" particles plus a variety of volatiles. Dust particles comprise graphite, metals, silicates, polysaccharides, and other organic polymers in submicron sizes. They can act as condensation nuclei for gases. Gases known to be present in such clouds, based on radio frequency absorption data, include both neutral molecules and ions (Table 3.2). Organic materials, including amino acids, are also present in significant quantities in one type of meteorite (carbonaceous chondrite). Although the origin of the chondrites themselves is still disputed, some are known to be older than the earth itself (4.6 billion years).

TABLE 3.2. *Molecules Observed in Interstellar Clouds and Comets.*

No. of Atoms	Interstellar Clouds	Comets
2	H_2, OH, NS, SO, SiO, SiS, CH, CN, CO, CS	CH, CN, C_2, CH, OH, CO, N_2
3	H_2O, H_2S, SO_2, CCH, CO_2, HCN, HNC, HCO, OCS	NH_2, CO_2, H_2O, HCN
4	NH_3, H_2CO, HNCO, H_2CS	NH_3
5	H_2CHN, H_2NCN, HCOOH, HC_3N	
6	CH_3OH, CH_3CN, $HCONH_2$	CH_3CN
7	CH_3NH_2, CH_3C_2H, $HCOCH_3$, H_2CCHCN, HC_5N	
8	$HCOOCH_3$	
9	$(CH_3)_2O$, C_2H_5OH, CH_7N	

3.5. PHYSICAL EVOLUTION: DIVERSITY, COMPLEXITY AND STABILITY

In the physical world, the two processes of which we have (or think we have) some understanding, that have taken place since the first microsecond after the Big Bang, and which appear to be valid examples of phylogenic (non-deterministic) evolution, are:

1. Nucleosynthesis: During the first 3 minutes after the Big Bang, the proton-neutron ratio shifted from 50:50 toward proton predominance (86:14) before stable atomic nuclei deuterons and helium nuclei formed. Had the cooling process (i.e., the expansion of the universe) been a little slower or a little faster, the ratio of neutrons to protons would have been different. Thus the trajectory of the proton-neutron ratio was governed by the linked non-linear equations of general relativity and of quantum electrodynamics. The early evolution of the proton-neutron ratio, in turn, determined the hydrogen-helium ratio that existed 700,000 years later at the time of radiation-matter decoupling.

2. Nuclear fusion created the heavy elements in stars and solar systems. All stars have a natural life cycle based on their size and initial composition. (The initial composition of "first generation" stars born soon after the Big Bang may differ significantly, for instance, from the composition of stars born later.) As the star's hydrogen is gradually converted to helium, and thence to heavier elements, the rate of energy generation power output eventually declines, and the star collapses. If the star was originally very large, the collapse can become explosive (nova or supernova) and some of the matter can be blown off into space. This can happen many times or only once, leaving a core (a white or black dwarf, or a neutron star). Stellar explosions result in rapid cooling of the mass blown off and redistribution

of the heavier elements into interstellar clouds from which dense planetary bodies and second-generation stars are formed by a condensation process analogous to the formation of raindrops. This set of linked creation/destruction processes has determined the structure (form) of the planetary and solar systems within galaxies and of the galaxies themselves. The wide variety of forms that we observe in the cosmos strongly suggests that the outcome of any given stellar explosion is strongly dependent on the initial conditions. This is a key characteristic of non-linear dynamical systems.

Each of the two types of evolutionary processes is clearly characterized by increasing **diversity of product** and increasing **complexity of form** together with increasing **stability** (or **persistence**) of complex forms. The existence and persistence of complex forms seems at first sight to contradict the second law of thermodynamics. In fact, it does not; rather, it exemplifies another characteristic of complex, non-linear systems, namely, the existence of multiple stable "attractors" far away from final equilibrium.

When the primordial plasma cloud of nucleons and electrons was very hot, all combinations of nuclei (and, for that matter, of atoms and molecules) were theoretically *possible,* but none could survive for more than an infinitesimal time before breaking up. Thus, the information embodied in such a plasma was almost entirely determined by the number of distinct components and the energy states (temperature) of each component.[12]

As the hot plasma cooled and began to condense into nuclei and then neutral atoms, the number of theoretically possible states did not change, but the probability of forming stable **bound** states such as deuterons (p-n), tritons (p-n-n), helium nuclei (p-p-n-n), and so on rose dramatically. Bound states are capable of embodying and preserving information over time. Heavy elements are more complex than light elements. As stellar fusion added more elements to the periodic table (working up from hydrogen), the range of possible chemical combinations and physical properties continuously increased.

The four dramatic phase changes that took place during the early evolutionary history of the universe (the GUT transition, or "freezing of the vacuum," the quark transition, the electro-weak transition, and the condensation of the baryons) were largely unsuspected until the 1970s. The details are still being worked out in conjunction with the gradual unfolding of the grand unification theory (GUT) of high-energy physics. The fifth phase change, condensation of neutral matter and the decoupling of matter and radiation, has been understood for several decades.

To recapitulate the key stylized facts: (1) the universe is expanding, and the radiation and neutrino components, at least, are cooling. On the other hand, (2) the atomic-material component has evolved in the direction of increasing quantities of still-heavier elements, including elements that are inherently unstable. A large palette of chemical elements with a variety of properties is a prerequisite of all chemical reactions. In addition, (3) physical order/structure appears to be increasing, both on the galactic scale and the atomic scale.

3.6. PHYSICAL EVOLUTION AND INFORMATION

As noted above, there has been a marked tendency toward structural differentiation on the very large (galactic) scale, as well as a trend toward condensation and chemical differentiation of matter into subsystems: galactic clusters, galaxies, stars, planets, etc. Differentiation of subsystems from one another, or from the universe as a whole, is tantamount to the embodiment of morphological and thermodynamic information in matter (as discussed in Chapter 2). Both of these trends are **phenotypic** in the Faber-Proops sense.

The second law of thermodynamics is a fundamental "covering principle" for physical evolution. It states that global entropy is increasing. This, of course, implies irreversibility on the cosmic scale and recalls all the controversies related to the apparent incompatibility of micro-reversibility and macro-irreversibility. The second law is, of course, a consequence of macro-irreversibility, which, in turn, is an extrapolation from observation and experiment. Yet, even as global entropy increases, physical and chemical differentiation have also continued. These processes signify an accumulation of "structural" information, or D-information.

It is worth emphasizing that the second law of thermodynamics is now generally acknowledged to be perfectly consistent with an increase in material order or structure, even in biological organisms. ("Entelechy" is dead. Few if any scientists today would argue that there are certain classes of phenomena to which the second law does not apply, as was frequently argued in the past.) On the other hand, nothing in classical near-equilibrium thermodynamics *per se* actually explains the increase in orderliness that we observe.

One final point of interest deserves mention here, in view of the relationship between order and entropy. The second law of thermodynamics states that global entropy can only increase. Yet, up to the time of the disappearance of free electrons (when the universe was less than 1 million years old), particles and photons were (roughly) in thermodynamic equilibrium. It would seem to follow that entropy must have already been at a maximum! How could it increase further? Has there been a resurrection of some sort from the "heat death" postulated by Nernst? According to some current views, the answer is essentially yes [Marx, 1984, 1987].

This apparent paradox is still being argued. It may be a question of semantics: If the ultimate equilibrium is a state "where nothing happens or can happen," then the early universe could have been in thermal equilibrium. But it was not in mechanical equilibrium precisely because it was (and still is) *expanding and doing work against gravity.* The rate of the expansion itself indicated the magnitude of the mechanical disequilibrium. (Only the *internal* degrees of freedom were in equilibrium.) Once one realizes that the macro-variables (volume, density, gravitational potential, and radiation temperature and pressure) must be considered, one can see that the transition from radiation-dominated to matter-dominated dynamics did, *in effect,* create thermal disequilibrium (and order) from a prior condition of thermal equilibrium. Or, in another sense, the transition may have simply changed the nature of the approach to thermal equilibrium. In this view, thermal equilibrium is still far away in the global sense.

On the other hand, since the residual neutrinos and antineutrinos, the residual black-body radiation, and matter effectively decoupled long ago, one can reasonably think of each component as being isolated and independent from the others, except to the extent that cooling matter radiates heat. The condensation of neutral matter from radiation enabled the latent order/structure (stars, galaxies, etc.) to emerge. In effect, the condensation processes discussed earlier in this chapter correspond to the unmixing of matter and radiation. Just as mixing corresponds to increasing entropy, so unmixing increases D-information (Chapter 2). It can plausibly be argued that "matter entropy" has decreased, along with an increase in "matter information," although "radiation entropy" continues to increase. Thus, it is possible for the matter component to become more ordered and D-information rich even as the radiation component becomes cooler and D-information poor.

The information needed to describe a complex object is greater than the information needed to describe a simpler one. The D-information needed to characterize complexity is, in effect, "embodied in" complexity. Thus, D-information seems to increase over time as complexity itself increases. The **material composition** of stars (and cooler matter) is, in turn, a function of the relative abundances of the heavier chemical elements relative to the abundance of hydrogen. (A relatively high abundance of heavy elements also favors chemical evolution, as will be seen.) Thus, complexity breeds complexity, and embodied D-information breeds embodied D-information.

As recent discoveries in high-energy physics have been applied to cosmology, a fairly remarkable fact has emerged. While no theory yet formulated is able to derive the numerical values of the basic physical constants,[13] it is becoming increasingly clear that, if these constants had significantly different values, life as we know it could not exist at all [Sussman 1985, Hawking 1988]. One of the most sensitive of these numbers is the charge of the electron. If its numerical value had been even very slightly different, stars could not "burn" hydrogen and helium to form heavier elements, or else they would have done it so slowly that there would have been no supernovae. Another very sensitive number is the original rate of expansion of the universe (in relation to the gravitational constant). If the expansion had been only slightly slower, the universe would not have reached its present age. It would long ago have reached its maximum volume and started a gravitational recollapse—allowing insufficient time for the evolution of intelligent life. If it had been very slightly faster the density of matter would have been too low for stars to have formed at all.

The fact that the conditions for life seem to depend so sensitively on the numerical values of the fundamental physical constants has led many physicists to wonder whether this could be an accident. Two versions of an **anthropic principle** have been put forward. The weaker version states, in essence, that "in a universe that is large or infinite in space and/or time the conditions necessary for the development of intelligent life will be met only in certain regions that are limited in space and time. The intelligent beings in these regions should therefore not be surprised if they observe that their locality in the universe satisfies the conditions that are necessary for their existence" [Hawking ibid p.124].

The stronger version of the principle resembles Copernican anthropocentrism: The universe is the way it is *in order to* provide the necessary conditions for intelligent life. Many scientists instinctively recoil from such a view (it is counter to the whole history of science), but it has its adherents.

Yet the very same facts that have led some physicists to ponder anthropic principles could be clues to something else. I have already noted that physical evolution is apparently characterized by increasing complexity, which is tantamount to an accumulation of D-information embodied in matter. This fact, together with the extreme sensitivity of evolutionary outcomes to starting conditions (including parametric values), suggests an alternative to the anthropic principle: *Evolution maximizes the embodiment of D-information (complexity).* In other words, from the beginning of the universe, it seems, the evolutionary choices that have been made were just those that permitted the most complex forms of matter, including ourselves, to exist.

I have characterized physical evolution in terms of increasing **diversity, complexity** and **stability.** It remains to express these concepts more rigorously in terms of **embodied D-information.** This task cannot be completed at present, however, because the first two of these three characteristics are not well defined in the mathematical sense. (Stability can be measured fairly well, in some physical systems, in terms of mean lifetime.) Until diversity and complexity are expressed adequately in quantifiable measures, I cannot proceed much further. Of course, I am not the first to rely on heuristics in the absence of quantification in this slippery field, and the same difficulty will arise again in the context of socio-economic evolution, where we shall also consider SR-information.

ENDNOTES

1. One of the most important discoveries of the past 40 years has been Lee and Yang's discovery in 1956 of nonconservation of parity in so-called "weak" interactions, which led to a theoretical understanding of β-decay processes. It also led, later on, to an appreciation of the importance of symmetry-breaking *per se.*

2. Einstein's failure to make this prediction himself was due to his commitment to the idea of a stationary universe. (This was an idea universally accepted at the turn of the century, and it was one of the sources of Boltzmann's conceptual difficulties in reconciling Newtonian reversibility with the second law of thermodynamics.) In Einstein's case, he had to introduce an artificial and non-physical term, which he called the *cosmological constant,* to prevent the universe from either collapsing or exploding in his relativistic field equations. Friedmann omitted this term and worked

out the implications. Later Einstein himself agreed that the term was inappropriate and called it his "biggest blunder."

3. The best-known challenger to the Big Bang theory is Hannes, Alfvén, the Nobel Prize-winning Swedish plasma physicist and cosmologist. Alfvén postulates an infinite universe in which large-scale structures and their dynamics are dominated by electromagnetic forces between vast currents of ionized matter and antimatter. It is the explosive interactions between matter-antimatter regions at the supercluster level of aggregation that accounts for the Hubble shift. See Alfvén [1966, 1981, and 1983]. For a recent exposition, see Lerner [1991]. A number of serious objections to the Alfvén hypothesis have been made, however, as well as some alternative theories to explain the Hubble expansion without a Big Bang. I cannot discuss all the possibilities here.

4. A second and possibly related problem with the Big Bang theory in its present form is the so-called flatness problem, first pointed out in the 1970s by Robert Dicke and Jim Peebles. In brief, unless space-time is very nearly geometrically "flat," one of two things would have happened by now, assuming Einstein's general theory of relativity to be valid. Either the universe would have gravitationally collapsed almost immediately, or (if space-time were actually flat) it would have expanded much faster than it has, leading to a much lower density than is actually observed. In shoft, both the lifetime of the universe and its present density are incredibly sensitive to the exact value of the density parameter in the early stages. Assuming the earliest time (10^{-12} sec after the Bang) for which we can reconstruct the composition, energy density, and temperature with some confidence, a density parameter consistent with stable Hubble expansion at the observed rate, within a factor of 10—without either premature collapse or extreme dilution—would have had to be fixed within 1 part in 10^{27}. Or, if we "start" the process with conditions that existed when the universe was 1 second old, the initial value of the density parameter had to be fixed to 1 part in 10^{15} to be consistent with the evolutionary trajectory.

5. In fact, there are several other reasons to postulate the existence of "dark matter." One is the fact that observed orbital velocities of various objects known to be in the gravitational fields of our own galaxy, the Milky Way, but outside the orbit of the earth are moving at speeds consistent with a mass at least ten times the luminous mass of the Milky Way. The luminous mass, in turn, can be estimated with reasonable accuracy from well-established relationships between star luminosity and mass. For a more thorough exposition, see Krauss [1989].

6. Einstein's is not the only theory of gravitation and spatial curvature; an alternative has been put forward by the physicist Huseyin Yilmaz [1985]. Yilmaz's theory predicts the bending of light in a gravitational field and the second-order correction to the orbit of the planet Mercury (the precession of the perihelion) that seemed to confirm Einstein's theory. However, Yilmaz's theory differs from Einstein's in several fundamental respects. First, unlike Einstein's theory, it explicitly includes the gravitational field self-energy in the mass-energy tensor. Second, unlike Einstein's theory, it is quantizable (Einstein's theory is thought to be inconsistent with quantum mechanics). Third, in contrast with Einstein's theory, the so-called Schwartschild radius is always zero: There is no finite density of matter such that light would not be able to escape. Thus the Yilmaz theory does not allow for "black holes."

7. These "standard" theories are:
   (1) Unified "electro-weak" interaction
   (2) Quantum theory of colored quarks
   (3) General theory of relativity (see previous note).

8. The distribution of the other, heavier chemical elements were subsequently determined by the slower processes of nuclear fusion within stars. But the stars themselves were only formed later on by a gravitational collapse of the early gas clouds. This process was first postulated by the British astrophysicist James Jeans. It is now known that at least eight different nuclear processes must have been involved in the synthesis of the atomic elements.

9. The observed inhomogeneities in the distribution of matter in the visible universe—up to the scale of superclusters at least—are far too great to be accounted for by random statistical fluctuations in the primordial-matter cloud that must have existed at the time of the transition from radiation-dominated to matter-dominated dynamics. James Jeans' original calculations assumed Newtonian gravity (no curvature of space) and concluded that the original "seed" inhomogeneities would grow exponentially over time. This would have allowed them to reach observed magnitudes. However, when the calculations were redone in the 1920s by the Russian physicist Evgeny Lifschitz, using Einstein's general theory of relativity (curved space), it was found that the inhomogeneities grow only in proportion to the expansion of the universe itself.

   Based on the Lifschitz calculations, the argument goes like this: Since the radiation-to-matter transition was finally completed when the universe was only about 1000 times smaller than it is now, the minimum "seed" size for a star of solar mass is about 1/1000 of a solar mass. By contrast, statistical fluctuations in a volume (of any size) containing enough atoms for a solar mass (10 to the power 57) would be of the order of the square root of that number, or less than 10 to the power 29—far, far less than the required 10 to the power 54! The obvious implication is that the primordial cloud must have already been extremely inhomogeneous.

   One proposed solution to the problem is the "expansion" theory suggested by Alan Guth of MIT [Guth and Steinhard 1984]. Crudely, the idea is that a sufficiently rapid rate of expansion in the first microseconds after the Big Bang could somehow "flatten" space-time to make the flat-space calculations relevant. Yilmaz's alternative theory of gravitation, mentioned in footnote 3, offers another way out of the dilemma. According to Yilmaz, "seed" inhomogeneities in his theory would grow at almost the same rate as in flat (Newtonian) space. In other words, Jeans' calculation may be roughly correct after all. If so, the difficulty disappears and the Big Bang theory survives.

10. In the absence of neutron bombardment (conditions on the earth's surface), there are inherently stable isotopes of all elements with atomic numbers up to 80, which is lead. All heavier elements are spontaneously radioactive. They belong to three series known at the actinides. The end product of each series is an isotope of lead. (The three branches are the uranium-radium series, which decays to Pb 206; the thorium series, which decays to Pb 208 and the actinium series, which decays to Pb 207.) For obvious economic reasons, nuclear explosives and nuclear (fission) power plants utilize only the energy available from fission of the actinides; indeed, only the most

unstable of the latter (U 235 and the artificial element P 240). Energy could theoretically be obtained from fissioning the stable heavy elements, like lead, but the process would be technically far more difficult (and less productive) than the fusion of light elements.

11. In fact, the state of decay of the radioactive elements in the earth's crust suggests that the supernova that created them must have occurred 4.55 billion years ago. This can be determined from the decay rates of various isotopes and the ratios of certain elements and their decay products. For instance, argon is almost entirely a decay product of potassium.

12. In actual fact, the plasma at the time in question contained photons, nucleons, electrons, and neutrinos and antineutrinos. The latter two components had already decoupled (ceased to be in thermal equilibrium with the plasma) at an earlier time, and cooled in direct proportion to the expansion of space. However, the electrons and nucleons at that time were still in thermal equilibrium with the photons, and were still being heated by electron-positron annihilation reactions and photon emissions (in effect, latent heat) from nucleon formation. Hence, the neutrino-antineutrino gas was about 30% cooler than the rest of the plasma [Weinberg 1978].

13. In fact, the inability of GUT to calculate these constants, especially the mass ratios, is counted as a major weakness [Fritzsch 1986; Motz 1986]. The values of the physical constants could indeed have also been in some sense accidental, "a result of God's playing dice (so to say, with Einstein)" [Sussman 1985].

REFERENCES

Alfvén, Hannes, *Worlds-Antiworlds: Antimatter and Cosmology*, W. H. Freeman and Co., San Francisco, 1966.

Alfvén, Hannes, *Cosmic Plasma*, D. Reidel, Holland, 1981.

Alfvén, Haanes, "On Hierarchical Cosmology, " *Astrophysics and Space Science* **89**, 1983 :313-324. references

Bekenstein, J., "Gravitation and the Origin of Large Structures in the Universe," in: *14th ICUS*, International Conference on the Unity of the Sciences, Houston, Tex., 1985.

de Vaucouleur, G., "The Case for an Hierarchical Cosmology," *Science* **167**, Feb. 27, 1970 :1203-1213.

Fritzsch, H., *The Creation of Matter*, Basic Books, New York, 1984.

Fritzsch, H., "Particle Physics and the Early Universe," in: *14th ICUS*, International Conference on the Unity of the Sciences, Houston, Tex., 1985.

Fritzsch, H., "Energy and Matter in the Early Universe," in: *15th ICUS*, International Conference on the Unity of the Sciences, 1986.

Guth, Alan and J. Steinhard, "The Inflationary Universe," *Scientific American* **250**(5), May 1984 :116-128.

Hawking, S., *A Brief History of Time: From the Big Bang to Black Holes*, Bantam Books, New York, 1988.

Horgan, John, "Universal Truths," *Scientific American* **263**(4), October 1990 :108-117.

Hoyle, Fred and C. Wickramasinghe, "Life Cloud," HarperCollins Publishers, New York, 1978.

Krauss, Lawrence M., *The Fifth Essence: The Search for Dark Matter in the Universe*, Basic Books, New York, 1989.

Lerner, Eric J., *The Big Bang Never Happened*, Random House, New York, 1991.

Marx, G., "Thermal History of the Universe," *Fortschritte der Physik* **32**, 1984 :185.

Marx, G., "Entropy of the Universe," *Acto Physica Hungarica*, 1987.

Motz, L., "Symmetry and the Laws of Nature," in: *15th ICUS*, International Conference on the Unity of the Sciences, 1986.

Sussman, G., "The Anthropic Principle and the Arrow of Time," in: *14th ICUS*, International Conference on the Unity of the Sciences, Houston, Tex., 1985.

Tully, R. Brent, and J. R. Fischer, *Atlas of Nearby Galaxies*, Cambridge University Press, Cambridge, England, 1987.

Weinberg, Alvin M., "Reflections on the Energy Wars," *American Scientist* 66(2), March/April 1978 :153-158.

Yilmaz, Huseyin, "New Theory of Gravitation," in: Ruffini (ed.), *Proceedings of the Marcel Grossman Meeting*, North Holland, Amsterdam, 1985.

Chapter 4

Geological and Biochemical Evolution

4.1. GEOLOGICAL EVOLUTION OF THE EARTH

As sketched briefly in the previous chapter, the more inhomogeneous the primordial matter, the greater the likelihood that different sizes and configurations of stars will develop. Greater inhomogeneity also increases the possibilities for relatively small and cool bodies (planets, moons, comets, asteroids) to form and survive for long periods of time in stable gravitational orbits around some star. This, in turn, results in the creation of many potential habitats near enough to a source of light and heat to drive various exothermic physical and chemical processes (such as the hydrological cycle) but not so close as to be burned to a cinder. In the evolution of the solar system, mainly phenotypic changes seem to be involved (in the sense defined by Faber and Proops, 1989). In geological evolution, however, there is also an element of ontogeny (predictive determinism) together with some unpredictable (phylogenic) changes.

As noted previously, the material of the earth seems to have been formed by a supernova 4.55 or 4.6 billion years ago. It is not known how long it took for the material, presumably already in orbit around the youthful sun, to agglomerate into planetary form. It is known from isotopic analysis (mainly of carbon isotopes 12, 13, and 14) that living matter did not appear on earth for another 600 million to a billion years thereafter.[1]

The first aeon (billion years) of the earth's history, known as the *Hadean* epoch, was a period of numerous collisions with meteors and planetoids. (Were it not for the smoothing effects of weather and living organisms, the earth would be as pockmarked as the moon.) Among the purely geophysical processes of importance were volcanic emissions of carbon dioxide, hydrogen sulfide, methane, and ammonia. Vulcanism was much more intense when the earth was young because the internal heat source that drives it (radioactive decay of uranium and thorium) was three times greater than it is now. Hot basaltic rock from volcanic eruptions, especially in the ocean, came into contact with ocean water, releasing hydrogen.

A second important geophysical process was the loss of hydrogen from the atmosphere into space. Under the intense ultraviolet radiation at the top of the atmosphere (and with no protection from the ozone layer) molecular hydrogen would be easily broken up and ionized to a plasma of free protons and free electrons. These particles can escape easily from the earth's gravitational attraction. This process of hydrogen loss continued as long as free hydrogen was being created by the breakup of such hydrogen-containing gases as H_4C (methane), H_3N (ammonia), H_2O (water), and H_2S (hydrogen sulfide) under the UV irradiation in the atmosphere. In chemical terms, the loss of hydrogen was partially responsible for the slow but continuous shift in the reduction/oxidation balance from "reducing" to "oxidizing." This shift has been accompanied by gradual acidification. The early oceans must have had a pH of around 11; currently it is closer to 8.3, while rainfall averages 5.5. In other words, the earth's chemistry shifted gradually from alkaline to acid. It is no longer shifting however, because any molecular hydrogen that is produced (for example by photolysis of water vapor in the atmosphere) is immediately re-combined with oxygen.

A third purely geophysical process of some importance was, and still is, the continuous removal of gaseous carbon dioxide by weathering of newly exposed basaltic rock, especially the conversion of orthoclase to kaolin (clay). Each year some CO_2 is permanently removed from the atmosphere by this mechanism, the precise amount being unknown but perhaps of the order of 0.01 percent of the amount in the atmosphere. While the weathering process is slow compared to the biological process of photosynthesis, it has operated for 4 billion years. The amount of mineral carbonate in the earth's crust is estimated to be about four times the amount of organic carbon in sedimentary rocks (shale), and perhaps 20,000 times the world's total coal reserves.

The balance between atmospheric gases in the early earth's atmosphere cannot be precisely known. It is evident that the temperature of the earth's surface must have been largely determined by that balance. One historical reconstruction puts the earth's temperature at the time of life's origin at about 23° C, just a few degrees warmer than the present [Owen et al., 1979]. This scenario assumes a carbon dioxide concentration of 200 to 1,000 times greater than the present level, depending on the amount of nitrogen present. In any case, the earth's average temperature could not have been outside the 0- to 50-degree range, or life could not have started.

Obviously, water is plentiful on the earth's surface. The standard geological theory is that the atmosphere of the early earth was probably created by the degassing of the magma (via vulcanism) and other physio-chemical processes that occurred *after* the agglomeration and solidification of the earth into its planetary configuration. In other words, according to this theory, the atmosphere was not left over from some primordial gas cloud. Any such water would have boiled off before the earth solidified. Unfortunately, this theory does not satisfactorily account for the quantity of water on the earth's surface.

The continued existence of liquid water on the surface of the earth is, in any case, not fully explained by purely geological theories, because some water is known to be lost from the atmosphere due to the breakup of the water molecule (induced by high levels of UV radiation) into H^+ and OH^- radicals. Of course, H^+ is nothing more than a free proton, which can escape fairly easily from the earth's gravitational field. Thus hydrogen mostly from water tends to "leak" from planetary surfaces. But it has been quite firmly established that the water in the oceans cannot be accounted for by the leaching or weathering (i.e., dehydration) of crustal rocks that has taken place since the earth was formed. Whereas the distribution of most chemical elements in seawater is consistent with such leaching or weathering, an excess of certain volatile elements, notably water itself, carbon dioxide, nitrogen and chlorine, cannot be accounted for. Unable to account for the oceans otherwise, Rubey [1951] postulated vulcanism as the source of the excess volatiles.

The most attractive alternative explanation has been put forward in the past few years by planetary physicist Louis Frank and his colleagues [Frank et al., 1986]. Based at first on NASA satellite observations of curious, short-lived "holes" in the stratosphere, Frank's theory postulates the existence of a hitherto unsuspected class of "small comets," each about 100 tons in mass, consisting mostly of ice, plus a little carbon and dust, which strike the earth's atmosphere on the average of twenty times per minute. After a bitter scientific controversy, a number of attempts have been made to detect these small comets. The measurements, and their interpretation, are still in dispute, but they have now been confirmed by astronomical observations [Yeates, 1989; Frank, 1990]. The weight of evidence now clearly favors Frank's theory of continuous bombardment of the earth's atmosphere by small comets.[2]

This continuing supply of comets would account for all the water in the oceans (and then some, but none of the data available so far would justify worrying over a mere factor of 3 or 4). Indeed, the rate of water accretion appears to be 1 inch per 10,000 years, which is undetectable. The composition of the early pre-life atmosphere is not known for certain. Likely components, in addition to water, would be carbon dioxide, hydrogen, ammonia, and methane. (The latter two are conspicuously present in the atmospheres of Jupiter and the outer planets, but not the inner planets.) Any free hydrogen present in the early atmosphere, or created by photo-chemical reactions, would have diffused outward into space fairly rapidly. The earth's gravitational field is not strong enough to retain gaseous hydrogen (or helium).

On the other hand, in the early stage of hyperactive vulcanism, gaseous hydrogen would have been regenerated continuously by chemical reactions between water and ferrous iron in volcanic magma or in the oceans.[3] In effect, the iron-rich magma grabs oxygen from the water molecules, releasing hydrogen. This reaction (and similar ones) would have tended to use up any free surface water or atmospheric water vapor during the Hadean epoch. It could help to account for a continuing supply of hydrogen in the early atmosphere, however.

The most generally accepted geological theory of the atmosphere today holds that the pre-biotic atmosphere of earth consisted mostly of carbon dioxide, with perhaps some gaseous nitrogen, traces of hydrogen, ammonia, and hydrogen sulfide, but no free oxygen [Holland, 1984]. It was what is known as a "reducing" atmosphere. Given the presence of free hydrogen, hydrogen sulfide, methane, and ammonia, abiotic processes can theorectically generate amino acids, the basic chemical "building blocks" of life. This topic will be considered later in this chapter.

The atmosphere and oceans maintain a sort of long-term balance partly through the hydrological cycle: evaporation, convection, precipitation, runoff. This cycle is possible only because the earth's temperature lies between the freezing and boiling points of water. In the long run, the stability of the cycle also depends on an extremely unusual property of water: It *expands* upon freezing. If it did not, ice would tend to sink to the bottom of lakes and oceans, and heat loss at the liquid surface would be much more rapid. Ice at the surface of the Arctic Ocean, for example, acts as a thermal blanket permitting a rich marine life to exist beneath the surface even in winter. If ice sank instead of floating, the arctic region would be as lifeless as the Greenland and Antarctic glaciers.

The hydrological cycle is, of course, an essential precondition for two important, irreversible geological processes: weathering and sedimentation. Weathering is the chemical/physical breakdown of rocks into fragments. Microscopic cracks in the rock expand as water seeps into them and freezes. Many cycles of expansion and contraction cause rock surfaces to crumble. Then wind erosion or water erosion removes the loose material. Eventually it accumulates in valleys and river bottoms, whence it migrates eventually to the oceans. Soluble mineral salts (such as sodium and potassium chlorides) accumulate in the oceans; insoluble minerals accumulate on the ocean floors. Chemical changes also occur at the surface of newly exposed rock faces. For instance, native metals or sulfur tend to oxidize, while some magmatic minerals react with carbon dioxide to form carbonates.

Oxidation depends, of course, on the availability of free oxygen in the atmosphere; this is a relatively recent event, geologically speaking. (The origin of the atmospheric oxygen will be considered later.) Another of the essential preconditions for life at least, terrestrial life is that there be *enough* free oxygen (for respiration), *but not too much*. The probability of a fire being ignited by lightning depends strongly on the fuel's water content. In general, each 1 percent increase in atmospheric oxygen content would increase the likelihood of ignition by 70 percent. If the oxygen concentration in the atmosphere were 4 percent higher than it is (25 percent instead of the present 21 percent), there could be no forests or other large accumulations of cellulosic material above ground. Fires would be too frequent and too hot to permit it [Lovelock, 1979].

Sedimentation has played an essential role in the maintenance of conditions permitting the sustenance of living matter. **Photosynthesis** depends on the existence of an atmosphere containing carbon dioxide, which must be replaced as it is removed. The waste product of photosynthesis is oxygen. Thus, another process to recombine carbon and oxygen is required to balance the cycle. In nature, this is accomplished by the **respiration** of plants, animals, and microbes. Respiration presupposes the existence of free oxygen, just as photosynthesis presupposes the existence of free carbon dioxide. But free oxygen is quite reactive. If all the chemical elements and compounds in the earth's crust were allowed to interact chemically, seeking the optimum partnership arrangements (the lowest energy state, or thermodynamic equilibrium), there would certainly be no free oxygen, nitrogen, carbon, hydrocarbons, sulfur, native metals, or metallic sulfides (the latter would oxidize to sulfates).

Yet, in fact, oxygen and nitrogen exist as gases in the atmosphere, while the rest except native metals are found in sedimentary rocks. This physical separation, essential to life, is basically due to sedimentation. Detritus from living organisms in the ocean falls to the bottom of the sea, where it is covered by mud and sediments and protected from contact with oxygen. Then, over geological aeons, the organic materials are "cooked" and carbonized at high pressures,[4] and carbon remains in the sedimentary rocks. In fact, on the average, about 0.1 percent of the carbon fixed annually by photosynthesis is left behind in this way buried in sediments. It was first suggested in 1951 (by Rubey), and is now generally accepted, that carbon burial is responsible for the existence of free oxygen in our atmosphere.

Actually, free sulfur is nearly as important. In fact, there is more than enough organic carbon in sedimentary rocks, together with elemental sulfur or sulfides (mostly iron pyrites, Fe_2S) to combine with all the oxygen in the atmosphere several times over. It is quite evident that a lot of the oxygen that might have been present in the pre-Cambrian atmosphere (combined originally with carbon or hydrogen) has become chemically unavailable. There is no mystery as to where it must have gone: It combined mainly with ferrous iron (FeO) and sulfur (as iron pyrite), yielding ferric iron (Fe_2O_3) and sulfates (SO_4). For each molecule of free oxygen now in the atmosphere there are four to eight molecules tied to iron and eleven molecules tied to sulfur.

Meanwhile, carbon dioxide has also been disappearing from the atmosphere. Only part of it (a small part) is accounted for as organic matter or its detritus. Most of it has been precipitated out on the sea floor as insoluble carbonates, mostly calcium. The ratio of organic to inorganic carbonate deposition in sedimentary rocks has apparently remained constant for a long time at about 1:4. That is, four CO_2 molecules are buried as inorganic carbonate for every one buried as organic material. The amount of CO_2 in the ocean (as CO_2, CO_3, and HCO_3) exceeds the amount in the atmosphere by a factor of more than 5,000. Thus, the amount that can be dissolved in ocean water is critical to the planetary carbon cycle. It depends, among other things, on water temperature and pH. The presence of such clay minerals as silica (SiO_2) and alumina (Al_2O_3) determines the carbonate formation and removal rate.

Some geologists [e.g., Holland, 1978] have argued for the existence of a non-biological feedback mechanism capable of stabilizing atmospheric CO_2. The equilibrating mechanism might work like this: Higher ocean temperatures would reduce the ability of the ocean to absorb CO_2 and might cause a net desorbtion, causing water temperatures to rise still higher because of the greenhouse effect. But this would increase rainfall, weathering, and erosion rates. Clay minerals washed into the oceans would tend to react with dissolved CO_2, precipitating insoluble $CaCO_3$. Eventually the level of dissolved CO_2 would fall to the point where reabsorption of CO_2 from the atmosphere begins. Of course, such an equilibration process would be very slow and could hardly operate on a time scale of less than thousands of years.

The origin of free nitrogen in the earth's atmosphere is not well known, at least in quantitative detail. We have some clues, however. Free nitrogen does not exist in other planetary atmospheres. Moreover, nitrogen is an essential element for living organisms, and living organisms appear to prefer nitrogen as ammonium (NH_4) or nitrate (NO_3) forms. Ammonia (NH_3) was probably present in the early earth's atmosphere (and it exists in large quantities in the atmospheres of the outer planets). Most likely, ammonia was present in early oceans as ammonium and hydroxyl ions, and early organisms utilized this form. Some primitive denitrifying bacteria probably used this reaction as a source of metabolic energy, yielding free nitrogen. (Similarly, some primitive marine organisms obtained metabolic energy by reducing hydrogen sulfide, leaving elemental sulfur deposits.) Later, as free oxygen was released by photosynthesizers, ammonia would have been oxidized to nitrogen oxides and nitrates. Higher plants have learned to utilize soluble nitrates, and more recently still to play a significant role in direct nitrogen fixation, from free nitrogen, thus closing the nitrogen cycle.

Another geological process, tectonic uplift, is important to the long-term habitability of the earth. In the first place, if the removal of carbon from the atmosphere continued with no mechanism for geological return, the atmospheric oxygen level would eventually rise to the point of excessive fire hazard. But, possibly even more important, one important biological nutrient, phosphorus, moves in the short run in one direction: from the land (via erosion and runoff) to the sea.[5] It can be returned to the land *only* by this means. There is no stable gaseous compound of phosphorus (analogous to carbon dioxide, methane, hydrogen sulfide, ammonia, or nitrogen oxides) that can carry phosphorus by atmospheric convection from the oceans back to the land. Fortunately, deposits of phosphate rock (from the bodies of marine organisms) have been uncovered by tectonic uplift and are being intensively exploited today as a source of fertilizer and chemicals.

It now seems clear that life itself must have played a major role in generating the oxidizing atmosphere that exists today (Table 4.1). Certainly, the present atmosphere is very far from thermodynamic equilibrium, in the sense that its constituents must be constantly renewed in some cases, or removed in others. For instance, the traces of methane in the atmosphere can be accounted for only

TABLE 4.1. *Planetary Atmospheres Before and After Life.*

	Venus	Earth A	Mars	Earth B	Earth C
Carbon dioxide	98%	98%	95%	99%	0.034%
Nitrogen	1.9%	1.9%	2.7%	0%	79%
Oxygen	trace	trace	0.13%	0%	21%
Argon	0.1%	0.1%	2%	1%	1%
Hydrogen					1 ppm
Temperature (°C)	477	290	−53	15	13
Pressure (bars)	90	60	.0064	1.0	1.0

Earth A Hypothetical lifeless planet interpolated between known conditions on Venus and Mars, reported by Lovelock [1979, Table 2].

Earth B Hypothetical chemical equilibrium, assuming uniform mixing of all planetary materials and allowing all reactions to go to completion at 15 °C. Calculated by Sillén, reported in Lovelock [1979, Table 1].

Earth C Actual earth, today.

by a continuous rate of emission, mainly from anaerobic decay processes. (Once in the atmosphere, methane is quickly oxidized by OH^- to CO, which subsequently oxidizes to CO_2 and H_2O.) And, as already pointed out, free oxygen can be produced only by photosynthesis; the existence of a large reservoir in the atmosphere is certainly a direct consequence of the disappearance by burial, over several aeons, of a large amount of fixed carbon.

The "distance from equilibrium" of the earth's atmosphere is indicated by Table 4.2. This distance can be regarded as a primary argument for the theory known as Gaia that life processes play an essential role in stabilizing the environment in conditions suitable for life [Lovelock, 1972, 1979]. As Table 4.1 indicates, a lifeless earth would almost certainly have an atmosphere consisting mainly of CO_2 and be too hot (because of an extreme form of the greenhouse effect) for water to exist in liquid form. Such an earth would therefore be too hot for life processes as we know them to have originated or survived. Unfortunately, the composition of the early atmosphere is not well enough known to calculate its radiation balance, hence its temperature, with high confidence. All we can really say is that the earth's temperature cannot have been very much hotter or colder than it is now, or life could not have existed.

In this context, Figure 4.1. shows the actual surface temperature on the earth's surface for the past several billion years. It has varied within very narrow bounds, despite the 30 percent increase in the intensity of incident solar radiation during that time. This, too, seems to support the Gaia hypothesis.

To summarize some key points: The geological and atmospheric structure of any planet is determined by its size, density, crustal composition, interaction with extra-planetary material, and its orbit around the sun. The more planets (and satellites) there are, and the greater their variety, the more likely some of them will be suitable for chemical and biological evolution. Based on what we

TABLE 4.2. *Some Chemically Reactive Gases of the Air.*

Gas	Abundance %	Flux in Megatons/ Year	Extent of Disequilibrium	Possible Function Under Gaia Hypothesis
Nitrogen	79	300	10^{10}	Pressure builder Fire extinguisher Alternate to nitrate in the sea
Oxygen	21	100,000	None; taken as reference	Energy reference gas
Carbon dioxide	0.03	140,000	10^3	Photosynthesis Climate control
Methane	10^{-4}	1,000	Infinite	Oxygen regulation Ozone regulation
Nitrous oxide	10^{-5}	30	10^{13}	pH control
Ammonia	10^{-6}	300	Infinite	Climate control (formerly)
Sulphur gases	10^{-8}	100	Infinite	Transport of sulphur-cycle gases
Methyl chloride	10^{-7}	10	Infinite	Ozone regulation
Methyl iodide	10^{-10}	1	Infinite	Transport of iodine

Note: Infinite in Column 4 means "beyond limits of computation." Source: Lovelock, 1979, Table 3, p. 68.

know of chemistry, the possibility of life on earth was absolutely dependent on two factors: the unique physical properties of water and the unique chemical properties of carbon (notably, its propensity to form polymeric chains). Thus, the conditions most propitious for chemical evolution seem to be: (1) availability of a number of light elements, especially H, C, N, O, P, and S, in the form of gaseous compounds such as methane, ammonia, hydrogen sulfide, carbon dioxide and di-hydrogen oxide (water); (2) moderate and relatively stable temperatures (roughly intermediate between the freezing and boiling points of water, but nearer the former than the latter), so that liquid-phase chemical reactions can occur at rates that are not too slow, but not too fast for their products to be stable, and (3) an atmosphere, oceans, and a solid and fairly stable crust. Crustal stability is obviously necessary to allow time for geological and chemical evolution.

Finally, on the planetary scale, morphological diversity and complexity are clearly preconditions to geo-chemical environmental complexity. (Again, the existence of elements heavier than helium in reasonable abundance, especially carbon is also a precondition of chemical evolution.) The more environments there are, and the more abundant the heavy elements are, the more likely some environment will be favorable to a chemical evolution leading to the synthesis of complex carbon-based organic molecules with self-replicating properties.[6]

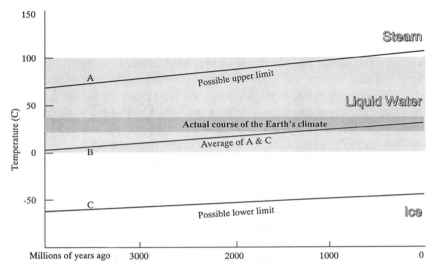

The average surface temperature of Earth has remained remarkably constant, approximately between 10 & 20 C [Lovelock 79]. Had the temperature been determined only by nonbiological subsystems, either line A or line C might have been realized. Even the average of these (B) would have made life improbable.

FIGURE 4.1. Surface Temperature of the Earth. Source: Reproduced with Permission from Lovelock, 1979.

4.2. CHEMICAL PRECURSORS TO ORGANIC SYNTHESIS

Following the period of active geophysical evolutionary change on the earth's surface was a period of genotypic chemical evolution. The exact triggers for the first organic synthesis are not known. As noted in the previous section, chemical evolution on the early earth probably required atmospheric conditions quite different from those currently prevailing. The Russian biochemist Alexander Oparin [1957] first suggested in the 1920s that primitive hydrocarbons were probably produced in a reducing atmosphere containing free hydrogen, methane (CH_4) and ammonia (NH_3). A few years later the British biologist J. B. S. Haldane [1985] also noted that the first organic molecules were most likely produced in an oxygen-free environment. Haldane and Oparin both suggested that ultraviolet radiation was the likely source of energy, and that the organic synthesis took place primarily in the oceans (the "hot dilute soup").

A very different explanation was put forward by John Óró [1961]. In brief, he suggested that at least some of the essential biochemical building blocks may have been brought to earth by comets. A similar notion seems to have been put forward by Francis Crick; the idea was worked out in some detail in the "life cloud" theory of Fred Hoyle and N. Wickramasinghe [1978].

I have already mentioned the presence, now confirmed, of gases like methane, carbon monoxide, carbon dioxide, formaldehyde, ammonia, hydrogen cyanide, and water in cold interstellar clouds. It is also now confirmed that such compounds can condense on tiny silica, alumina, or iron particles, forming a kind of icy coating. Exposed to various kinds of radiation, these particles can undergo chemical reactions, producing more complex hydrocarbons, including polymers. In fact, large comets (such as Halley's) are now known to be black and hard on the surface, and not "dirty snowballs" as had previously been thought. The most plausible explanation for the hard black surface is that it is a kind of tarry carbon-polymer, up to half an inch thick. (The tail of the comet is from occasional jets of escaping water ice or other chemicals released by punctures of the carbon mantle by collisions with small meteorites.)

The main problem with this theory is not how organic molecules could exist in comets—there seems little doubt that they can and do—but how the organic chemicals could have reached the earth's surface without burning up. Oxidation *per se* is not the problem: The lack of oxygen in the early atmosphere takes care of this objection. However, frictional heating alone, not to mention collisions with the ground, would seem (at first glance) to be energetic enough to dissociate, if not ionize, all complex molecules. Further analysis, however, yields some possible explanations of how organics could survive passage through the atmosphere. To begin with, a carbon-polymer "mantle" may have provided significant protection to the interior of a meteorite. Such a mantle would have much the same properties as the nose cone of a re-entry vehicle (which gets hot, but sheds its heat by shedding some of its surface layers).

NASA scientists Kevin Zahnle and David Grinspoon have recently proposed a "non-destructive" mechanism that swept up cometary dust containing organic molecules (such as nucleotides) and "sprinkled" it gently onto the surface of the earth over thousands of years. The low rate of this process would help explain why so few of the molecular possibilities have been adopted, a point that arises again later. A denser atmosphere might also have had a "cushioning" effect on incoming comets. Finally, the "small comet" theory of Frank [Frank *et al.*, 1986], mentioned earlier, seems to explain other problems at the same time, including the origin of the excess volatiles in the oceans themselves.

The subsequent stages of chemical evolution are not at all well understood, either, and there are competing hypotheses [e.g., Miller and Orgel, 1974; Shapiro, 1986]. The range of possibilities is wide:

> "Must the organic medium be condensed and 'hot' (something like a 'warm, thick and salty primordial bouillon') [Oparin, 1957] or rarified and cold [Goldanskii, 1979] as in dark clouds. Must the organic matter be subjected to corpuscular and electromagnetic radiation or must it be protected from them? Answers ... are absent. Moreover, the principles such answers can be based on are not clear" [Morozov and Goldanskii, 1984].

Several categories of organic molecules essential to life must somehow be accounted for (Figure 4.2). These are as follows:

- **Amino acids**, from which **peptides** and eventually **proteins** are constructed. Proteins are simply long chains of amino acids, folded over in complex

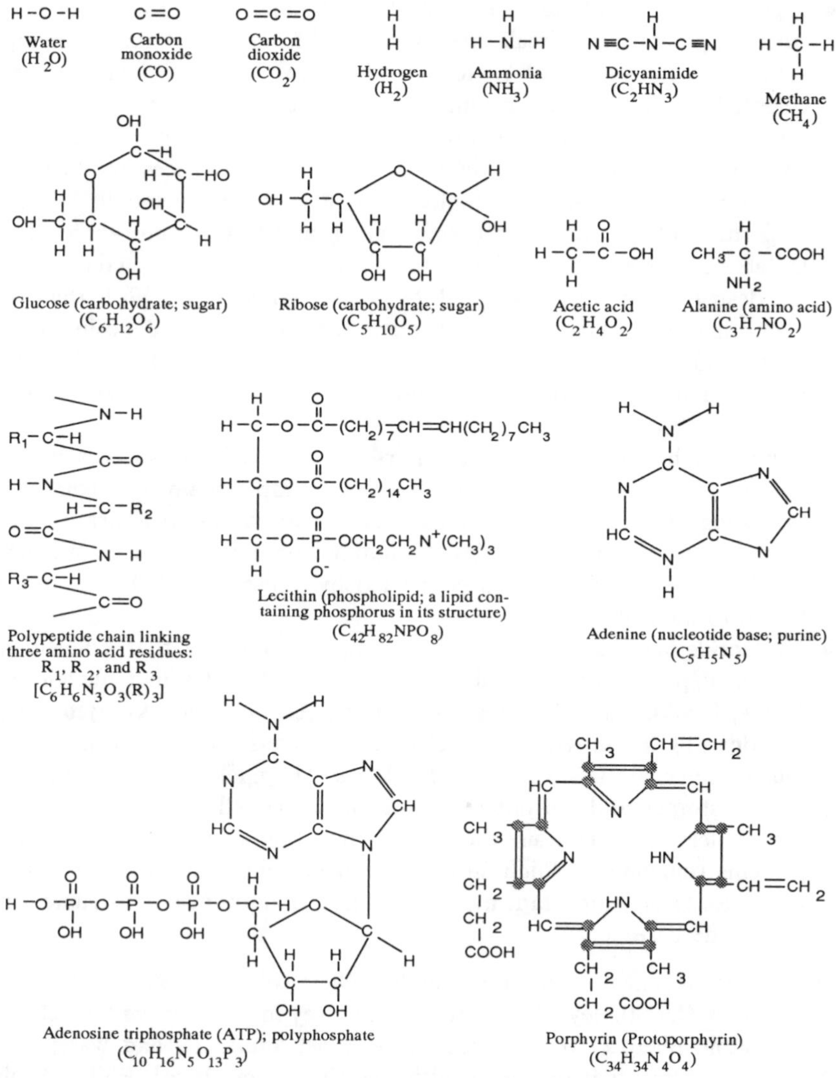

FIGURE 4.2. Basic Chemical Building Blocks of Life.

ways. Proteins play many roles in higher organisms, but, within a simple primitive cell, they are needed primarily as catalysts, known as **enzymes**. Enzymes are needed to facilitate every chemical reaction within the cell, including photosynthesis and nitrogen fixation. There are twenty amino acids in proteins, although the laws of chemistry would permit several hundred or even thousands. Thus, the limited number of amino acids found in nature is one of the things to be explained.

- **Lipids** (fatty acids), are chains of H-C-H units. **Fats**, the major energy stores of plants and animals, are constructed from lipids linked by complex alcohols, such as glycerol. Lipids linked to phosphate (PO_4) groups are the essential building blocks for semipermeable membranes. The essential function of cellular wall membranes is to maintain a controlled interior (**autopoietic**) environment. Such an environment is necessary to contain and protect the large macro-molecules that carry out metabolic functions, to permit nutrients to enter (osmotically), and to force unwanted (toxic) metabolic wastes out. Without this ability, life processes would quickly self-destruct.[7] It was suggested by Peter Mitchell in 1961 that these membranes also play an essential role in the synthesis of adenosine triphosphate (ATP), the universal energy-carrier molecule found in every known living organism. Mitchell's hypothesis (the "chemiosmotic theory") is now generally accepted.[8]
- **Sugars**, such as lactose, glucose, and ribose, are needed as sources of energy for metabolic processes. Carbohydrates are simple chains of sugars, cellulose being a familiar example. Sugars also constitute the structural backbone of many important molecules, including the nucleotide chains, which consist of five-carbon sugars connected by phosphate (PO_4) groups and linked to nucleotide bases.
- **Nucleotide bases** are the nitrogen-containing molecules that, connected to chains of phosphates and sugars, produce **nucleotides** and **nucleic acids (DNA, RNA)**. The five bases are: **guanine (G), adenine (A), cytosine (C), thymine (T)** and **uracil (U)**. Nucleotides are constructed from these bases, plus some sugars (above) and phosphates. The nucleic acids are the information storage and reproduction systems of the cell. Among other things, the nucleic acids contain the "programs" that control the synthesis of proteins (enzymes), which, in turn, govern all metabolic and reproductive activities. Other important molecules, like ATP, are also constructed from nucleotide bases.

An important milestone in understanding chemical evolution was the laboratory synthesis (by Stanley Miller and Harold Urey in 1953) of **amino acids** in a reducing atmosphere of free hydrogen, methane, ammonia, and water vapor, possibly resembling that of the primitive earth [Miller, 1974]. Philip Abelson later pointed out the possibility of amino acid synthesis in an atmosphere with very little free hydrogen. The rules of chemistry favor certain combinations, and it is not too difficult to imagine how some of the biologically important molecules could have been created in a "hot soup," especially in the absence of free oxygen. For instance, adenine, one of the nucleotide bases, is nothing more than five copies of the inorganic hydrogen cyanide molecule HCN, linked to itself as $(HCN)_5$. ATP consists of a five-carbon sugar molecule (ribose) linked to adenine and three phosphate groups. Carbohydrates and lipids might have formed spontaneously in such an environment.

In the 1930s, Alexander Oparin experimented with a sequence beginning with suspensions of oily liquids in water. Tiny droplets of oil in water are called **coacervates**. Oparin [1938, 1957] demonstrated that such droplets analogous to cell membranes can convert soluble sugars to insoluble starches (**polysaccharides**) and "grow," given the availability of appropriate enzymes. He could not, of course, explain the origin of enzymes (which are proteins). Nor could he explain how such droplets acquired the ability to replicate themselves, the last and most critical characteristic of life.

Polymers such as polysaccharides, polypeptides, and polynucleotides are an essential component of all known organic life. Fatty acids, cellulose, proteins, and DNA/RNA are all examples. At some stage in chemical evolution, simple organic compounds (monomers) must have been concentrated and combined into chains to form polymers.[11]

The preferred theory of polymerization today seems to depend on some sort of autocatalysis. The concentration of polymers in the primitive ocean was subsequently built up by some, as yet unknown, mechanism. One possibility has been explored by Sydney Fox [1980, 1988], who has shown that, when high concentrations of amino acids are heated under pressure, a spontaneous polymerization occurs, yielding "proteinoids." This has suggested the possibility that, instead of life originating in shallow ponds, organic life may have originated deep in the ocean in superheated regions near volcanic vents. One might account for polysaccharides in a warm organic soup by a mechanism such as that postulated by Oparin, given the availability of protein catalysts (enzymes).

A high-temperature/high-pressure polymerization mechanism near undersea hot springs, along the lines of Fox's ideas, just might account for the proteinoids, hence the catalyst, though this is speculative. As explained already, however, the conditions for nucleotide formation and peptide polymerization were quite different. And it seems highly unlikely that either could have occurred naturally in seawater, due to the high rate of decomposition by hydrolysis. Most likely, polymerization occurred either in cold clouds (e.g., in comets), as postulated by Oro, Crick, and Hoyle, or around the edges of shallow, evaporating ponds, or on the surface of clay-type minerals, as suggested recently by Cairns-Smith [1984, 1985]. Both the cold, dark "life cloud" scenario and the Cairns-Smith scenario are unproven. But these two hypotheses are, to date, the only plausible explanations of the co-evolution, and co-existence, of nucleotides with amino acids.

A small digression on proteins seems worthwhile here. The number of different proteins involved in living processes has been assumed to be incalculably enormous. Yet, surprisingly, there is new evidence that all proteins are constructed from a much smaller number of component elements, called **exons**, corresponding roughly to shape modules [Dorit et al., 1990]. An analogy would be that paragraphs are constructed from words, words are constructed from syllables, and syllables from letters. These components, or exons, consist of sequences of 20 to 120 (or so) amino acids, which correspond to specialized information-carrying segments of DNA. The exons are usually separated or sometimes connected by much longer sequences of amino acids, essentially

TABLE 4.3. *The Genetic Code.*

Nucleotide Bases	Amino Acids	Codons for DNA
Guanine (G)	Phenyl alanine	UUU, UUC
Cytosine (C)	Leucine	UUA, UUG, CUU, CUC, CUA, CUG
Adenine (A)	Isoleucine	AUU, AUC, AUA
Thymine (T)	Methionine	AUG
Uracil (U)	Valine	GUU, GUC, GUA, GUG
	Serine	UCU, UCC, UCA, UCG, AGU, AGC
	Proline	CCU, CCC, CCA, CCG
	Threonine	ACU, ACC, ACA, ACG
	Alanine	GCU, GCC, GCA, GCG
	Tyrosine	UAU, UAC
	Histidine	CAU, CAC
	Glutamine	CAA, CAG
	Asparagine	AAU, AAC
	Lysine	AAA, AAG
	Aspartic acid	CAU, GAC
	Glutamic acid	GAA, GAG
	Cysteine	UGU, UGC
	Tryptophan	UGG
	Arginine	CGU, CGC, CGA, CGG, AGA, AGG
	Glycine	GGU, GGC, GGA, GGG

Note 1: U can be replaced in any codon by T, since U appears only in RNA and T appears in the corresponding codons in DNA.

Note 2: Three codons (UAA, UAG, UGA) do not correspond to any amino acid, but are "punctuations" in the sequence.

chemical punctuation marks, called "introns." The introns apparently carry no useful genetic information and are sometimes called "junk DNA" (although some biochemists think they once had an evolutionary role). A **gene** consists of a sequence of 15 to 20 exons. Dorit's evidence suggests that the "universe" of exons may be as small as 1,000 to 7,000 [*ibid.*].

Notice the following hierarchy:

1. There are twenty amino acids, constructed from the four basic elements carbon (C), hydrogen (H), oxygen (O), and nitrogen (N). Three of the amino acids also contain atoms of sulfur (S). Each amino acid corresponds to one or more DNA codons, consisting of sequences of the five nucleotides (Table 4.3).
2. There seems to be a limited number (1,000-7,000) of different exons.
3. All proteins are apparently constructed by "shuffling" such modules.

The evolutionary implications are very important. If all combinations of amino acids were equally probable, then a protein molecule consisting of 200 amino acids would have an inherent improbability of 20 to the power 200. If all exons were equally probable, the number of possibilities for an exon of length 40 would be 20 to the power 40 (or 10 to the power 52). Yet the actual number of exons appears to be less than 10,000 (10 to the 4th power), and if all exons were equally probable, a five-exon protein would be one of (less than) 50,000 possible combinations. While this seems like a large number, it is an inconceivably tiny fraction of the number of theoretically possible combinations of amino acids.

This line of argument has powerful implications of another sort. The proteins in living organisms represent the outcome of a search for the best of all possible ways of accomplishing each metabolic function, *within a given domain*. Until recently, it was thought (or assumed) that the search encompassed all possible combinations of amino acids. It now appears that the domain actually canvassed must have been far, far smaller. This leaves open the possibility that alternative, equally probable, evolutionary processes (on other planets, for instance) could have discovered much more efficient functional solutions.

4.3. DIGRESSION: OTHER UNSOLVED MYSTERIES

One of the most interesting features of biochemical evolution is the persistent violation of **mirror symmetry** in biomolecules. *Ceteris paribus*, asymmetric molecules should occur with equal probability in either of their two mirror-symmetric isomeric forms.[10] In organic-synthesis experiments like Stanley Miller's this is what happens. Yet L-amino acids and D-sugars are overwhelmingly predominant in organic matter, a fact first noted more than a century ago by Louis Pasteur. This phenomenon was named **chirality** or **chiral purity** by the British physicist William Thomson (Lord Kelvin).

The natural mirror symmetries of bio-molecules are, of course, direct consequences of intermolecular forces. These, in turn, are electromagnetic in origin (subject to the rules of quantum mechanics). However, the underlying forces exhibit no preference for chirality. The problem of evolutionary **symmetry-breaking** is thus a fundamental one (as it is in high-energy physics). There are many links between the symmetries of simple organic molecules, such as amino acids and sugar, and the properties of macro-molecules (polypeptides, nucleic acids and proteins).

Many important biochemical interactions probably depend on structural details, which depend strongly on symmetry elements. For example, some proteins interact with single-stranded nucleic acids; others with double-stranded nucleic acids, and still others with globular structures. In some cases the protein recognizes specific sequences of nucleotides; in others it recognizes the details of the sugar-phosphate "backbone" of the nucleic acid [Pincheira, 1986]. Thus, evolutionary discrimination between isomers at the monomer level could be an indirect consequence of discrimination at the macro-molecular level. In other words, macro-molecular symmetry preferences may have evolved later, leading indirectly to L-D symmetry discrimination at the micro-level.

Another extremely interesting "stylized fact" of chemical evolution is the

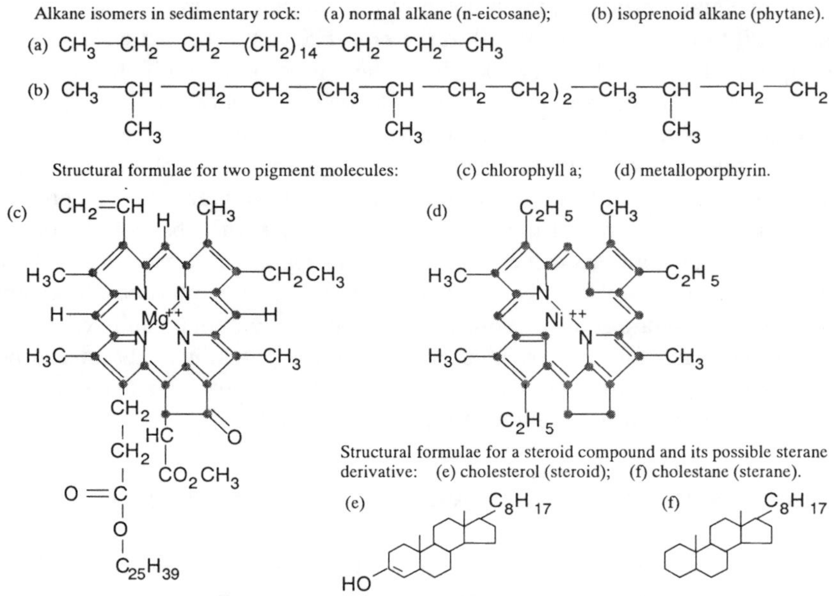

Alkane isomers in sedimentary rock: (a) normal alkane (n-eicosane);   (b) isoprenoid alkane (phytane).

(a) $CH_3-CH_2-CH_2-(CH_2)_{14}-CH_2-CH_2-CH_3$

(b) $CH_3-\underset{\underset{CH_3}{|}}{CH}-CH_2-CH_2-(CH_3-\underset{\underset{CH_3}{|}}{CH}-CH_2-CH_2)_2-CH_3-\underset{\underset{CH_3}{|}}{CH}-CH_2-CH_2$

Structural formulae for two pigment molecules: (c) chlorophyll a;   (d) metalloporphyrin.

Structural formulae for a steroid compound and its possible sterane derivative: (e) cholesterol (steroid);   (f) cholestane (sterane).

FIGURE 4.3. Characteristic Chemical Structures.

early and continued predominance of a few characteristic chemical structures among a great many possibilities. Most bio-molecules are constructed from simple units like OH, CO, CH_2, CH_3, HCN, NH_2, NH_3, etc. These, in turn, are linked by carbon atoms (or, in some cases, phosphate groups) that constitute the "glue." Carbon atoms, and CH_2 groups, are uniquely inclined to link to one another in chains. Given the large number of simple units that can be linked to carbon atoms, it is clear that the number of possible chemical combinations for a chain of even five or six carbon atoms is enormous. When the number of carbon atoms increases to twenty or so, the number of possibilities is truly astronomical. Yet remarkably few structural types are actually observed in nature. As mentioned already, only twenty different amino acids are found in proteins. Why so few? The most obvious explanation is that the evolutionary search process, being random and myopic (not systematic), has not extended over the entire range of chemical possibilities.

Consider another instance: The most stable decomposition residues of all lipids (fatty acids) tend to be of stable saturated hydrocarbons called **alkanes**. A typical alkane has the composition $C_{20}H_{42}$, consisting of (possibly branched) chains. Normal alkane would be a simple chain (polymer) of 18 CH_2 units with a CH_3 at either end, as shown in Figure 4.3a. Allowing for branches, however, an enormous number of alternative structures (**isomers**) is possible. Yet, of this array of possibilities, only one other structure—the five-branched isoprenoid structure shown in Figure 4.3b—is found in fossil rocks. As far as we can tell today, unsaturated relatives of these two isomers (and no others) have been the predominant building blocks of lipids for at least 1.5 billion years.

Another ancient structure that has a key role in living systems is the tetra-pyrrole ring structure. Porphyrin (Figure 4.2) is an example. It is basic to both **chlorophyll** (the green pigment in plants, Figure 4.3c) and **hemin** (the red pigment in blood). In the case of chlorophyll, a magnesium atom sits in the center of the ring; in the case of hemin, it is replaced by iron. Traces of these ring structures are found in old sedimentary rocks as metalloporphyrins, as shown, for instance, in Figure 4.3d. Very ancient rocks also contain traces of steranes, like those shown in Figure 4.3f. These are degraded relatives of steroids (Figure 4.3e), the class of structures common to most organic catalysts or enzymes. Again, the evolutionary search process seems to have been limited to a very small region of configuration-space. This is consistent with the notion that, at the chemical level, the process is fundamentally myopic.

4.4. SELF-ORGANIZATION AT THE CHEMICAL LEVEL

The evolution of complex molecules was obviously a necessary condition for life. But the phenomenon of **self-organization** was still the biggest step in chem-ical evolution. Somehow, populations of polymers interacted among themselves, and a crude sort of chemical self-organization emerged in the form of a self-sustaining chemical process displaying spatial differentiation of parts, or **struc-ture**. This spatial differentiation required a flow of free energy; it was thus a dissipative process, far from thermodynamic equilibrium. Apart from the diffi-culties noted above, it is impossible to explain this evolutionary step on the basis of classical thermodynamic concepts:

> "The probability that at ordinary temperatures a macroscopic number of molecules is assembled to give rise to the highly ordered structures and the coordinated functions characteristic of living organisms is vanishingly small. The idea of *spon-taneous* genesis of life in its present form is therefore highly improbable, even on the scale of the billions of years during which prebiotic evolution occurred" [Prigogine *et al.,* 1972; emphasis added].

Theories of chemical evolution generally fall into two classes. The first class is sometimes associated with Oparin [1957, 1961]. It can be termed the **continuous** model, in contrast to **discrete** or **catastrophic** models (as defined by René Thom). The best-known example of the continuous model is Manfred Eigen's **hypercycle** theory, discussed later. In such a model, evolution consists of a chain of small incremental improvements constituting a stable dynamical trajectory without discontinuities or singularities in the space of proper variables (pheno-typic attributes).

In the past two decades an alternative class of theories has emerged, featuring such non-linear dynamical phenomena as **bifurcations** and **catastrophes**. In this class of models, major discontinuities (analogous to **phase transitions** in physics) can occur when the previous dynamical trajectory becomes unstable. Such models of biochemical evolution have been suggested by Prigogine and his colleagues, in particular. They emphasize the importance of **non-linearities**, which create a multiplicity of possible trajectories for the system. They also emphasize the role of **fluctuations**, which cause the system to choose one of the possible paths and to "lock into" it. The theory of open-system thermodynamics has produced a variety of simple model chemical systems exhibiting self-organizing characteristics, including the so-called "Brusselator" [Nicolis and Prigogine, 1977], and the "Oregonator" [Field and Noyes, 1974; Tyson, 1978, 1983; Field, 1985].[13]

The choice between continuous and discontinuous theories remains unresolved. A significant straw in the wind, however, may be the recent discovery that the prevalence of **chiral purity**, the conspicuous violation of mirror symmetry with regard to the "right-handedness" or "left-handedness" of organic molecules, can be explained only by a discontinuous or catastrophic type of model [Morozov and Goldanskii, 1984].

The first laboratory example of self-organization in the chemical domain may have been the discovery by the Russian chemist B. P. Belousov in 1958 of a simple reaction that exhibits both periodicity in time and the kind of large-scale spatial differentiation and complexity that had previously been observed only in biological systems [Zhabotinsky, 1973]. The so-called Belousov-Zhabotinsky (B-Z) reaction,[14] named for its discoverers, shows evolving geometric forms (concentrational inhomogeneities) beginning at one or more "leading centers" and propagating as traveling waves until the waves collide, but do not penetrate each other. As the system evolves, many forms can result, including "target patterns" and rotating spiral structures with one, two, or three arms [Agladze and Krinsky, 1982]. No explanation of the B-Z reaction is possible without invoking the theory of open-system (non-equilibrium) thermodynamics [Winfree, 1972, 1973]. The B-Z reaction can therefore perhaps be regarded as a direct confirmation of Prigogine's ideas.

In recent years a number of researchers have discovered other reactions with such properties [see, for example, Epstein *et al.,* 1983]. More to the point, so far as this book is concerned, is that some scientists are beginning to see simple, self-organizing chemical reactions (such as the B-Z system) as a possible evolutionary link between simple organic chemical synthesis and self-replicating structures.

4.5. SELF-REPLICATION

It now appears that the major difficulty in evolutionary theory is explaining the synthesis of nucleotides, the essential building blocks of nucleic acids. Admittedly, interesting synthesis experiments have been done in laboratories. Leslie Orgel, for instance, has demonstrated the spontaneous formation of a sequence of fifty nucleotides in a "DNA-like" molecule, starting from simple carbon compounds and lead salts. What can be done in a test tube under laboratory conditions, however, would not necessarily happen under natural conditions. A fundamental difficulty with the "hot soup" hypothesis is that unprotected nucleotides, and chains, decompose very rapidly in water due to hydrolysis. This difficulty is one of the most persuasive arguments for an extraterrestrial (cometary) origin of organic synthesis, along the lines discussed above.

A more "down-to-earth" theory of the origin of nucleotide replication is due to A. G. Cairns-Smith [1984, 1985]. He postulates that life started, not with carbon-based organic materials, but with self-replicating silicon-based crystals. The idea is that it is much easier to imagine a dissipative chemical structure persisting in the presence of a solid "template" than merely floating in a liquid solution. The complex surfaces of certain natural minerals offer ideal templates for the formation of complex carbon-based compounds. According to the theory, replication of the crystals led to replication of the associated organics. Gradually, by evolutionary selection, the carbon-based compounds "learned" to replicate without the help of the silicon crystals. This scheme has been dubbed "the genetic takeover."

A possible mechanism for direct protein replication without the DNA machinery has been postulated by Robert Shapiro [1986]. It requires that there be an external "substrate" for the protein molecule to be reproduced, and an enzyme to recognize and match amino acids at both ends. It simplifies the cycle shown in Fig. 4.4. There is little direct evidence for this hypothesis, but it does suggest a reason why such a small number of amino acids (or the next building block, exons) are found in proteins.

A generalized, non-linear "catastrophe theory" model generating order from disorder without assuming any specific chemistry has been presented by Freeman Dyson [Dyson 1982, 1986]. Dyson postulates a self-reproducing, virus-like macro-molecule. He then considers the number a of different amino acids or nucleotides (monomers) in the "cloud" or "soup," the number b of different types of catalytic synthesis the macro-molecule can perform, and the number N of molecular units in the chain. Larger values of a are needed to accommodate larger values of b.

Dyson's model allows for both disordered (dead) and ordered (living) states. With some combination of the three parameters only dead states are possible; with other combinations all states are living. However, there is an interesting region where both states can occur. The simplest such case corresponds to 7 distinct monomers and about 50 types of catalytic synthesis. The region of 8-11 monomers and 60-100 types of catalytic synthesis is of particular interest.

Insofar as the Dyson model has implications for the transition from non-life to life, it seems to favor the priority of proteins, or at least protein-modules (*exons*) on the grounds that nucleotides *per se* have insufficient diversity to exhibit an order-disorder transition leading to a condition where life can survive. It also suggests that a transition from disorder to order is much more likely to occur for a non-replicating system (i.e., one with a very high probability of copying

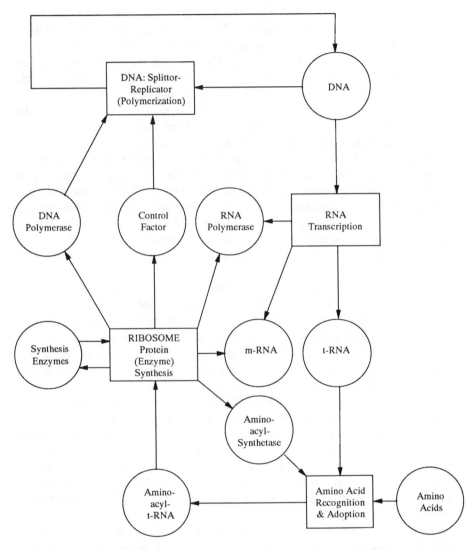

FIGURE 4.4. DNA Replication and Protein Synthesis. Source: Adapted from Eigen, 1971; Rich and Kim, 1978.

errors) than for a true replicating system. In fact, the conditions under which replication could occur with reasonable likelihood are such that the probability of transition is

$$(1+a)^{-N}.$$

For $a = 9$ this implies that only quite small values of N $(N < 100)$ could be consistent with accurate replication. This presumes, of course, that copying errors are random.

4.6. GENETIC REPRODUCTION AS AN INFORMATION PROCESS

The most salient feature of life is the ability to reproduce, not just at the molecular level, but at the cellular level. The minimum elements of a self-reproducing system (of any kind) were first identified in 1948 by the mathematician John von Neumann, viz:

- A **blueprint** (i.e., stored information).
- A **factory** to fabricate the components and assemble the units.
- A **controller** to supervise the operations of the factory.
- A **duplicating machine** to copy the blueprint.

Of course, von Neumann [1961-1963] characterized all these items in fairly abstract mathematical language, but the biological counterparts of these elements are easily recognized.

As expressed in biological organisms today (Figure 4.4), the "blueprint" is the DNA, the cytoplasm and its enzymes are the "factory," the specialized replicase enzymes play the role of "controller," and the DNA-replication process is the "duplicating machine." The mechanism for information replication need not be described in detail here, though the unraveling of the universal **genetic code** was one of the greatest milestones of 20th century biology (see Table 4.3).

The basic information storage unit **DNA (deoxyribonucleic acid)** consists of a kind of twisted ladder (the famous double helix, first discovered in 1953 by Francis Crick and James Watson). The "rungs" of the ladder are pairs of nucleotide bases ($A \leftarrow \rightarrow T, G \leftarrow \rightarrow C$). Each base is anchored to a ribose (sugar) molecule on the helical frame. The sugar molecules are linked longitudinally by phosphates. Information is stored in the longitudinal sequence of bases. Each (overlapping) set of three bases corresponds to a three-letter sequence of the four letters A, T, G, C.

Note that one of the five possible bases (U) never appears in double-stranded DNA. It appears only in the single-strand nucleic acid RNA, as a direct substitute for T. The equivalent pairings between corresponding strands of RNA and DNA are therefore ($U \leftarrow \rightarrow A, A \leftarrow \rightarrow T; G \leftarrow \rightarrow C$).

The standard core hypothesis of molecular biology is that genetic information is transferred from DNA to RNA to proteins by the generic scheme

$$DNA \ transcription \rightarrow RNA \ translation \rightarrow Protein$$

The standard hypothesis does *not* admit any flow of information from proteins back to DNA. (Such a reversal would, in principle, permit *learned* information to be retained permanently in the genes. This process was once postulated by the 19th century biologist Lamarck who didn't know about genes—and was later revived by the discredited Soviet biologist Lysenko. But there is no experimental evidence for such a process.)

The actual DNA replication is simple in principle: A special enzyme (protein molecule) "unwinds" the double helix and separates the two strands. Because of the strict pairing rules (A $\leftarrow\rightarrow$ T; G $\leftarrow\rightarrow$ C), each strand can then be used as a template to construct an exact duplicate of the original. This is done by other special enzymes.

The translation of DNA information into protein structures begins with the synthesis of single strands of nucleic acid (called messenger RNA, or mRNA) by the appropriate base-pairing rules. This process is called *transcription*. The mRNA contains the same coded information stored in a segment of the DNA (except that U replaces T in each codon). The mRNA is then "decoded," codon by codon, by a protein molecule, called a **ribosome**, roughly analogous to the assembly worker in a factory. The materials (amino acids) for constructing new proteins have to be brought to the assembler by carrier molecules called **transfer RNA**, or tRNA.

Suppose, for instance, the ribosome is decoding the sequence AUC in the mRNA. From Table 4.3 it can be seen that this particular codon corresponds to the amino acid **isoleucine**. The ribosome needs a tRNA molecule that exactly fits the AUC codon, but in reverse order. This is an **anti-codon**. By the pairing rules, one of the anti-codons for isoleucine is GAU. When the ribosome encounters GAU (or one of the others that match), it detaches the amino acid from the tRNA and reattaches it to the protein "workpiece." When the ribosome encounters a "stop" instruction on the mRNA chain, it releases the complete protein.

The foregoing summary does not cover all the key synthesis processes, but it does summarize what is known of the most important ones. It is an inescapable fact that nucleic acids and proteins are now interdependent, as clearly indicated by Figure 4.4. From an evolutionary perspective, therefore, it is necessary to explain not only how both proteins and DNA were generated from non-living matter, but also how such a *process* came into existence. Hence there is a "chicken and egg" problem. How could DNA be generated in the first place without protein replicase enzymes? How, on the other hand, could proteins be produced without DNA, RNA, and ribosomes?

In the context of biological evolution, the nucleic acids building blocks of polynucleotides were probably a precondition for the first self-replication event, because of their ability to catalyze reproduction through template action. However, a number of template-catalyzed reactions have to be coupled via chains or self-reproductive catalytic **hypercycles** as postulated by Manfred Eigen [1971]. Eigen, like Oparin, assumes a primordial soup, already containing peptides or small proteins and fatty acids, as well as nucleotides. (In other words, he assumes away the difficulties noted previously.) He also assumes a prototype RNA molecule formed by accident, possibly assisted by the proteins in the soup. The RNA molecule is a molecular self-replicator (but not a gene).

But how can such a molecule "learn" to synthesize proteins? The problem is to explain replication without replicase enzymes, which detect and correct errors. Under such conditions one would expect a high natural-error rate. The high copying-error rate limits the maximum number of nucleotide bases that can be copied successfully to the order of 100 [Casti, 1989]. That is about right for a single amino acid. But it is nowhere near enough to encode a protein complex enough to be an enzyme. In other words, without having enzymes in the first place, the replication-error rate is too high to reproduce enzymes.

Eigen's hypercycle is one scheme to get around the difficulty. It postulates that the genetic message for a complex protein can be *modularized* and that each module (an exon, presumably) is reproduced, and evolves, independently. But independent evolution also implies competition for resources. This, in turn, implies that some genes could outcompete and eliminate others. There is no obvious reason why they should cooperate.

Eigen and Schuster [1979] suggested a simple explanation. Suppose, for purposes of illustration, that the genetic entity, or "message," to be reproduced is a word. As an example, let us take the word DARWIN. (Here, each exon is analogous to a letter, not a word.) Each letter corresponds to a distinct population of exons, with one (or more) catalytic functions. The exons are encoded by single strands of RNA. One of the functions of each exon assists in the formation or recognition of the information carrier RNA for the *next* exon in the sequence. It turns out the relationships between populations are stable if (and only if) each population catalyzes the next one in the sequence, and if the sequence is closed on itself, i.e.:

$$\leftarrow D \leftarrow A \leftarrow R \leftarrow W \leftarrow I \leftarrow N \leftarrow$$

This is the hypercycle. Not only is the system reasonably stable, but it can evolve and grow in complexity.[13] Thus, in effect, Eigen postulates Darwinian evolution at the molecular level.

A different scheme for RNA replication in the absence of replicase has been suggested by Walter Gilbert [1986]. He postulates autocatalysis by the RNA molecule, itself, leading to proliferation and evolutionary improvements, then to protein synthesis, and finally to DNA synthesis. There is some experimental evidence of autocatalysis by RNA [e.g., Orgel, 1983]. But for both Eigen and Gilbert the original synthesis of RNA remains unexplained. From the chemical point of view, the process was genotypic (in the sense of fixing possibilities), even though genes had not yet appeared. Biological evolution which coincided, to a large extent, with chemical evolution began with the appearance of DNA (along with other associated chemical species) or, more properly, with the emergence of the **DNA-replication cycle** (Figure 4.4) roughly 4 billion years ago.

Whatever the detailed mechanism for self-reproduction at the chemical level turns out to be, it is increasingly clear that the processes can best be described in terms of information storage and transfer. In all likelihood, the final theory will be expressed in the language of information theory. Eigen emphasizes that the theory of selection can be derived from classical (or quantum) statistical mechanics by adding a parameter representing *a selective value* for each information state. In fact, he characterizes his own approach as *"a general theory which includes the origin or self-organization of ('valuable') information,* thereby uniting Darwin's evolution principle with classical information theory" [Eigen, 1971, p. 516; italics in original].

These words are generally applicable to modern molecular biology, as well. It seems almost superfluous to add that Eigen's notion of "selectively valuable" information is essentially identical to what I have previously called D-information. Perhaps also needless to say, the notion of "value" takes on different meanings as we move later in the book into the realm of social science and economics.

ENDNOTES

1. For reasons that are not well understood, the photosynthetic carbon-fixation process mediated by the enzyme rubulose-1,5-bisphosphate carboxylase discriminates against the heavier isotopes of carbon, so that ^{13}C is depleted by about 25% in organic carbon, as compared to inorganic carbon. Thus the ratio of the isotopic composition of carbon becomes a definite indicator of autotrophic activity.

2. The source of the small comets remains speculative. Frank postulates a belt of small icy comets orbiting the sun beyond the belt occupied by large known comets. He also postulates a large unknown planet X with an extremely eccentric orbit that regularly intersects the outer region occupied by small comets. Once every 26 million years, he believes, the planet's orbit intersects the large comet belt, causing swarms of large comets to be deflected into the inner solar system. Frank believes a swarm of comets, some of which collided with the earth, caused the extinction of the dinosaurs 65 million years ago. The next such swarm would arrive in about 12 million years. The existence of this orbiting belt of small comets, and the planet X, have not been confirmed by direct observation. For a more complete discussion, see Frank *et al.* [1990].

3. Recall that the core of the earth is mostly molten iron and nickel—the same combination found in many meteorites. This explains the presence of metallic iron or incompletely oxidized (ferrous) iron in basalt rock or volcanic magma.

4. Most organic material has the approximate chemical formula CH_2O. Aerobic oxidation (respiration being a version of this process) is the usual route for decomposition: The products are CO_2 and H_2O. Anaerobic decomposition (accomplished by bacteria) yields one molecule of CO_2 and one molecule of CH_4 for each two molecules of CH_2O. In both of these cases the cycle is closed; there is no mechanism for carbon accumulation. However, in conditions of high temperature and pressure there is a third route: CH_2O simply decomposes to carbon and water. The water combines with other minerals or escapes, leaving the excess carbon behind. Natural deposits of coal, gas, and oil presumably result from some combination of the latter two routes.

5. A minor exception to this rule is the guano left on some coastal islands and cliffs by sea birds living on fish.

6. At one time it was thought that silicon (one row below carbon on the periodic table with the same valence structure) might also be a possible basis for long-chain molecules. However, it now appears fairly certain that carbon is the only element capable of playing this role due to its uniquely well-balanced electronic shell-structure $(15)^2(25)(29)^3$. Without this fine-tuned electronic shell structure, it seems, organic life would not be possible [Sussman 1985].

7. The only "living" organisms that can exist without a protective cell wall are viruses. Even viruses outside a cell are inactive, however. They are activated only when they penetrate a cell.

8. Semipermeable membranes are constructed from phospho-lipids lined up parallel to each other like blades of grass or the pieces of yarn in a rug. Each phospho-lipid molecule has a phosphate group, which is electrically charged at one end. The charged (phosphate) end attracts water molecules, while the uncharged end repels water. Thus, phospho-lipids tend to form a film on water, with the phosphate groups on the inside. Surface tension makes the films "fold" into droplets. Such droplets may have been the first "proto-cells." Cell walls tend to maintain both electrical (voltage) gradients and acidity (pH) gradients. If electrons are removed from the phosphate groups at one side of the membrane by a multi-step process, H^+ ions—which are simply protons—can penetrate the membranes to restore electrical neutrality. This is the basic electrical mechanism that underlies both photosynthesis and respiration [Hinkle and McCarty, 1978].

9. Of course, this required a reasonably high concentration of organic molecules. It is generally assumed that the concentration occurred mainly around the edges of ponds, by evaporation and freezing of the water.

10. Basically, these can be thought of as "right-handed" and "left-handed." There are at least two common terminologies. In referring to sugars and amino acids it is usual to denote right-handedness and left-handedness by D and L, where these letters stand for the Latin "dextrogyre" and "levogyre," respectively. To confuse matters, organic chemists also refer to isomers as para- and ortho-, respectively, with similar but not identical connotations.

11. Indeed, the need for a new open-system non-equilibrium thermodynamics was perhaps first expressed by the Belgian physical chemist Theophile De Donder in the 1930s. Its importance for understanding biological systems was recognized as early as 1940 by the biologist Ludwig von Bertalanffy and the physicist Erwin Schrödinger [1945].

   For most chemical systems (at constant temperature and pressure), the appropriate thermodynamic potential function—or measure of distance from equilibrium—is the Gibbs' free energy

   $G = U + PV - TS,$

   where U is the internal energy, P is the pressure, V is the volume, T is the temperature, and S is the entropy. The change in free energy at constant temperature and pressure is given by

   $dG = dU + PdV - TdS.$

The analog of "force" driving the system toward equilibrium is definable (in principle) in terms of G. In general, the governing relationships are non-linear, though they can be linearized as the system approaches close to equilibrium (in the limit as G is small). Non-linear dynamic equations can have multiple discrete solutions (or even continua of solutions), only one of which reduces to the solution of the linearized equation. This has been called the "thermodynamic" branch [Glansdorff and Prigogine, 1971].

Far from equilibrium, however, the non-linearity of the system is critical to describing its behavior. Some of the solutions correspond to situations where the Gibbs free energy G is at a local minimum, but not a global minimum. Such a local minimum may persist at finite temperatures and entropy because of the term $-TS$ in the G function. In fact, it can correspond to a localized maximization of S, resulting from a high rate of localized energy dissipation, which can continue as long as the necessary free energy is supplied by some external source. It is this behavior that has led Prigogine to coin the term "dissipative structure.
" The persistence of such a dissipative structure is itself a kind of coherent order, or organization. The organized state can be created from a disorganized state by a random fluctuation [Nicolis and Prigogine, 1971]. (Transitions of this sort, between different branches or trajectories of a dynamic non-linear system, have been classified topologically and called catastrophes by the French mathematician René Thom [1972]).

12. The reaction involves oxidation of cerium ions catalyzed by bromate ions but inhibited by bromide ions, in a citric acid/sulfuric acid medium.

13. Unfortunately, the hypercycle is not stable against all possible perturbations [Niessert 1987]. Computer simulations have uncovered three types of possible discontinuities that can result from unfortunate mutations (copying errors), viz:

   (1) One RNA molecule learns how to replicate faster than others but "forgets" to catalyze the production of the next one.
   (2) One RNA molecule changes its role and skips one or more steps in the chain, leading to a "short-circuit."
   (3) One population dies out due to statistical fluctuation

It turns out that the first two types of catastrophe increase in probability with population size, while the third type has a high probability if the population is low. This implies a finite lifetime for a hypercycle, unless, of course, the hypercycle succeeds in evolving a protective mechanism (such as replicase).

REFERENCES

Agladze, K. I. and V. I. Krinsky, Title missing, *Nature* **269**, 1982 :424.

Cairns-Smith, A. G., *Genetic Takeover and the Mineral Origin of Life*, Cambridge University Press, Cambridge, England, 1984.

Cairns-Smith, A. G., "The First Organisms," *Scientific American* **252**(6), 1985.

Casti, John L., *Paradigms Lost: Images of Man in the Mirror of Science*, W. Morrow and Company, New York, 1989.

Dorit, R. L., L. Schoenbach and W. Gilbert, "How Big is the Universe of Exons?" *Science* **250**, Dec. 7, 1990 :1377-1382.

Dyson, Freeman J., "Origin of Life," *Journal of Molecular Evolution* **18**, 1982 :344-355.

Dyson, Freeman J., *Origins of Life*, Cambridge University Press, Cambridge, England, 1986.

Eigen, Manfred, "Self-organization of Matter and the Evolution of Biological Macro-molecules," *Naturwiss* **58**, October 1971.

Eigen, Manfred and P. Schuster, *The Hypercycle: A Principle of Natural Self-Organization*, 1979.

Faber, Malte and John L. R. Proops, *Evolution in Biology, Physics and Economics: A Conceptual Analysis*, Discussion Paper (131), University of Heidelberg Department of Economics, Heidelberg, Germany, January 1989. [Mimeo]

Field, R. J., "Chemical Organization in Time and Space," *American Scientist* **73**, March/April 1985 :142-149.

Field, R. J. and R. M. Noyes, Title missing, *Journal of Chemical Physics* **60**, 1974 :1877.

Fox, Sydney, "New Missing Links," *Science*, January 1980 :18-21.

Fox, Sydney, *The Emergence of Life*, Basic Books, New York, 1988.

Frank, Louis A., J. B. Sigwarth and J. D. Craven, "On the Influx of Small Comets into the Earth's Upper Atmosphere, II: Interpretation," *Geophysical Research Letters* **13**, April 1986 :307.

Frank, Louis A., J. B. Sigwarth and C. M. Yeates, "A Search for Small Solar-System Bodies near the Earth Using a Ground-based Telescope: Technique and Observations," *Astronomy and Astrophysics* **228**, February 1990 :522.

Gilbert, Walter, "The RNA World," *Nature* **319**, 1986 :618.

Glansdorff, P. and Ilya Prigogine, *Thermodynamic Theory of Structure, Stability and Fluctuations*, Wiley-Interscience, New York, 1971.

Goldanskii, V. I., Title missing, *Nature* **279**, 1979 :109.

Haldane, J. B. S., *On Being the Right Size and Other Essays*, Oxford University Press, London, 1985.

Hinkle, Peter C. and Richard E. McCarty, "How Cells Make ATP," *Scientific American*, March 1978.

Holland, H. D., *The Chemical Evolution of the Atmosphere and the Oceans*, Princeton University Press, Princeton, N.J., 1984.

Hoyle, Fred and N. Chaudra Wickramashinghe, *Life Cloud*, Harper and Row, New York, 1978.

Lovelock, James E., "Gaia as Seen Through the Atmosphere," *Atmospheric Environment* **6**, 1972 :579-580.

Lovelock, James E., *Gaia: A New Look at Life on Earth*, Oxford University Press, London, 1979.

Miller, Stanley L., "The First Laboratory Synthesis of Organic Compounds under Primitive Conditions," in: Neyman (ed.), *The Heritage of Copernicus* :228-241, MIT Press, Cambridge, Mass., 1974.

Miller, S. and L. Orgel, *The Origins of Life on The Earth*, Prentice-Hall, Englewood Cliffs, N.J., 1974.

Morozov, L. L. and V. I. Goldanskii, "Violation of Symmetry and Self-Organization in Pre-biological Evolution," in: Krinsky (ed.), *Self-Organization: Autowaves and Structures Far from Equilibrium*, Chapter 6 :224-232, Springer-Verlag, New York, 1984.

Nicolis, Gregoire and Ilya Prigogine, "Thermodynamics of Structure, Stability and Fluctuation," *Interscience*, 1971. [New York]

Nicolis, Gregoire and Ilya Prigogine, *Self-Organization in Non-Equilibrium Systems*, Wiley-Interscience, New York, 1977.

Niessert, V., "How Many Genes to Start With? A Computer Simulation About the Origin of Life," *Origins of Life* **17**, 1987 :155-169.

Oparin, Alexander I., *Origins of Life*, Macmillan, New York, 1938. [Translated by S. Margulis]

Oparin, Alexander I. (ed.), *Origin of Life on Earth*, Publishing House of the U.S.S.R. Academy of Sciences, Moscow, 1957. [in Russian]

Orgel, Leslie, "The Evolution of Life and the Evolution of Macro-molecules," *Folia Biologica* **29**, 1983 :65-77.

Óró, J., "Comets and the Formation of Biochemical Compounds on the Primitive Earth," *Nature* **190**, 1961 :389.

Pincherra, Guido, "Symmetry in Biomolecules: Its Physiological Meaning," in: *15th ICUS*, International Conference on the Unity of the Sciences, 1986.

Prigogine, Ilya, Gregoire Nicolis and A. Babloyantz, "Thermodynamics of Evolution," *Physics Today* **23**(11/12), November/December 1972 :23-28(Nov.) and 38-44(Dec.).

Rich, Alexander and Sung Hon Kim, "The Three Dimensional Structure of Transfer RNA," *Scientific American*, January 1978.

Rubey, William W., "Geologic History of Seawater: An Attempt to State the Problem," *Bulletin of the Geological Society of America* **62**, September 1951 :1135.

Schrödinger, Erwin, *What is Life? The Physical Aspects of The Living Cell*, Cambridge University Press, Cambridge, England, 1945.

Shapiro, Robert, *Origins: A Skeptic's Guide to the Creation of Life on Earth*, Summit, New York, 1986.

Sussman, G., "The Anthropic Principle and the Arrow of Time," in: *14th ICUS*, International Conference on the Unity of the Sciences, Houston, Tex., 1985.

Thom, R., *Structural Stability and Morphogenesis: General Theory of Models*, 1972.

Tyson, J. J., Title missing, *Journal of Mathematical Biology*, **5**, 1978 :351.

Tyson, J. J., *Oscillations and Travelling Waves in Chemical Systems*, John Wiley and Sons, New York, 1983.

von Neumann, John, *Collected Works*, Macmillan, New York, 1961-63.

von Neumann, John, "The General and Logical Theory of Automata," in: von Neumann (ed.), *Collected Works* **5** :288-328, Macmillan, New York, 1963.

Winfree, A. T., Title missing, *Science* **175**, 1972 :634.

Winfree, A. T., Title missing, *Science* **181**, 1973 :937.

Yeates, C. M., "Initial Findings from a Telescopic Search for Small Comets near Earth," *Planetary and Space Science* **37**, October 1989 :1185.

Zhabotinsky, A. M., *Concentrational Oscillations*, Moscow, 1973. [In Russian]

Chapter 5

Biological Evolution

5.1. PRIMITIVE ORGANISMS

The first cellular organisms seem to have appeared 3.5 billion years ago. They were **prokaryotes,** which means "cells without nuclei." These first living cells comprised several varieties. It is widely believed that the direct precursors of most forms of life now existing obtained the free energy needed to sustain their metabolism and reproduction cycle by a process known as **anaerobic fermentation** of organic molecules (e.g., sugars) present in the early oceans. However, it is by no means clear where such molecules might have come from, or even whether they could have been produced by any natural process on the surface of the earth.

There is evidence that primitive organisms were able to exploit certain hydrogen-rich gases as "food." Some early proto-bacteria extracted energy from hydrogen sulfide, releasing hydrogen gas and depositing elemental sulfur. The underground sulfur deposits of Louisiana and Texas are leftovers from these organisms. Bacteria that metabolize sulfur and metallic sulfides are still found in mineral ores. It is plausible—even likely—that other early proto-bacteria were able to obtain metabolic energy from ammonia, reducing it to hydrogen gas and molecular nitrogen. Denitrifying bacteria that descended from these organisms still abound in the soil and play a major role in closing the nitrogen cycle. In fact, the existence of free nitrogen in the earth's atmosphere is perhaps attributable to these primitive but hypothetical organisms.

Proto-bacteria functioned *anaerobically,* i.e., in the absence of oxygen, since at the time in question the atmosphere contained neither free oxygen nor free nitrogen. Actually, if the early atmosphere had contained any free oxygen, there could have been no buildup in the concentration of organic molecules in the first place, because such molecules could not have survived long enough. Indeed, without a very efficient (non-linear) synthesis mechanism—such as Eigen's hypercycle—macro-molecules would have broken up faster than they could have been created [Wald, 1954]. On the other hand, no ozone layer could form in the absence of oxygen. Hence, the level of UV radiation at the earth's surface must have been much higher than it is at present. This would have resulted in a much higher level of photolytic chemical reactivity in the atmosphere (and the surface waters) than we now observe.

In anaerobic cellular fermentation, conventional theory holds that the food supply was a stock of glucose (of unclear origin) dissolved in the oceans (or ponds). In the proto-cells this was enzymatically split into two molecules of pyruvate. This splitting is called **glycolysis.** Energy released by glycolysis is mostly lost as low-grade heat, but some is transferred to the molecule adenosine triphosphate, or ATP, the energy-carrier in all known living systems.[1] Excreted waste products included lactic acid, ethyl alcohol, and carbon dioxide. Note that the ecosystem based on fermentation was inherently unstable: It would eventually either use up its food supply of glucose or poison itself with its waste products. The conventional theory has another fundamental problem already alluded to: It assumes the prior existence of enzymes (proteins) to catalyze the reactions. How were the proteins created in the first place? For that matter, how can we account for the existence of ATP? The honest answer is that we don't know.

What we do know is that the next great innovation in biological evolution (3 billion years ago) was **anaerobic photosynthesis,** the process by which organisms used solar energy to synthesize glucose from carbon dioxide, without oxygen. In fact, these organisms, known as **prokaryotic photobacteria,** replenished the glucose supply but produced oxygen as a waste product. For the next billion years or so, the atmosphere remained free of oxygen, because all the oxygen generated by photosynthesis was used up as fast as it was produced. It combined with dissolved ferrous magmatic iron in the ocean water. A specialized type of organism evolved to utilize it efficiently. These were the stromatolites, which laid down the great hematite (iron oxide) deposits currently being exploited as iron ore.

The dissolved iron in the oceans was apparently used up about 1 billion years ago. Meanwhile, free carbon was gradually being removed from circulation by the slow but steady rain of organic detritus to the bottom of the oceans, where it was covered by sediment and sequestered. This carbon is distributed throughout all forms of sedimentary rock, such as shale. To balance the disappearance of the carbon in sediments, the level of free oxygen in the atmosphere began to rise. While oxygen in the atmosphere created a shield (the ozone layer) against destructive UV, it also increased the rate of oxidative dissolution of all macro-molecules. The anaerobic photobacteria were in trouble. At this stage, the system as a whole (the **biosphere,** to use a term that has become popular), was still unsustainable. A use for the oxygen was badly needed. The direct oxidation of iron by the stromatolites was a precursor of the next major evolutionary step, direct oxidation of glucose, or **respiration.** A particularly fortunate mutant acquired this capability.

Respiration begins with glycolysis. But in this case, the intermediate products lactic acid and ethyl alcohol are further oxidized by a sequence of ten reactions collectively known as the **citric acid cycle,** which yields only carbon dioxide as

TABLE 5.1. *Energy-Conversion Processes during Various Evolutionary Eras.*

Form of Evolution	Era	Environment	Energy Source	Structure and Outcomes
Chemical	I	Anaerobic: methane, ammonia, hydrogen sulfide Loss of most free hydrogen	Ultraviolet light; heat	Acetate, glycine, uracil, adenine, other organic molecules in aqueous medium
	II	Anaerobic: traces of gaseous oxygen	Ultraviolet light; heat	Polyphosphates, porphyrins, peptides, porphyrin catalysis of photo-reduction
		Loss of most ultraviolet light	Visible light	& oxidation
Biological	III	Anaerobic: traces of gaseous oxygen, free nitrogen & carbon dioxide Loss of many free organic molecules	Visible light	Replicating organic molecules; photochemical reactions
	IV	Anaerobic: gaseous carbon dioxide & traces of gaseous oxygen Loss of most anaerobic environmental regions	Photo-reduction; fermentation	Organelles; cells; two-step light-energy conversion
	V	Aerobic: anaerobic pockets	Photosynthesis; respiration	Free-living cells; organs & organisms

Source: Adapted from Gaffron, 1965.

a waste.[2] Respiration is 18 times more efficient, in energy terms, than its predecessor, fermentation. For each molecule of glucose "food," it produces 36 molecules of the energy-carrier ATP, as compared to only 2 for the first stage (glycolysis) alone.[3] This enormous competitive advantage (as compared with anaerobic fermenters) led to a gradual displacement of the fermentation organisms from all habitats where oxygen was available.

Meanwhile, the anaerobic photobacteria were displaced by aerobic photosynthesizers (**cyanobacteria,** or blue-green algae) that were less sensitive to oxygen. Sometime between 1.5 and 1 billion years ago, the basic energy metabolism and the atmospheric **carbon cycle** stabilized. Table 5.1 summarizes the various evolutionary stages in energy metabolism. In due course, the oxygen content of the atmosphere arrived at its present level (more or less), and the biosphere as a whole achieved a reasonable ability to sustain itself.

Whereas some early organisms obtained energy by inefficient anaerobic processes akin to fermentation, later aerobic organisms utilized the more efficient photosynthetic processes. Again, it is interesting, and characteristic, that the newer, more efficient oxidation processes (e.g., the citric acid cycle) are more complex and involve more intermediate stages and more enzymes than the older fermentation process. The more complex process requires far more information embodied in its structure than the older one.

The evolution of the modern atmosphere also reflects evolutionary changes in the **nitrogen cycle.** Early proto-cells probably obtained nitrogen from dissolved ammonium (NH_4) ions or an iron-ammonium complex. As the iron was oxidized and atmospheric oxygen began to build up, however, free ammonia was (and is) rapidly oxidized or removed by other reactions. One group of oxidation reactions yields nitrogen oxides, NO and NO_2. Reacting with ozone, NO is oxidized in the atmosphere to NO_2, which is further oxidized to N_2O_5. N_2O and N_2O_5, in turn, react with ammonia to form ammonium nitrites and nitrates, which are redeposited (wet or dry) on land or ocean surfaces, where the nitrogen becomes available to plants.

Some reaction paths for the direct oxidation of ammonia and for denitrification of organic materials can lead to molecular nitrogen and water vapor. Molecular nitrogen is extremely stable and quite unreactive (except when reactions are caused by lightning), and, as a consequence, the atmosphere is a nitrogen "sink." As the rate of input of free ammonia to the atmosphere from volcanos gradually decreased while the need for nitrogen by living organisms increased, additional mechanisms were needed to recapture some of the atmospheric nitrogen in soluble and biologically usable form.[4]

5.2. THE INVENTION OF SEXUAL REPRODUCTION

Prokaryotic single-cell organisms reproduce by division (**mitosis**), but they have an ability that was not fully appreciated until recently: They free-exchange bits of genetic material. This ability gives them enormous adaptability. Nevertheless, single-cell organisms are inherently limited in terms of capability, since each individual cell must carry out all the functions of life. The appearance of **eukaryotes** (cells with true nuclei) was the next major development in evolution. This step toward increased specialization, which occurred 1.4 billion years ago, seems to have resulted from the first (successful) evolutionary merger. Lynn Margulis [1970] has shown that, in all probability, the cellular nuclei were originally independent organisms (bacteria) that were either ingested by prokaryotes or invaded them and evolved a symbiotic relationship.[5] In effect, prokaryotes are multi-cell organisms consisting of a nucleus and mitochondria associated with a "parent."

Eukaryotic organisms differ from their predecessors in having a specialized information storage and processing center, the nucleus. Free of the need to carry out other functions, it was able to specialize more effectively on this one particular function. The specialized gene-carrying nucleus opened the way to sexual reproduction.

Eukaryotes were followed by the first truly multi-cell organism roughly 700 million years ago. This was almost certainly another and more complex form of merger involving a number of eukaryotes working and reproducing together as a unit. Probably the earliest multi-cellular organisms evolved from successful symbioses. For instance, symbiotic cooperation between photosynthesizers (algae) and nitrogen-fixing or other nutrient-concentrating bacteria could have evolved into the earliest multi-celled plants. This would explain why such organisms appeared so late and lack recognizable precursors in the fossil record.

Calcium began to be used for protective shells (and subsequently for the internal skeletons of animals) between 0.8 and 0.6 billion years ago. Simultaneously, the free-oxygen level in the atmosphere began to rise sharply, along with the "sweetening" of the oceans. As respiratory metabolism took over, much of the dissolved carbon dioxide (CO_2) in the oceans was gradually taken up by either photosynthetic plants (in the carboniferous era) and later deposited as coal, or by calcium carbonate ($CaCO_3$), deposited on the ocean floor as chalk or diatomite. Some of these deposits were ultimately converted by heat and pressure into limestone. Calcium carbonate deposition in the ocean is a major part of the global carbon cycle. Carbon is regenerated on land by weathering of the silicate mineral $CaSiO_3$, which reacts with atmospheric CO_2 to form silica and bicarbonate.

The net effect of the advent of multi-cellular organisms seems to have been an increase in the efficiency of the organism and its ability to compete and survive. For instance, specialized sense organs developed, along with skins, shells, skeletons, limbs, central nervous systems, specialized digestive systems, specialized reproductive systems, specialized breathing and circulatory systems, warm blood, physical and chemical weapons for attack and defense, and—a new feature—the life cycle, from birth to death. Multi-cell organisms (variegated by mutation) also permitted more diversity of form and function. Sexual reproduction, together with a finite life span, resulted in the rapid proliferation of species with widely different functional specializations, including grazing, predation, parasitism, and decay. Specialization of function also permitted the gradual invasion of the polar regions, the land, and the air.

Biological evolution seems to have favored increasing *intelligence*. The term is difficult to define, but no one can deny that it exists and that it appeared fairly late. Intelligence comes in various degrees, of course. Mild forms include learning from experience, learning by imitation, tool-using behavior, and verbal communication. Written communication, abstract thought, and certain other types of organized social behavior appear to be unique to humans. This fact has tempted many biologists (and others) to argue that intelligence is the *object* of evolution. This idea was propounded in the writings of Herbert Spencer, the "father of sociology," as early as 1872. In recent years, however, this teleological view has been sharply attacked.[6] The most significant contribution of humans to biological evolution is probably that humans have, in fundamental ways, changed the rules of the evolutionary game itself. I will return to this point later.

5.3. EVOLUTIONARY MECHANISMS AND DISCONTINUITIES

In the examples of truly unpredictable (phylogenic) physical and chemical evolution described briefly above, random fluctuations appear to have played a fundamental role. The original Darwinian theory described no specific mechanism for the acquisition and inheritance of characteristics, but it was already clear that the introduction of new characteristics by some means (**mutation**) was necessary. The genetic discoveries of the remarkable monk, Gregor Mendel, in the late 19th century, opened the way to an explanation. The next step was the elucidation of the fundamental nature of genes and chromosomes by H. J. Muller and others. The theoretical work of the physicist Erwin Schrödinger (based in part on the X-ray diffraction work of Max Delbrück) provided a much deeper insight into the physical nature of the genetic substance and its ability to repair and replicate itself with extraordinary accuracy. Schrödinger called attention to the essentially quantum mechanical nature of the phenomenon.

Meanwhile the development of the field of population genetics by R. A. Fisher, J. B. S. Haldane, and Sewall Wright and empirical studies of natural populations by Theodosius Dobzhansky provided a deeper scientific basis for understanding—and better mathematical tools for analyzing—the interaction between variation and selection. As of the early 1960s there was a general consensus among geneticists as to the essentials of the process. In brief, mutations (or "copying errors") create variants of genes, called **alleles,** which are combined and recombined in chromosomes and which spread through the species by sexual reproduction. It was thought (mainly on the basis of phenotypic studies) that each new allele was either preferentially selected within a few generations, or just as quickly disappeared.

A third possibility—**selective neutrality**—was suggested by 1967 by the Japanese biologist Motoo Kimura [1979]. The so-called neutral theory holds that some mutations can spread through a population by random drift, without possessing any selective advantage. An important implication of this finding is that many — perhaps most—molecular variants are more or less functionally equivalent. Evidence for this theory is growing. It is of interest because it emphasizes, once again, the importance of random processes (or deterministic chaos).

It is important to note, for future reference, that the adaptation of species by mutation and selection from a distribution of genetic variants is not the only kind of adaptation mechanism observed in biology. Cooperation and/or merger—as exemplified, in all probability, by the evolutionary invention of the cell nucleus (eukaryotes) and later by the multi-cell organisms, are cases in point [Margulis, 1981]. The specific mechanisms by which accidental symbiosis evolved into true genetic mergers are not well understood, although the advantages to the organism are clear enough.

Later, the advent of the central nervous system and, still later, an evolved brain in animals made it possible for individuals themselves to react to and avoid or overcome environmental threats. Among animals, "lower" forms are lower because they can react only instinctively; their reactions are pre-programmed. Higher animals with bigger and better brains gradually developed an ability to adapt by learning from trial and error, or (in other words) information feedback and **self-modification**. Learned responses are not passed on genetically, but they can be taught to offspring. Indeed, this mechanism of adaptation by learning is much faster and more efficient than mutation and breeding mechanisms.

In this regard, the selection mechanisms in the evolution of species are similar to those in economics, which I discuss in Chapter 6. In making economic decisions, people do not behave like *Homo economicus*. Rather, they use heuristic "rules of thumb." This kind of behavior has been termed **satisficing** (in contrast to **optimizing**). I suggest that the same thing is effectively true for phylogenic evolution. Regarding the population as a kind of (collective) decision-maker, it is clear that many of the selection decisions are essentially arbitrary; that is, they do not result in any significant improvement. (Kimura's "neutral theory" of natural selection is based on this fact.)

Each time an allele is introduced into a population, a kind of unfocused, random competition takes place, but the winner is not always, or necessarily, the most efficient. A good deal of luck is involved, especially when there are only a few copies of the allele in the population. If an infinite number of trials were to be carried out, in all possible combinations of circumstances—and the results recorded—the results could be said to constitute perfect information about the performance of the allele. But this never happens. Usually, either the allele disappears fairly quickly after a few trials (which might be due to bad luck), or it eventually takes over. The latter process involves a far larger, but still finite, number of trials. Thus both selection and (especially) rejection often occur on the basis of imperfect information.

At the morphological level, the same point has been emphasized by Stephen Jay Gould. He comments:

> "Any replay of the tape [of life] would lead evolution down a pathway radically different from the road actually taken. But the consequent differences in outcome do not imply that evolution is senseless, and without meaningful pattern; the divergent route of the replay would be just as interpretable and just as explainable after the fact, as the actual road" [Gould, 1979, p. 47].

In advanced animal societies, learning (and teaching) activity is itself largely instinctive or accidental. Among the most advanced animals, however, culture is a factor. A wide variety of higher animal species, from birds to lions, have developed learned patterns of social behavior in which the mother assumes responsibility not only for protecting the young, but also for teaching them what they need to know to survive. (The very low survival rates for young animals raised among humans and later released to the wild attests to this.) In some cases (such as elephants, walruses, and anthropoid apes), a larger social group takes on this responsibility.

It is difficult to identify a satisfactory Darwinian selection mechanism to explain the evolution of such social behavior. In pure Darwinian selection, the genes of fitter individuals are passed on (and gradually penetrate the population), because fitter individuals are able to compete more effectively for food, mates, and other requisites of survival and are therefore able to produce more offspring than their less fit rivals. The fighting between males of many species for breeding rights exemplifies the standard model. How can a tendency toward cooperation and self-sacrificing behavior enable an individual to compete effectively in the breeding stakes? This question has not yet been fully answered.[7]

Humans have carried the adaptation mechanism another stage forward. We have developed the societal learning-cum-teaching phenomenon much further than have other animals. Several "quantum leaps" can be identified. The first was the development of spoken language, with grammatical structure and a significant vocabulary of words, to facilitate interpersonal communication in general and teaching in particular. (Other species, such as whales and dolphins, wolves, monkeys, apes, and elephants, obviously also communicate to some non-trivial degree, but decades of research have not yet uncovered anything resembling a true language among any other species.) In any case, humans began supplementing spoken language with pictographs, which developed about 4,000 years ago into written language and written records. Social history, as we know it, began with that breakthrough.

To anticipate a point that will be discussed later, it is likely that the pure Darwinian selection mechanism sketched above *cannot* account for social evolution. In particular, it is unlikely that inheritable differences dominate the process, or that fitness, in the social context, results in higher reproduction rates. In fact, the most superficial observation suggests, to the contrary, that fitness in the social context is reflected in economic, not reproductive, performance. There is an excellent correlation, in fact, between high levels of economic performance and low rates of human reproduction and population growth. The spread of Western technology and industrialization throughout the world is clearly driven by social learning and imitation, not by genetic inheritance.

In summary, biological evolution began 4 billion years ago and proceeded for almost all of that time via a slow (Darwinian) mechanism in which information contributing to survivability was accumulated accidentally and passed on by inheritance, through the genes. The selection process in species evolution is like a "random walk"; it is not goal driven. In behavioral terms, the process more closely resembles "satisficing" than "maximizing" or "minimizing." It also is important to emphasize (in view of later discussion) that all the known evolutionary selection mechanisms in the natural (i.e., non-human) world are essentially *myopic*.

Human history, very brief by comparison, exemplifies a much faster (Lamarckian) process based on brain learning and teaching, and the passing on of information from individual to individual. The formalization of this process by society, and the recent formalization of the related processes of knowledge acquisition (scientific research), could well be the most significant evolutionary event of all. In contrast to the myopia of natural selection, humans and human institutions and societies are beginning to make use of learning and forward vision (presbyopia) in the evolutionary selection process.

5.4. INFORMATION AND THE EVOLUTIONARY MECHANISM

Living organisms relate to one another within a larger framework, called an **ecosystem,** which consists of photosynthesizers, grazers, higher predators, parasites, and decay organisms. An ecosystem is a self-organizing entity at the macro-scale. It strives to grow (i.e., to capture all available resources), and it competes against other ecosystems on its boundaries. It can also increase its chances for survival by modifying its environment in favorable ways or by incorporating homeostatic elements.

Of course, the behavior of an ecosystem is determined not only by the environment but also by the characteristics of its various species. But the latter, in turn, depend on the former. The ecosystem as a whole is like a social club, in that it "chooses" which species shall be allowed to belong to it, while, at the same time, it is defined by its membership. The energy-flow relationships within an ecosystem reflect the various roles of its component species. As a general rule, the more complex an ecosystem is—the more interaction within it—the more efficiently the ecosystem utilizes the available resources (sunlight, water, minerals) and the more biomass it can support. Admittedly, this statement is close to a tautology, since the most natural way to define ecological efficiency—or, at least, to recognize it—is in terms of the ability to support biomass. The relationship suggested by Robert Ulanowicz between energy flows and competitiveness, or *ascendance,* was discussed briefly in Chapter 2.

Viewed from the perspective of entropy, three generalizations are defensible. First, a living organism is a **self-organizing dissipative structure** far from equilibrium, which captures and stores negative entropy (information) from its environment for future use, in the form of more complex *organization* or *structure.* The information thus captured and stored for future use, either in genes or in brains, is SU-information.

Second, **evolutionary level** can be defined, tentatively, as the ability of a living system to capture, store, and utilize SU-information. At the lower levels, information is stored in genetic structures and utilized at the cellular level. At higher levels, individuals acquire and utilize information via the sensory organs and the central nervous system; they accumulate and store information in brains. Genetic information is probably predominant in all forms of life (at least quantitatively), but brain information is increasingly important—and the nature/

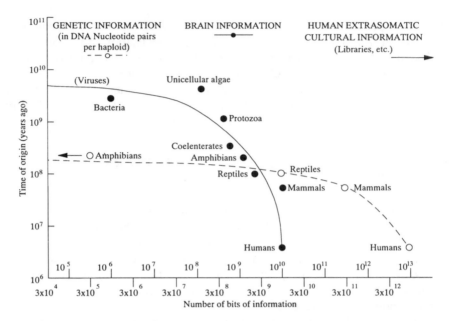

FIGURE 5.1. Genetic Information VS. Brain Information. Source: Reproduced with Permission from Sagan, 1977.

nurture balance is declining—in the higher forms (as indicated in Figure 5.1). All living organisms are characterized by the ability to store and utilize SU-information. The same is probably true of organized communities or ecosystems [MacArthur, 1955; Ulanowicz, 1986]. In the case of human social systems, one could make a plausible argument that brain information and social communication are now more important than genetic endowment.

Third, **intelligence** can be defined as the ability to learn, to modify behavior, and/or to modify the external environment, i.e., to create external structures (cultures and cultural artifacts) for information storage. The ability to modify the external environment is apparently possessed to some extent by all higher animals, but the ability to create external structures explicitly for information storage is possessed (as far as we know) only by humans. Carl Sagan [1977] has called this *extra-somatic* information. Whereas A. J. Lotka's principle (discussed in the next section) suggests that evolution seeks to maximize the ability to process free energy, it is suggested below that evolution tends to correlate with the ability to capture free energy and, most important, *to process it, utilize it, and convert it to morphological information embodied as structure or organization. To the extent that intelligence enhances this ability, evolution seeks to maximize intelligence.* This perspective differs from Lotka's in that it implies maximizing stocks and flows of SU-information rather than stocks and flows of free energy.

Consistent, at least in part, with these three generalizations, Werner Ebeling and Rainer Feistel [1984] have defined evolution as a "historical irreversible process consisting of an unlimited series of *self-organization* steps. Each step is triggered by a critical fluctuation ... leading to an *instability* of the original semi-stable state" (emphasis added). These authors have also suggested the necessary conditions for evolution, as follows:

1. The system must be far from thermodynamic equilibrium, though it can (and must) be in or near a local stable state.
2. It must have a *structural reserve* (of free energy) to permit infinite chains of self-organizing processes, including self-reproduction.
3. It must have the ability to *produce, store,* and *use* information, by means of *multi-stability, self-reproduction, finite lifetimes,* and the *selection of subsystems.*
4. It must be subject to *fluctuations* that vary the stored information.

That organisms capture and utilize disembodied information as negentropy from the environment and store it in physical structures has already been pointed out. The related hypothesis that living organisms retard the local—if not the global—increase of entropy has also been put forward many times in the biological literature. [See, for example, Johnstone, 1921; Breder, 1942; Needham, 1943; and Blum, 1955.] Prigogine, Nicolis, and Babloyantz [Prigogine *et al.,* 1972] raised the question of how living systems have acquired the ability to dissipate intensely. The role of fluctuations, as emphasized by Prigogine and his colleagues, and by Ebeling and Feistel, remains less clear than one would like.

The basic idea seems to be that fluctuations (like mutations in the Darwinian paradigm) constitute an important element of the evolutionary search process. Information losses, for instance, would normally have negative utility in most situations. Occasionally, however, random deletions will eliminate critical pieces of *mis*-information, or otherwise open the way to completely new evolutionary directions. It is often said of young scientists that "they haven't had time to learn what can't be done." Usually it wastes time and resources to keep challenging conventional wisdom—witness the never-ending stream of attempts to patent perpetual motion machines—but, while most such attempts fail, the occasional success may outweigh the many failures.

The Austrian physicist Erwin Schrödinger [1945] was one of the first to point out that the gene itself is nothing more nor less than a packet of information stored compactly in molecular form. The gene contains both **morphological** and **functional** information needed by the organism. The information embodied in genes tells cells how and when to divide, how and when to differentiate, how to manufacture various enzymes, hormones, etc. It specifies the many coupled biochemical pathways (**hypercyles**) that carry out the metabolic processes. It also tells the organism as a whole how to react to various stimuli, what food to eat, when and how to mate, where to lay eggs, etc. This information storehouse results from a long evolutionary learning process, described by Darwin as natural selection. The cumulative nature of the process is evident from the fact that the higher organisms, arriving later on the evolutionary scene, carry far

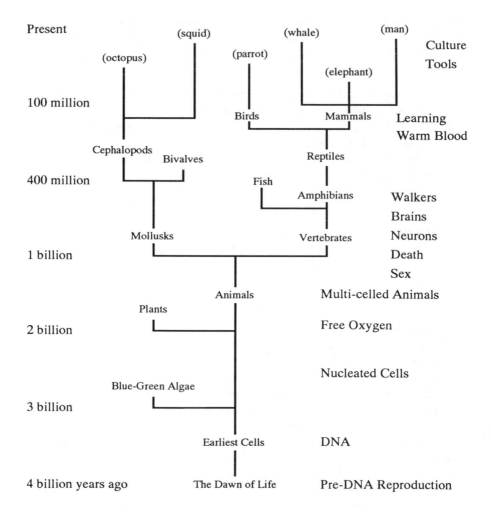

The diagram gives timing, family relationships and significant innovations in
the development of terrestrial intelligence. It is likely that very early evolution
occurred in an information carrier other than DNA. With the advent of learned
behavior in mammals and birds DNA lost a significant part of its job. More
than half of what makes a modern human being is passed on culturally.

FIGURE 5.2. Intelligence on Earth. Source: Reproduced with Permission from
Moravec, 1988.

more genetic information than the simplest, earliest organisms (Figures 5.1 and
5.2). The ability of higher organisms to accumulate and reproduce this information by extragenetic means can also be regarded as evidence of increasing
intelligence in the sense defined above.

Steven Polgar [1961] has identified four key evolutionary mechanisms, namely, *persistence, self-replication, environmental modification,* and *social evolution. Self-modification* (i.e., brain learning) might be added to Polgar's list, although it occurs by different selection mechanisms and on a much shorter time scale.

At the molecular (DNA) level, Schrödinger [1945] showed that **persistence,** as reflected by the low rate of mutation, can be explained only in terms of quantum mechanics. The critical difference between classical and quantized systems is that the former are characterized by a continuum of possible states, whereas the latter are characterized by discrete states with discrete transitions between them. At the level of chemical systems, Glansdorff and Prigogine [1971] and Nicolis and Prigogine [1977] have given an explanation of the persistence of structure—as they define **self-organization**—in terms of non-equilibrium thermodynamics.

Replication is, of course, one of nature's basic mechanisms for long-term survival (and growth). As noted previously, a detailed understanding of replication at the cellular level may still be years away. However, replication alone cannot explain evolution. A mechanism to ensure variability and selection is needed. The variability can result from random processes—for example, by mutation, or *copying errors* in the replication process as it occurs at the molecular level, as postulated by Eigen and co-workers [Eigen, 1971; Eigen and Schuster, 1979; Eigen *et al.,* 1981], Dyson [1982], and Ebeling and Feistel [1982, 1984]. It is the selection mechanism that, somehow, converts a small (but critical) proportion of nature's mistakes into evolutionary winners.

The variation-selection mechanism at the genetic level can be understood as an algorithm that operates on a genetic message (allele). The first step is the introduction of random copying errors, or "noise." It might be thought that copying errors (or mutations) decrease the information content of the genetic package. This is true from the perspective of communication. That is to say, if the replication process is viewed as the sending of a coded DNA message to posterity, the information equals "truth." Copying-error mutations could only contaminate the message by losing information and reducing its "truth" content.

The second, more critical, step processes that erroneous information through a sequence of decoding transformations and filters. (This is the same process that the correct message, or allele, undergoes as the organism develops ontogenically.) In the vast majority of cases, the variant allele is either neutral or harmful to the organism, whence it is not successfully reproduced at all, or is reproduced with low frequency, and dies out. Once in a long while, however, a variant allele turns out to be beneficial, in that the organism gains an enhanced capability of some sort. As the new allele replaces the old, over a period of generations, the new "message" also replaces the old. The interesting thing is that, when this happens, the message also tends to become more complicated, presumably because its enhancement of the organism's performance requires additional instruction.

Clearly, biological evolution is not really like communication. The selection process weeds out most copying errors as inferior, but it does not simply reward accuracy of replication. It does not, in this sense, resemble a test for typists or telegraphers. The highest prize in the evolutionary contest is given for *improvements* on the original message. The selection process has been compared (and modeled) as "hill-climbing in phenotype valuation landscapes" [Ebeling and Feistel, 1984]. As noted in Chapter 1, a number of biophysicists and computer scientists have built "hill-climbing" models of evolutionary selection.[8] Such models have the useful feature of abstracting the dynamics of the selection process from the specific attributes being selected for, or from the general evolutionary principles describing the overall trajectory.

By comparison, environmental modification has received less attention as an evolutionary mechanism despite its immense importance. As discussed in Chapter 3, living organisms have massively modified the atmosphere and the ocean over the past 2 billion years. The earth's primordial atmosphere was mostly carbon dioxide (CO_2) and water (H_2O) plus methane (CH_4) and ammonia (NH_3). It had no free oxygen or nitrogen, and the primitive oceans were much smaller and saltier. Life as we know it today could not survive in such an environment. As pointed out earlier, the present oxygen-rich environment is quite unsuitable to the spontaneous formation of organic molecules by known mechanisms.

Environmental modification is still practiced today at the micro-level by viruses (on their host cells), by wasps that paralyze their prey and lay eggs in them, and by every nest-building species. At the meso-level one sees the consequences in beaver ponds and coral reefs. On a larger scale, forests and savannahs create their own macro-environments. (Deforestation in the tropics, for example, can lead to irreversible desertification.) Of course, such human activities as agriculture and fossil-fuel consumption have enormous environmental impacts.

Another way in which negentropy (i.e., information-as-structure) in the biosphere cuts down on the free-energy flux per unit of biomass (and, thus, retards the rate of increase in global entropy) is through species diversity and specialization. This happens because each species of plant or animal having a different function (i.e., a different niche) from the functions of other species will also have a different structure to enable it to perform that function. Hence, it will have a different micro-composition (different skeleton, different digestive system, different enzymes, etc.). It follows that each species can be expected to require a somewhat different combination of environmental conditions (as Justus Leibig noted in 1876 in the context of agriculture [Lotka, 1950]). Thus, the total biomass of two species exploiting the same resource base can generally be larger than that of any one alone. (The genetic information embodied in the ecosystem will be somewhat greater, if not twice as great.) A third species, with a different specialization, can add still more to the total biomass without exhausting the available resources (but requiring still more embodied genetic information), and so on.

The lowest-level trophic animals obtain their food mainly as carbohydrates or cellulose directly from photosynthetic plants. Higher animals, in turn, are able to exploit lower animals and obtain their food mainly as proteins and fats, which are more useful and easier to digest. The top predators are therefore able to eat and digest much less bulk than they would otherwise need, by consuming foods more similar to their own tissues and more easily broken down into sugars and amino acids for reassembly. This enables active carnivores to consume more food (free energy) but spend much less time eating and digesting than the herbivores they prey on.

Significantly, the amount of genetic information required to reproduce an organism tends to increase with its trophic level in the predator-prey hierarchy. So, in general, does the information-processing capacity of the organism itself: The processing capacity of the brain and central nervous system tends to increase both with trophic level and with the passage of time since the earliest life, as Figures 5.1 and 5.2 show. For higher mammals, the theoretical D-information storage capacity of the brain significantly exceeds the storage capacity of the DNA, based on rough counts of neuron connections and their possible "states."[9] Figure 5.2 reflects the increased importance of **adaptive response** and **learning** in evolution as organisms and ecosystems become more complex.

5.5. THE DIRECTION OF PHYLOGENIC EVOLUTION

Biologists since the early 19th century have conjectured that the evolution of species follows a definite path. Among the stylized formulations of evolutionary direction are the following:

1. The evolution of species generally proceeds from simpler to more-complex forms. Dollo's "Law of Irreversibility" (1893) asserted that biological evolution is irreversible and tends toward increasing complexity and structure/organization, culminating in the human brain. Dollo's law was reformulated by Julian Huxley [1956] and has been taken up again in modern form by Carl Sagan [1977] (see Figure 5.1). Sagan notes that, significantly, the information processing and storage capacity of the brain and central nervous system of advanced animals increasingly exceeds their capacity to store genetic information.

2. The evolution of species proceeds from fewer, less-specialized forms to a greater number of more specialized ones, enabling the biosphere to exploit increasingly diverse physical habitats. This is conventional wisdom in biology and ecology. Of course, the statement is strictly qualitative and difficult to prove. It is theoretically plausible, however, that greater diversity (via specialization into "niches") increases the overall efficiency of a community in its use of available resources,[10] and observation appears to bear this out. The relationship between complexity and stability in ecosystems remains an open question. Conventional wisdom used to hold that

"complexity begets stability," that resilience is a *consequence* of complexity. Recent discoveries in the behavior of non-linear dynamical systems have raised a suspicion, yet to be confirmed, that complexity and stability may, in fact, be somewhat antithetical to each other.[11] Even if this is true, however, it would not be incompatible with an evolutionary tendency toward both greater organism complexity and system stability.

3. Species evolution appears to coincide with increasingly efficient capture and use of energy from the environment. "The life struggle is primarily a competition for free energy" [Boltzmann; cited by Lotka, 1945, p. 179]. And Bertrand Russell [1927, p. 174] stated: "Every living thing is a sort of imperialist, seeking to transform as much as possible of its environment into itself and its seed."

4. The biosphere* appears to *retard* the global increase in entropy [e.g., Johnstone, 1921; cited in Lotka, 1945, p. 184], even though individual life processes (which occur far from thermodynamic equilibrium) actually *maximize* the local production of entropy. Many biological writers have noted this entropy-retarding "thermodynamic imperative" of living organisms, including Breder [1942], Needham [1943], and Blum [1955], all cited by Polgar [1961]. Prigogine and his colleagues have put the emphasis on the second part of the proposition. They ask how living systems acquired the ability to "dissipate intensely" enough to sustain order/structure in conditions far from thermodynamic equilibrium. (As more and more free energy is captured and embodied in living organisms, the ability of the biosphere to retard global entropy seems to be increasing.)

5. "Evolution is entropy." The evolutionary theories of Daniel Brooks and E. O. Wiley [1986] are essentially an elaboration of the theory of self-organization in dissipative systems far from equilibrium. [See Prigogine *et al.*, 1972; Nicolis and Prigogine, 1977; and Prigogine and Stenger, 1984.] Brooks and Wiley essentially argue along the lines of (2) above that evolution is moving toward diversity, complexity, specialization, and organization. They go even beyond Prigogine and his colleagues in arguing that this trend is a necessary consequence of the laws of thermodynamics: i.e., that entropy *causes* evolution. Critics argue that their theory is tautological.

6. The biosphere* tends toward zero net material waste (perfect recycling), at least over the long run. In most cases the material cycle is partly closed biologically and partly closed geologically. The continuous burial of elemental carbon and sulfur is necessary, for instance, to permit the existence of free oxygen in the atmosphere. The burial of calcium and phosphorus (in skeletal material and shells) must be compensated for by geological exposure of materials buried in the past.

The asterisk (*) after *biosphere* means that in this case the term explicitly excludes the activities of industrialized humankind. (These can be called the **industrial metabolism,** which operates in the **anthroposphere.**) The extent to which any statement about the biosphere can have meaning if humankind is excluded must be discussed in due course. As I have noted, however, humans have changed the rules of the evolutionary game, and human evolution can probably be best considered as a separate problem.

The foregoing list of stylized formulations of evolutionary direction is hardly complete. I have excluded, for instance, statements so broad as to be meaningless, such as characterizations of evolution in terms of "progress" or "increasing organization." (To be sure, its characterization in terms of increasing complexity could be criticized on similar grounds, although I think the meaning is clear and complexity potentially measurable.)

There is no absolutely clear distinction between a stylized formulation and a "covering principle." Lotka's principle, in particular, blurs the distinction. Building on the observations in (3) above led Lotka [1922] to conclude:

> "The operation of natural selection tends to increase the biomass and embodied free energy in the biosphere[*]. It also tends to increase the rate of circulation of both matter and [free] energy flux through the system."

Clearly, Lotka's principle is in the tradition of physics: It is a heroic attempt to find a simple extremum principle (analogous to the famous "principle of least action" in mechanics) from which the extremely complex phenomenon of biological evolution can be derived by deductive logic. Unfortunately, Lotka's principle, while more explicit and more easily tested, suffers the same fate as Dollo's earlier "law." Both explain many of the facts, but neither explains all of them, and (worse) there are exceptions to both. With regard to Dollo's law, Lotka himself has pointed out the difficulty:

> "Dollo's Law of Irreversibility, even in its modern phrasing: 'Evolution is reversible in that structures or functions once gained may be lost, but irreversible in that structures once lost can never be regained,' is not borne out by the facts" [Cave and Haines, 1944; cited by Lotka, 1945].

Lotka's principle, in turn, fails to account for all the facts of biological (or social) evolution. In particular, it fails to shed any light on the evolutionary roles of complexity, diversity, stability, order, and intelligence. In contrast, the theory of Brooks and Wiley, "evolution as entropy," emphasizes the fact that evolution is an irreversible process characterized by increasing self-organization, complexity, specialization, and organization, although it does not make the proposed causal connection very clear.

5.6. A DIGRESSION ON THE EVOLUTION OF CONSCIOUSNESS

Given the intense interest these days in artificial intelligence, and whether computers can "think" like humans, I want to briefly address the question "Is consciousness the end-point of evolution"? Undeniably, many people believe this. One of the axioms of those who discount the very possibility of artificial intelligence is that computers can never be "conscious" as we humans are. At bottom, such questions can be discussed seriously only on the basis of an accepted definition of **consciousness.**

There may be wide disagreement on such a definition, but I shall start from the premise that consciousness is awareness of the independent existence of a *self* that is alive and aware of its surroundings. In other words, *consciousness* implies an ability to discriminate between *self* and *not-self,* and to observe the changing relationships between these two entities. Consciousness does not necessarily require any particular degree of intelligence. Intelligent organisms can clearly be unconscious (when they sleep without dreaming, for instance). The most confusing aspect about consciousness is that it can lead, recursively, to thinking about thinking about thinking—a good way to get a headache. The potential for confusion is so great that many people have concluded that consciousness actually belongs to an independent entity, "the soul," which normally resides in our bodies (but occasionally goes off on "out-of-body" adventures) and finally departs for good when the body dies.

It is tempting to ascribe life-after-death to the soul. To accept the notion that consciousness is a phenomenon of the physical brain itself, one must give up the idea of the soul's immortality. This is undoubtedly a source of deep-seated resistance to the notion that computers might eventually be able to think. If computers can think (and be conscious) without having a soul, why then.... The implications are painful. On the other hand, the soul is, by definition, a non-physical entity that cannot be explained by science. The only proof of its existence is that it seems to explain the mystery of consciousness. Actually, however, the idea of the "soul" is not very satisfactory, even if one sets aside the question of its supposedly non-physical nature.

To explain the phenomenon of self-awareness in terms of a soul, it is necessary to imagine the soul as a kind of "homunculus" sitting in the "central processor" of the brain, directing its awareness to this sensory input or that, and formulating responses. But, to act in this way, the soul (or homunculus) would have to have its own central processor, with another homunculus sitting in it, and so on and on. Infinite regress.

One way out of this logical difficulty is to postulate either a hierarchy of homunculi of decreasing intelligence at lower levels. Another is to postulate a large number of semi-independent homunculi, each in charge of its special function (eyes, ears, legs, etc.) and each communicating with the others in a sort of broadcast fashion. In other words, it may not be necessary to assume any central switchboard at all. Each of the major functions (like vision) can be further broken down. There may be "person detectors"—perhaps further distinguishing males, females, children, "our tribe," and "enemies." There may be animal detectors (dangerous predator, possible prey, etc.), tree detectors, rock detectors, etc. Reporting to these may be little processors in charge of identifying horizontal lines, vertical lines, circles, bilateral symmetry, and so on. The details are too complex to describe here. Suffice it to say that recent work in computer science and cognitive psychology makes this view look more and more like the correct one. Consciousness, in this view, is a kind of prioritizing meta-function that selects from the cacophony of internal signals ("voices") and creates a linear real-time narrative of "what is going on" moment by moment.

In short, consciousness appears to be a natural consequence of information exchange among elements of a distributed processing system. In this sense, consciousness has resulted from the evolutionary tendency toward the embodiment and processing of more and more D-information.

5.7. SUMMARY

To draw the various themes together, **phylogenic** evolution is essentially the accumulation of genetic information that has been screened for *relevance* and *usefulness.* Learning can be categorized as a corresponding accumulation of relevant information about the environment and the behavior of other organisms. But relevant learning results in effective adaptive behavior. The interaction of organisms with the environment can be interpreted in information-processing terms. The organism sends messages to the environment and receives messages (threats and opportunities) in return. Successful adaptation on the organism's part can be regarded as correct interpretation of the environmental message. Alternatively, the environment as information-processor can be regarded as the "selection algorithm" for phylogenesis.

The key characteristics of biological evolution are similar in some respects to the characteristics of physical evolution, namely, *increasing diversity of products,* with *increasing complexity of form,* and *increasing stability of complex forms.* As the next chapter will discuss in more detail, all these trends are consistent with the increasing quantity of *structural information* and a parallel tendency toward an increasing *capacity to acquire, process, reduce, and transmit information.* Most people will not find these generalizations either particularly surprising or profound.

One key feature of biological evolution, however, is the evolution of cooperative behavior (symbiosis) and genetic mergers. In fact, this feature may well be responsible for the observable increase in biological complexity and stability. At the primitive level, "cooperation" may be nothing more than a fortunate combination of mutually beneficial behaviors. The combination is reinforced by increased competitiveness and survivability on the part of the combination. At more advanced (i.e., more recent) levels, cooperation has been linked to altruism and "far-sightedness." It is unclear which of these notions is the more fundamental.

Indeed, any self-regulating system can be regarded (by a small stretch of the imagination) as a sort of cooperation. Thus, it is not too far-fetched to think of cooperation between biological organisms and inanimate geochemical or geophysical systems. This kind of cooperation increases overall stability—or homeostasis — another way of describing *survivability* for a system. It is worth mentioning that, in the context of this sort of self-regulating system, the apparent conflict between ecosystem stability and complexity seems to disappear.[12] To mention one simple case in point, tropical rain forests are adapted to high humidity and frequent rainfall, which are themselves largely determined by evapo-transpiration from the forest. Thus the rain forest contributes to itself a large part of the moisture it needs to survive as a tropical rain forest.

I believe, and tried to show explicitly in Chapter 2, that the general trends toward diversity, complexity, and stability can be most usefully expressed in the language of information theory. It seems clear, however, that it will not be possible to explain all the phenomenology underlying these trends in terms of a simple maximization or minimization principle analogous to Lotka's. In effect, the evolutionary system on the macro-scale does not seek to maximize anything, least of all a single attribute or measure.

The most general characterization of this process I can think of is the following: Biological evolution tends to increase the ability of organisms to survive and grow by various means. These include sensing the environment, "learning" useful behaviors—ranging from parasitism and predation to symbiosis and cooperation—and storing the information in retrievable and reproducible form.[13]

ENDNOTES

1. Glycolysis is actually quite a complex process, involving a series of 12 enzyme-induced reactions.

2. It is obvious that biochemical metabolic cycles, such as the citric acid cycle, are close relatives of Eigen's hypothetical protein-reproduction hypercycle. In fact, all metabolic cycles could well be evolutionary descendants of some primitive reproduction cycle.

3. The ATP is recycled. It contains high-energy phosphate bonds, which can easily yield their energy to drive various organic synthesis processes inside the cell, while the depleted form of the carrier molecule, known as adenosine diphosphate or ADP, makes its way (by diffusion) back to a recharging point where it is reconverted to ATP.

4. This ability was acquired, eventually, by specialized micro-organisms that live symbiotically on the root structures of certain plants (the legumes). Symbiosis—a very close form of cooperation, in biological terms—is probably a fairly recent evolutionary development.

5. Margulis [1981] has presented strong evidence that organelles, mitochondria, and cellular flagellae also evolved from symbiotic relationships.

6. Harvard biologist Stephen Jay Gould [e.g., 1979] has strongly debunked what he calls "the iconography of the ladder," by which he means the idea that evolution is an inevitable progression along some definite goal-directed trajectory—toward intelligence or whatever.

7. Recently, Herbert Simon [1990] has presented a simple model (intended to apply to humans, but applicable perhaps to other advanced social species, such as apes and elephants) in which he postulates that an inherited quality of docility (meaning "disposed to be taught") contributes to social fitness via learned skills and behaviors. One of the kinds of behavior that can be taught (and learned) is, of course, altruism. Simon's theory may not be "true"—it has yet to be tested empirically—but it is, at least, evidence that the spread of altruism is not inconsistent with Darwinian selection rules.

8. See, for example, Holland [a, 1986, b, 1987, 1988], Kaufmann [1989], and Schwefel [1988, 1989a,b].

9. Unfortunately an exact comparison is impossible. Though we understand the basic genetic coding scheme (Table 4.3), we have no comparable understanding of the information-coding scheme used by the brain.

10. The argument can be sketched as follows: Assume an ecosystem consisting of several interacting species in an open habitat. (If the habitat were closed, as in a spaceship, in principle a finite number of species might recycle all minerals and utilize only sunlight). *Open* means that conditions are variable and there are some non-energy inputs and outputs of a random nature. Efficient use of the resources under a range of different external conditions—say, rainfall—therefore implies a combination of large, efficient, long-lived dominant species (analogous to a base load electric power supply) and smaller, faster-growing, short-lived species able to multiply fast and take up the slack quickly when conditions are favorable (analogous to peak load supply). To be available when needed, of course, the latter type must be represented in some number, even when conditions are unfavorable for it. As the number of constraints increases (sunlight, nitrogen, phosphorus, potassium, trace minerals, etc.) the number of possible tradeoffs favoring different specialists also increases.

11. According to one mathematical ecologist: "A wide variety of mathematical models suggest that as a system becomes more complex, in the sense of more species and a more rich structure of interdependence, it becomes more dynamically fragile ... as a mathematical generality, increasing complexity makes for dynamic fragility rather than robustness." [May, quoted by Lovelock, 1988, p. 51]

12. Gaia theory is concerned with the regulatory effects of the biosphere as a whole on the earth's climate and environment, especially the atmosphere. It is a hypothesis of Gaia theory that the earth is maintained and stabilized in a condition to support life by life itself. With respect to the supposed conflict between complexity and stability, no such problem seems to arise when non-biological controls, such as temperature, rainfall or nutrient availability, are incorporated in the models [Lovelock, 1988, p. 52 *ff.*]

13. The brief discussion of reversibility and irreversibility in computers in Chapter 1 made the point that one aspect of irreversibility is the loss of information. The key to evolution—as well as computation—is the discrimination and selection of useful information from the mass of useless information.

REFERENCES

Blum, H. F., *Time's Arrow and Evolution*, Princeton University Press, Princeton, N.J., 1955.

Breder, C., "A Consideration of Evolutionary Hypotheses in Reference to the Origin of Life," *Zoologica* **27**, 1942 :132-143.

Brooks, Daniel R. and E. O. Wiley, *Evolution As Entropy: Towards a Unified Theory of Biology*, University of Chicago Press, Chicago, 1986.

Cave, A. J. E. and R. Wheeler Haines, "Dollo's Law of Irreversibility," *Nature*, November 1944 :579.

Dyson, Freeman J., "Origin of Life," *Journal of Molecular Evolution* **18**, 1982 :344-355.

Ebeling, Werner and Rainer Feistel, *Physik der Selbstorganization und Evolution*, Akademie-Verlag, Berlin, 1982.

Ebeling, Werner and Rainer Feistel, "Physical Models of Evolution Processes," in: Krinsky (ed.), *Self-Organization: Autowaves and Structures Far from Equilibrium*, Chapter 6 :233-239, Springer-Verlag, New York, 1984.

Eigen, Manfred, "Selforganization of Matter and the Evolution of Biological Macro-molecules," *Naturwiss* **58**, October 1971.

Eigen, Manfred et al., "The Origin of Genetic Information," *Scientific American* **244**, 1981 :88-118.

Eigen, Manfred and P. Schuster, *The Hypercycle: A Principle of Natural Self-Organization*, 1979.

Gaffron, H., "The Role of Light in Evolution. The Transition from a One-Quantum to a Two-Quantum Mechanism," in: Fox (ed.), *The Origins of Prebiological Systems and of Their Molecular Matrices* :450, Academic Press, New York, 1965.

Glansdorff, P. and Ilya Prigogine, *Thermodynamic Theory of Structure, Stability and Fluctuations*, Wiley-Interscience, New York, 1971.

Gould, S. J., "Another Look at Lamarck," *New Scientist* **84** (1175), 1979 :38-40.

Holland, John H., "A Mathematical Framework for Studying Learning in Classifier Systems," *Physica* **22D**, 1986 :307-317.

Holland, John H., *Evolution, Games and Learning*, North Holland, Amsterdam, 1986a.

Holland, John H., "Genetic Algorithms and Classifier Systems: Foundations and Future Directions," in: Grefenstette (ed.), *Genetic Algorithms and Their Applications*, Lawrence Erlbaum, Hillsdale, N.J., 1987.

Holland, John H., "The Dynamics of Searches Directed by Genetic Algorithms," in: Lee (ed.), *Evolution, Learning and Cognition*, World Scientific, Singapore, 1988.

Huxley, J. S., "Evolution, Cultural and Biological," in: Thomas (ed.), *Current Anthropology* :3-25, University of Chicago Press, Chicago, 1956.

Johnstone, James, *The Mechanism of Life in Relation to Modern Physical Theory*, Longmans, London, 1921.

Kauffman, Stuart A., *Origins of Order: Self-Organization and Selection in Evolution*, xxx, xxx, 1989. [unpublished manuscript]

Kimura, Motoo, "The Neutral Theory of Molecular Evolution," *Scientific American*, November 1979.

Lotka, Alfred J., "Contribution to the Energetics of Evolution," *Proceedings of the National Academy of Sciences* **8**, 1922 :147.

Lotka, Alfred J., "The Law of Evolution As a Maximal Principle," *Human Biology* **17**(3), 1945 :167-194.

Lotka, Alfred J., *Elements of Physical Biology*, Dover Publications, New York, 1950. [First published in 1925]

Lovelock, James E., *The Ages of Gaia: A Biography of Our Living Earth*, Oxford University Press, London, 1988.

MacArthur, Rilt, "Fluctuations of Animal Populations and a Measure of Community Stability," *Ecology* **36**, 1955 :533-536.

Margulis, Lynn, Origin of Eukaryotic Cells, Yale University Press, New Haven, Conn., 1970.

Margulis, Lynn, *Symbiosis in Cell Evolution*, W. H. Freeman and Co., San Francisco, 1981.

Moravec, Hans, *Mind Children: The Future of Robot and Human Intelligence*, Harvard University Press, Cambridge, Mass., 1988.

Needham, Joseph, *Time, the Refreshing River*, Allen and Unwin, London, 1943.

Nicolis, Gregoire and Ilya Prigogine, *Self-Organization in Non-Equilibrium Systems*, Wiley-Interscience, New York, 1977.

Polgar, Steven, "Evolution and the Thermodynamic Imperative," *Human Biology* 33(2), 1961 :99-109.

Prigogine, Ilya, Gregoire Nicolis and A. Babloyantz, "Thermodynamics of Evolution," *Physics Today* 23(11/12), November/December 1972 :23-28(Nov.) and 38-44(Dec.).

Prigogine, Ilya and I. Stengers, *Order Out of Chaos: Man's New Dialogue with Nature*, Bantam Books, London, 1984.

Russell, Bertrand, *An Outline of Philosophy*, 1927.

Sagan, Carl, *Dragons of Eden: Speculations on the Evolution of Human Intelligence*, Random House, New York, 1977.

Schrodinger, Erwin, *What is Life? The Physical Aspects of the Living Cell*, Cambridge University Press, London, 1945.

Schwefel, Hans-Paul, *Collective Intelligence in Evolving Systems*, Discussion Paper, Informatics Center, University of Dortmund, Dortmund, Germany, 1989a.

Schwefel, Hans-Paul, *Evolutionary Learning Optimum-seeking on Parallel Computer Architectures*, Technical Report, University of Dortmund Department of Computer Science, Dortmund, Germany, 1988. [P.O. Box 500500, D-4600, Dortmund 50, Germany]

Schwefel, Hans-Paul, *Collective Learning in Evolutionary Processes*, International Symposium on Evolutionary Dynamics and Non-Linear Economics, University of Texas, Austin, Tex., April 1989b.

Simon, Herbert A., "A Mechanism for Social Selection and Successful Altruism," *Science* 250, December 1990 :1665-1668.

Ulanowicz, Robert E., *Growth and Development: Ecosystems Phenomenology*, Springer-Verlag, New York, 1986.

Wald, George, "The Origin of Life," *Scientific American*, August 1954.

Chapter 6

Evolution in Human Social Systems

6.1. THE EVOLUTION OF COOPERATIVE BEHAVIOR

Social systems are, in one respect at least, extraordinarily complex when even compared to biological organisms and communities. This reason alone makes the key evolutionary (or developmental) changes difficult to identify. The language and typology (phylogeny, ontogeny; genotype, phenotype) that come from biology seem to fit physical evolution fairly well. But it is not entirely clear that this typology is even applicable to the social case.[1]

The inherent difficulty of social science can be attributed in very large part to one fundamental difference between social and biological systems, the importance of which cannot be overestimated: In biological systems, natural selection is a myopic, random-walk process, requiring many generations or life cycles. Thus, at each moment in time the selection criteria confronted by an evolutionary innovation (**mutation**) are essentially static. In social systems, selection is a learning process that can (in extreme cases, at least) be nearly instantaneous. Social systems are inherently dynamic.

Feedback between evolutionary changes at the organism level and the environment does exist. But it is slow enough, in almost all cases, to be ignored. Thus a static-equilibrium concept (known, incidentally, as **evolutionary stability**), plays a central role in most current theories of biological evolution.[2] In the language of game theory, the organism is "playing a game against (unchanging) nature."

The co-evolution of viruses and bacteria in an epidemic presents an obvious counter-example. A parasite that kills its own host will not survive long unless it can go dormant or transfer to another organism. Thus the survival of parasitic micro-organisms depends on the evolutionary development of mutually benign relationships with the host. The appearance of a virulent (mutant) strain of microbes induces the host to evolve more and more effective defenses. The return to evolutionary stability, to be sure, requires many generations of microbes, but only a single generation (or two or three) of hosts. (This situation resembles a

124

game in which there are no winners, only losers.)[3] Notwithstanding the case of viruses and microbes, feedback between the evolving entity and its environment is normally much slower and chancier in biology. So much so, that it is often easy to overlook. Nevertheless, co-evolution is a historical fact.

On the other hand, co-evolutionary feedback is normally much more efficient in social systems (whether the evolving entity is an individual, group, community, or organization). As discussed in the previous chapter, several forms of passive cooperation (symbiosis) are observed in biological systems. One organism can find a useful niche for itself by providing some service for another, and vice versa. The role of bees and butterflies in pollinating plants illustrates one form of symbiosis. The relationship between pilotfish and sharks is another. The domestication of dogs, sheep, cattle, and other animals by humans is still another example. Or consider the nitrogen-fixing bacteria living on the roots of legumes or the digestive bacteria living in the guts of almost all higher animals (notably the bacteria that enable ruminants to digest cellulose).[4]

Active cooperation among individuals within a colony—bees, wasps, ants, termites, beavers, antelopes, cattle, sheep, horses, zebra, reindeer, elk, wolves, hyenas, monkeys, baboons, elephants, and so on—is a familiar phenomenon. (Because of it, we call them "social" species.) But beehives and anthills, or even beavers or antelopes, have limited relevance to human behavior. Social cooperation among closely related kin sharing a common genetic heritage emerged among insects (ants, termites, bees) hundreds of millions of years ago. This phenomenon has been satisfactorily explained within the Darwinian evolutionary paradigm by W. D. Hamilton [1964].

Cooperation and altruistic behavior beyond the kinship level, however, is a much more recent and essentially different kind of phenomenon, although not unique to humans. The more advanced the species (i.e., the more recently it emerged on the evolutionary scene), the more likely it is that cooperation is not only active, but also to some extent voluntary. Voluntary cooperation is, in many cases, indistinguishable from altruism: actions taken for the good of others without prospect of a return. By contrast, there is nothing voluntary about the behavior of insects. A genetic basis for social cooperation among higher animals cannot be excluded, of course [see, e.g., Wilson, 1975]. Instinct, genetically programmed, obviously plays a role among higher species, even man, but it is not a complete program for behavior. Indeed, the evolutionary selection mechanisms involved in human society are almost certainly cultural, notwithstanding Herbert Simon's recent proposal of a Darwinian model for the preferential selection of a propensity for "teachability" (*docility*), resulting, in turn, in socially induced altruism [Simon, 1990].

For higher animals with advanced central nervous systems and well-developed brains, behavior involves real choices and, therefore, more- or less-conscious decisions.[5] Decisions, in turn, are based on the acquisition and processing of information about the environment. Human societies create institutions that tend to induce individual decisions in one direction or another, but free will exists even in the most coercive societies. The evolution of voluntary behavior, conscious decision-making, far-sightedness (presbyopia), and cooperation are facts. In effect, both individual and evolutionary choice are becoming less myopic and more far-sighted. In this connection, Robert Trivers [1971] showed in the early 1970s that **reciprocal altruism** (i.e., *quid pro quo*) is not necessarily inconsistent with myopic (Darwinian) genetic selection mechanisms. The inherited ability to identify individuals and to remember their past behavior, and the outcomes of that behavior, seems to be the critical factor in cultural evolution.

6.2. GAMES AND RATIONAL CHOICE IN SOCIAL SYSTEMS

In social systems the responses and adjustments between decision-making entities whether an individual, firm, or nation are relatively fast. On the time scale of biological evolution such responses are instantaneous. A firm announces a price change, a bank increases its prime rate, or the Federal Reserve announces a change in interest rates, and the response occurs within hours or days. Responses to moves on currency, stocks, and commodity markets can occur in minutes (or, with computerized program-trading, in seconds). Obviously some social responses are much slower than these examples. But, clearly, social systems can and do respond very rapidly to some stimuli.

This quickness of response implies that **selection** in social systems, at most levels, relies largely on conscious and explicit decision-making. Indeed, the best model of social decision-making is **game-playing**, where the game itself is the system. The usual conception of game-playing is one of pure competition, subject to rules, whose objective is to gain points and, ultimately, to **win**. Each player makes moves based on his or her assessment of the current situation and future prospects. Such prospects depend on the past and present behavior of the other players. (In chess, for example, a good player tries to work out the options as far ahead as six to eight moves.) But, in the real world, the number of possible moves and strategies, never mind the number of options, is far, far greater than that in a simple recreational game. In particular, competition is not the only strategy in the real world. **Cooperation**, while not important in most recreational games,[6] is critically important in social systems.

Many recreational games, especially those involving physical activity, have another misleading aspect: The only **rationality** required on the part of players is defined by the rules. In contrast, the "games" of social life depend on human behavior. The notion of rationality arises, in some sense, *ex post*, as an attempt to derive general rules explaining human behavior in one domain (say, economics) from simple axioms. The prevailing notion of economic rationality was formulated axiomatically by John von Neumann and Oskar Morgenstern [1944, pp. 26-27]. In brief, an individual is rational if, when faced with alternatives, his preferences are **reflexive**, **transitive**, and **complete**; that is, they can be put in a definite, unique order. The authors state:

"We have practically defined numerical utility as being that thing for which the calculus of mathematical expectations is legitimate" [*ibid.*, p. 28].

It is interesting that the axiomatic basis of utility-maximization theory as set forth by von Neumann and Morgenstern has remained almost unchallenged until recently.[7] Yet the realism hence the applicability of utility maximization to economics has faced strong challenges from behavioristic and organizational perspectives.[8]

Given a criterion for rational choice, games can be analyzed by defining consistent decision-making rules, called **strategies**, and systematically comparing the outcomes of different strategies. In most cases this comparison can be done only probabilistically, which requires one to take into account the possibility of failure. This, in turn, necessitates an approach that assesses not only the probability but the disutility associated with a loss.[9] A systematic compilation of the outcomes of all possible strategies by one player against all possible strategies by other players is called a **payoff matrix**. (In principle, such a matrix will be multidimensional, depending on the number of players. It is a square matrix only if there are just two players.)

A perfectly rational player is one who has computed the payoff matrix for the game (based on his or her preferences for certain gains and losses) and who selects a strategy on the basis of expected payoffs (higher payoffs being preferred to lower ones). If cooperative behavior is possible, all possibilities for alliances and coordination between and among players are implicit in the payoff matrix. In reality, the computational requirements for such an approach are so impossibly enormous that only the simplest games (like tic-tac-toe) can be analyzed in this way.

Perfect rationality in the von Neumann–Morgenstern sense is only a theoretical abstraction. Real players are forced to play real games with only the dimmest knowledge of the payoff matrix. The much fuzzier decision-making rules employed by real players have been called **bounded rationality** [Simon, 1977, 1979, 1982, 1984] to distinguish them from perfect rationality. Bounded rationality, or BR, is simply characterized by a preference for gains over losses and the application of "common sense."

In the game-playing perspective outlined above, there are two limiting cases of theoretical interest. First is the case where the individual is so insignificant compared to the number of similar or comparable individuals in the collectivity (e.g., the "competitive free market") that his actions generate no specific response. This case is a mathematically convenient assumption partially justified by "the law of large numbers." It is the intellectual basis for much of standard neo-classical economic theory, as will be noted later. In this case, the individual actor is really "playing a game against nature," as in the biological case. In the economic context, he is a "price-taker."

The second limiting case is opposite and even less realistic. The player (usually an institution) is assumed to be so massive and dominant that it can be assumed to be the **only** player in the game. In economics, it is a monopolist if a producer and a monopsonist if a consumer. The classic monopolists were Standard Oil, I.G. Farben, and AT&T (all of which have been broken up). The classic monopsonists are the Pentagon (for military hardware) or NASA (for spacecraft). In either case, it is a price-maker in the economic context.

In either of these cases, the **utility function** of the player is well defined, and the game becomes trivial as defined in game theory, though not necessarily easy to compute. With perfect information, the player can, in principle, determine his or her optimum strategy. With imperfect information, this is no longer true, and all sorts of speculative behavior and instabilities can, and do, result.

Real social and economic life, of course, consists almost entirely of intermediate cases, where active "gaming" takes place among individuals, groups, and economic units. The crucial fact is that, in a "zero-sum" setting (or an approximation thereto), there must be some losers to accommodate winners.

Because each individual player or economic unit has a unique utility function, the notion of "social optimum" cannot be consistently defined in terms of individual utility functions, as Kenneth Arrow [1967] proved with his "impossibility theorem." The best one can do, it seems, is to define some aggregate measure of social welfare—the usual choice is Gross National Product, or GNP—and try (as a central planner) to maximize it. (In effect, this assumes that money is truly the measure of all things.) So the question arises: Is there any consistent way to define an optimum? Or, failing that, an equilibrium (or stationary) state toward which an interacting aggregation of economic units is drawn?

Vilfredo Pareto, an Italian economist with an engineering background suggested an answer around the turn of the century. The so-called **Pareto optimum** is a state in which trading has halted because no individual can achieve a gain except at the expense of another. Voluntary exchanges occur only if all parties benefit. Once all possible voluntary exchanges have taken place, the benefits of such exchanges are exhausted. Any one individual can benefit further only via exchanges in which another individual loses, thereby violating the condition for a Pareto optimum.[10] The Pareto optimum does not necessarily allocate resources **equitably**. (Equity is, of course, an ethical concept.) Nor is it a unique state. There can be several, or even many, Pareto-optimal states.[11]

A more complicated example, which illustrates many features of real-world social and economic phenomena, is called "the prisoner's dilemma." Assume there are several prisoners accused of a joint crime. If all prisoners refuse to cooperate with the authorities, there will be insufficient evidence to prosecute and all will be released from prison immediately. If any of the prisoners confesses and testifies against the others, however, he will get a short prison term while the others will stay in prison for a long time. The optimum result for all players requires perfect cooperation and trust. But each prisoner is much better off than the others if *he* is the first one to confess. In other words, there is a strong motive for betrayal. In these circumstances, the optimal result (release for all) is much less likely than the suboptimal result (all lose).

Here we have a competitive situation in which even the Pareto optimum is difficult to achieve, to say nothing of any sort of global optimum.

In the real world, many system-level disasters probably do occur because of difficulties in arriving at cooperative solutions. One of the best-known (generic) examples has been named the "tragedy of the commons" [Hardin, 1968]. This situation arises with respect to the use of so-called common-property resources, such as fisheries, communal pastureland or woodland, fresh water, wild animals, national parks, and the assimilative capacity of the environment. Again, the socially optimal solution is a cooperative one that limits the total annual harvest to a level that can be sustained indefinitely. (Within that sustainable yield, the allocation among individual users is less critical.) But, absent a cooperative solution, when many people have the right of access to such a resource each is motivated to capture as much of the resource as possible for his or her own use. Nobody has any motive to conserve the resource, because voluntary restraint by one user simply makes more resources available for the next user. For this reason, common-property resources—unless they are effectively infinite in magnitude—are extremely vulnerable to overuse and exhaustion, to the detriment of all users.

The absence of cooperation in common-property situations accounts for overgrazing by sheep and severe erosion over much of the African Sahel, the cutting of communal woodlands and deforestation in the Himalayas and much of the tropics, the rapid decline in the water table throughout much of the U.S. Midwest, the decline and near-disappearance of some fish species (such as the North Sea herring) and many whale species, the approaching disappearance of the rhinoceros (hunted for its horn), the elephant (hunted for its tusks), and so on and on.

As I shall discuss in more detail later, a relatively simple social invention ameliorates this problem. This invention is *private ownership* of the resource. Private individual ownership eliminates the damaging effects of free competition for the resource, while permitting competition between owners in their role as producers of products for the marketplace. But the institution of private ownership, itself, requires a high level of social cooperation to protect the rights of owners. Also, it is ineffective when owners are myopic.

I should also note that ownership is not always possible. The reader will instantly realize that ownership implies the possibility of both exclusive possession and exchange. Land or minerals or physical objects can be possessed (and, therefore, exchanged). Air cannot be possessed or exchanged in this manner, at least on earth. (In a spaceship, the situation would clearly be different.) Nobody can own the sunshine or the scenery, or the biosphere, or the climate, because it is not physically possible to absolutely exclude others from access to them. Nor can anyone exclusively own information (in the sense of data) once it has been published. By the same token, universal access to public goods opens the possibility of causing indirect harm to other users. The institutions of ownership and markets do not apply in such cases.

Another example of how optimal behavior by an individual can lead to the poor performance of a system level can be found on the factory floor. Frederick Taylor's [1911] theories of scientific management advocated the most extreme possible division of labor, with each task carefully analyzed and designed from a human-factors (ergonomic) perspective to maximize individual output. Taylor and his followers thought that maximizing the output of each worker would "obviously" maximize the product of the whole organization.

Sometimes the obvious conclusion is incorrect. It is now known and mathematically provable that, for an interdependent system, optimization at the unit level is actually a recipe for disaster at the system level. Maximizing the output of each machine in a job shop (called "line balancing" by manufacturing engineers), based on the theory that "what is not needed now can always be used later," inevitably results in a buildup of inventory. Worse, it creates bottlenecks that interfere with filling actual orders. This kind of counterproductive result can and often does occur if factory foremen are instructed to maximize machine utilization [Goldratt and Cox, 1986].

Arms races offer a last illustration of how competition without cooperation can lead to suboptimal results. Each actor tries to gain military security by piling up more armaments, which the others inevitably perceive as a threat, resulting in further escalation [see, e.g., Kahn, 1962; Schelling, 1978]. Thus overall security is not increased while enormous resources are diverted to essentially unproductive purposes. Moreover, if the competitive system is unstable and one of the actors perceives that it has a temporary window of advantage, it might actually initiate a conflict that would not have occurred at all, had there been no buildup. (World War I seems to have been an example of this process.) Evidently, the destructiveness of any conflict will be greater, the longer the arms race has been going on.

Theorists have had some difficulty until recently in understanding how cooperative strategies could evolve in a multi-player game. Such strategies pose particular difficulties from a static myopic game-theory point of view. Based on the biological analogy [Maynard Smith, 1982], the question can be posed in the following way: Under what circumstances is a given distribution of inheritable behavioral strategies in an animal population stable against perturbation, i.e., invasion by a radically different mutant strategy?

In genetic selection, a variant reinforces its success by an ability to survive and reproduce. In iterative game situations, competitive or cooperative strategies are reinforced by success and undermined by failure, to the extent that this feedback mechanism continuously modifies the decision-making algorithm to incorporate elements of successful strategies and discard unsuccessful ones. In "prisoners dilemma" games, for instance, it has been shown that no universally superior decision-making rule emerges if each iteration is independent (i.e., there is no memory and no learning).

Various decision-making rules, such as "tit-for-tat," have been shown by repeated simulations to be evolutionary-stable under myopic conditions. This means that, if all players use that rule consistently, a "mutant" (i.e., a single player or small group adopting some other strategy) cannot succeed in taking over the game. Thus, assuming myopia, a strategy involving cooperation could not evolve from a strategy of "constant betrayal." The introduction of *learning from experience* (from previous encounters with the players) changes the situation drastically, however [Axelrod, 1984, 1987; Axelrod and Dion, 1988].

When players are allowed to (1) play the same game over and over and (2) remember their previous encounters with other players, especially (3) when they expect to be playing in the future, a dominant cooperative strategy begins to emerge. The strategy will be based, roughly, on the presumption of cooperation to achieve global optima, combined with "tit-for-tat" to punish defections [*ibid.*]. This phenomenon has already been identified with presbyopia, or far-sightedness,[12] in contrast to myopia, or short-sightedness.

The mechanism for achieving far-sightedness or altruism in the system itself remains unclear, although Robert Trivers' seminal work on "reciprocal altruism," cited in the previous section, appears to point in the right direction. Interesting research in this direction has been initiated by computer scientist John Holland. He calls his core idea *adaptive non-linear networks* (ANNs). These are intentional analogs of the central nervous system. The key feature of ANNs that Holland's research is intended to elucidate is their ability to *anticipate*, not merely react to stimuli. (Anticipation is, of course, another word for presbyopia.)[13]

The point was made earlier that individually rational—even optimal —strategies can lead to collective disasters. Conversely, under some conditions, non-optimal strategies may lead to collective success, e.g., economic growth. Among the crucial questions that need to be answered is this: How do individual actions based on limited information and bounded rationality lead to either collective progress or retrogression? How can the balance between competition and cooperation itself be optimized? Is there a maximization principle for the social system?

6.3. STATIC EQUILIBRIUM IN ECONOMICS: THE FREE MARKET

It may seem strange at first to discuss static-equilibrium-based theories in the context of evolution, since static-equilibrium theories do not (and, in fairness, were never intended to) explain evolutionary change. Nevertheless, quasi-static-equilibrium theories still constitute the central paradigm (in T. S. Kuhn's sense) of economics. The equilibrium mechanism (and one of the core ideas of economics, along with private property) is the free competitive market.

From the central perspective of so-called classical and neo-classical economics, the economic system represents a kind of orderly behavior at the societal level arising from a large number of essentially random encounters at the individual level. The organizing element is that the social institution the market through which goods and services are offered for sale, priced, and exchanged.

Thus, both supply and demand are mutually adjusted. Adam Smith coined the phrase "invisible hand" for this regulating process, which appeared quite remarkable to the 18th century observer.

Since then the idea of "the free competitive market" has become so familiar to economists, especially of the present generation, that it is taken for granted. Its deep significance as a control mechanism may not always be as fully recognized as it should be. The fact that the market is, in itself, an example of cooperative behavior, which must have evolved somehow, is even less well appreciated.

To better appreciate the crucial role of the market, consider an idealized primitive society of people living in an isolated valley and trading with one another. One can reasonably assume some occupational specialization, based on an individual's experience, location, and other comparative advantages. For example, there are grain farmers who till the soil of the river bottoms; herdsmen who graze sheep and goats on the higher pastures; fishermen; hunters who bring back meat, skins, and horns from distant forests; woodsmen who collect wood and provide charcoal; tanners who scrape and prepare hides; potters who make baked clay containers; miners who dig and collect pieces of native copper or nuggets of silver; copper- and silversmiths who make metal implements and decorative objects. There may also be wood carvers and carpenters; stone masons, spinners and weavers, bakers, and so on.

This idealized picture assumes a social order of some sort, which means it assumes away the basic facts of life where societies do not exist: the exploitation of the weak by the strong. Even in modern societies, we are constantly reminded of what happens when law and order disappears, even briefly, as in the aftermath of a natural disaster.[14] In the absence of social order (continuously and visibly enforced), goods can be acquired by means other than manufacture or trade. They can be "taxed" or "confiscated" by authority (meaning those with power) or simply stolen. It is perfectly clear that an anarchic society of thieves is likely to be much less productive than a well-ordered society, and as I have already pointed out, optimal solutions are rarely achieved by myopic evolutionary selection mechanisms.

Assume, however, an ordered society under rule of law, at least to the extent that no theft or involuntary exchange is allowed. In the absence of any organized auction market (or equivalent institution), exchanges will occur only in two-way barters. The exchange rate for each transaction (units of commodity x_i per unit of commodity x_j) will depend on the current availability of each commodity and the relative preference (marginal utility) of each commodity to each of the two parties at the time.

For instance, suppose the fisherman after a good catch has 10 fish more than he can consume, so he goes to the baker's hut and offers to exchange his 10 fish for 10 loaves. But the baker has only 10 surplus loaves that day (having already exchanged 5 loaves for a basket of charcoal and 5 more for an urn of wheat flour). He wants to save his extra loaves of bread to trade for a piece of cloth and a clay pot. Finally they agree to exchange 3 fish for 2 loaves. The fisherman does not need anything else that is available that day, so he decides to preserve some of this remaining 7 fish by drying them over a fire. For this he needs charcoal. The charcoal burner has already disposed of most of his supply to the baker (1 basket) and the potter (for a cooking pot), so he drives a hard bargain with the fisherman: 3 fish for 1/2 basket of charcoal.

At the end of the day, the following exchanges have occurred, among others:

$$1 \text{ basket of charcoal} = 5 \text{ loaves,}$$

$$1/2 \text{ basket of charcoal} = 3 \text{ fish,}$$

$$3 \text{ fish} = 2 \text{ loaves.}$$

The price of 1 basket of charcoal is 5 loaves in one transaction and 6 fish in a second. But in a third transaction the price of a loaf was $1\frac{1}{2}$ fish. Thus there were two different exchange rates between loaves and fishes: a **direct** rate (2 loaves = 3 fish) and an **indirect** rate, in terms of charcoal (5 loaves = 6 fish). If we were to catalog all the transactions in the course of a month, many more such inconsistencies would appear. In fact, it would be most unusual for the direct and indirect exchange rates for any two commodities to coincide.

The situation characterized by this simple example is one of economic anarchy (disorder) in the specific sense that there is little or no correlation between direct pairwise exchange rates and indirect exchange rates. **Order**, in this context, implies (1) that, at all times, direct and indirect exchanges are consistent, i.e., that each commodity commands a unique exchange price in terms of every other commodity and (2) that all goods offered for sale at that price or below are actually sold, while all goods desired at the market price or above are actually purchased. (In other words, the market **clears**.) The question of how a market actually approaches this state of (static) equilibrium is discussed later.[15]

It is sufficient that exchange rates be well defined between every commodity and a medium of exchange with universally acknowledged value. The medium of exchange in a given society is often an historical accident. A wide variety of items (e.g., seashells) have served as the equivalent of coins in various places. Gradually, bronze, silver, and gold coins became widespread. "Paper money" is merely another commodity (an abstract one, in some respects) with its market price relative to other currencies.

In brief, economic order emerges from economic anarchy when consistent prices are defined for each good or service. Prices can be determined, in principle, either by government (i.e., by central planners) or by impersonal, competitive bidding between producers and consumers. In practice, the failures of central planning are only too painfully clear. The functioning of an impersonal market depends on accurate and timely information about the availability and demand for each good or service offered for sale. It is axiomatic that, in a money economy, the distribution of individual preferences and incomes throughout the population, assuming each actor (individual or firm) is in some sense **rational**, shapes the characteristic relationships between producers and consumers.

The notion of the market as an impersonal, collective balancing mechanism was first articulated by Adam Smith in his evocative phrase "the invisible hand." It was conceptualized more precisely in terms of supply and demand by the French economist Jean Baptiste Say. Say's law is usually stated simplistically as "supply creates its own demand."[16] Lacking any detailed phenomenological theory of how or why this balance should occur, Say's law was nonetheless taken as axiomatic by such turn-of-the-century luminaries as Alfred Marshall. Marshall's interpretation of Say's law was that output (to meet demand) was limited only by the factors controlling supply: labor, land, natural resources, and capital.

The French engineer-economist Leon Walras [1874] formulated Say's law in mathematical terms. The standard Walrasian (neo-classical) model of an economic system consists of a set of production activities with cost functions; a set of independent resources, commodities, and services, and a final demand function.[17] It is assumed for convenience that all resources, commodities, and services can be produced entirely from linear combinations of others in the system (i.e., the system is closed), with an appropriate expenditure of labor. The wages of labor constitute the income available to satisfy demand. In the original version, no joint products or co-products are permitted and (by assumption) there are no "free goods" or waste.

Walras postulated a unique set of equilibrium (market-clearing) prices such that supply would exactly balance demand for each good or service. Even though Walras was not able to prove this conjecture, it had an enormous influence on subsequent developments in economic theory.[18]

The proof of the existence of a static supply-demand equilibrium was one of the great achievements of neo-classical economics because it seemed to provide a theoretical explanation for Adam Smith's price-setting "invisible hand." Walras's grand conceptualization was called "the Magna Carta of economics" by Joseph Schumpeter, and Milton Friedman has been quoted as saying, "We curtsey to Marshall, but we walk with Walras" [in Hoos, 1983, p. 33]. There can be no question that the operation of a money-based, free competitive market generates a kind of coherence, or long-range order, somewhat analogous to so-called cooperative phenomena in physics—in contrast to the unstable price/ wage anarchy that prevails, for instance, in a barter society. (Central planning attempts to introduce order by another means.)

But, however elegant the Walrasian general-equilibrium model may be, it does not represent the real world at all well. It has been properly criticized as "a system devoid of human beings" [Seligman, 1962, p. 385] and a spuriously deterministic one, at that. Friedman once described it as "a form of analysis without much substance" [ibid.]. Janos Kornai [1973] has written a detailed and exhaustive critique titled, simply, "Anti-Equilibrium." Most of these criticisms remain valid.[19] In particular, Walrasian theory does not explain economic growth; it is theoretically compatible only with a special (and unrealistic) kind of growth. Walrasian equilibrium is inherently static. At best, it accommodates a kind of smooth, proportional expansion bearing little resemblance to the kinds of growth that are actually observed.

Unquestionably, however, neo-classical equilibrium theories have focused attention on an important phenomenon: the emergence of a collective economic order (something that has happened only within the past few hundred years). The significance of this order is that it preceded and probably caused, or at least facilitated, a major increase in productivity and output. Some of the underlying mechanisms are not hard to identify. The emergence of a relatively reliable and impersonal price-setting mechanism made long-range planning and investment feasible. This, in turn, enabled merchant-entrepreneurs (such as ship owners) to increase their capital base by selling equity shares in their ventures. Indeed, it may not be too far-fetched to suggest that capitalism was actually a consequence of the coming of the "invisible hand."

While collective phenomena and long-range order are not unique to economics, the market-price system as such has no close analogy in physics or biology. The unique feature of the price system is not that it automatically processes environmental information to generate control signals all living organisms do that but that it functions with **symbolic** information. I will return to this point in Chapter 8, where I discuss the economy as an **information processor**. Finally, the market system is, in some (not quite clear) sense, self-regulating and capable of recovering, for instance, from a perturbation in demand. It therefore displays **homeostatic** and **self-organizing** attributes.

6.4. QUASI-STATIC (HOMOTHETIC) GROWTH IN ECONOMICS

It is intuitively clear that growth and evolution are closely related. The tendency of all evolutionary processes to exhibit something like irreversibility or "time's arrow" has been noted many times. What remains unclear, however, is the source of this irreversibility in economics. The questions arise: Does the simple expansion of aggregate output or wealth give economic growth its irreversible character? If not, what does? The first question is the easier: The answer is no! But, in keeping with the traditions of science, it makes sense to see how much can be explained with very simple assumptions before going on to more-complex models. The second question can be answered only by degrees.

Although they relied to some extent on physical analogies, the early neo-classicists (Walras, Pareto, and their successors) did not fully face the fundamental contradiction of growth in an equilibrium state. The simplest way out of the difficulty is to rely on a physical analogy: **adiabatic** change, which is (in essence) change driven by external factors at a rate slow enough to permit continuous thermal equilibration.

In a similar manner, one can assume that economic equilibrium (in the Walrasian sense) is constantly maintained and attribute growth entirely to external factors.[20] If "supply creates its own demand" (Say's law) then growth in equilibrium can be explained by an exogenous increase in the factors of production, namely, labor, land, resources, and capital. Population growth automatically takes care of the first factor labor supply and the exploitation of the new lands in the Western hemisphere did, for a while, increase the availability of land and natural resources. But by the late 19th century, there was little more empty land to be exploited. Yet wealth per capita was increasing rapidly even in Europe, where all available land was in use and other natural resources were being exhausted.

Neo-classicists, therefore, had to try to explain equilibrium growth in terms of accumulation of capital. This put the emphasis on **savings** as the engine of growth. In the hypothetical Walrasian equilibrium state, net saving by wealth-holders was assumed to be exactly balanced by net investment in additional production capacity. "Normal" profits of enterprises were assumed to replace depreciated capital. "Supernormal" profits (which might finance growth) existed only in imperfect markets and depended on the existence of barriers to entry, i.e., an effective monopoly.

The leading economists at the turn of the century, such as Alfred Marshall, tended to deny any economic role for "supernormal" profits from monopolistic enterprises. They pointed out that, in a truly competitive free market, where monopolies cannot exist (by definition), no such supernormal profits are possible. It was generally assumed in neo-classical theory, as in Marxist theory, that, in a competitive labor market, the necessities of life would entirely consume workers' wages. Hence, workers could not be expected to save voluntarily. Therefore, in the ideal neo-classical perfect market, as conceived at that time, accumulation (i.e., saving) by wealthy property owners was the only acknowledged engine of economic growth or wealth creation.

This idealized turn-of-the-century worldview was undermined by several recalcitrant facts. One of them was that supernormal profits from technological innovations in such industries as telephones, electrical equipment, autos, and chemicals did exist and were demonstrably financing rapid growth and job creation in those industries. Neo-classicists often dismissed these as examples of "monopolistic abuse," overlooking both fierce competition and the sharp and continuing price reductions that greatly improved the general standard of living.

Another awkward fact was that workers' wages tended to rise in real terms, even in competitive labor markets. The unmistakable implication of this was that skilled and experienced workers are worth more than unskilled and inexperienced workers. (Hence, there is no such thing as a meaningful universal unit of labor.) In effect, skilled workers can claim some of the benefits of a monopoly even without organizing themselves into labor unions.

But the neo-classical theory survived until it was finally made untenable by the prolonged British slump of the 1920s and the catastrophic worldwide depression of the 1930s. Both events demonstrated painfully that supply does *not* necessarily bring forth matching demand. Economic orthodoxy in the 1920's asserted that reduced profits and excess production capacity would automatically lead to reduced wages and reduced investment, until the natural Walrasian equilibrium of supply and demand was restored.[21]

As early as 1930, John Maynard Keynes and his Cambridge followers argued forcefully that lower wages and investment actually lead to a vicious circle. Reduced demand leads to reduced revenues, which justifies still-lower wages and less investment. Similarly, if savings exceed investment, money lies idle and demand falls. Thus, non-inflationary equilibrium is possible at *any* level of production short of full capacity. In other words, contrary to the neo-classicist's axiom (Say's law), there is *no* natural tendency to fully utilize available labor and capital. (Hindsight suggests that the contraction in demand brought on by the Federal Reserve's tight money policy may have converted a normal recession into the Great Depression.) This was the background for Keynes's proposal that government should become the demand-creator of last resort ("digging holes and filling them in," if necessary). In due course, Keynes's view was widely accepted and became the new economic orthodoxy.

The Keynesians, having discredited the supposed exclusive role of savings as the engine of growth and wealth creation in a competitive market economy, proceeded to put more emphasis on investment behavior. But, still under the influence of the static neo-classical paradigm, they offered no serious explanation for why investments were ever made beyond the replacement level. Keynes himself [1965] attributed investment in new enterprises to the "natural animal spirits" of entrepreneurs. In any case, he seems to have disregarded growth as an important or enduring phenomenon. He expected economic growth to come to an end in two or three generations at most, when the industrial countries would reach a plateau of consumption (called "bliss" in some of the literature of the period) as the declining marginal utility of consumption approached zero. Fundamentally, however, Keynes was unconcerned with problems of economic growth. Appropriately for his time, he focused his attention on policies for achieving full employment in a relatively static economy.

Nevertheless, beginning with Frank Ramsey [1928], the Cambridge school of economists devised savings-driven theories of development (i.e., growth) using Keynes's notions of "multipliers" and "accelerators."[22]

Homothetic growth, that is, growth without structural change, can be modeled theoretically in several ways—for example, as an optimal-control model with aggregate consumption (or welfare) as the objective function. The control variable is the rate of savings diverted from immediate consumption to replace depreciated capital and to add new capital to support a higher level of future consumption. The rate of growth in a simple model of this kind is directly proportional to the rate of savings, which, in turn, depends on the assumed depreciation rate and an assumed temporal discount rate by which one can compare present versus future benefits. Note that assumptions about the operation of the market play almost no role in this type of growth model. Savings, in this type of model, can be voluntary or enforced by government. Technological change is strictly exogenous.[23]

Models of this sort are extremely elegant but reflect almost none of the important dynamic characteristics of the real economic system. In fact, they have almost nothing worthwhile to say about economic policy, and, perhaps for that reason, they have fallen out of favor in recent years.[24]

To be sure, homothetic-growth models allowed some adjustments for technical change, but they nevertheless made capital formation the primary driver of growth. This was strictly for convenience. In this regard, these models were fundamentally erroneous and misleading to the point of perversity, in that the deficiencies of the model were used as an apology for central planning. Given the prevailing fixation on capital formation, it was argued that the economy could grow only when capital and labor flowed to various sectors in fairly fixed proportions.[25] Development economists interpreted a lack of growth as evidence of chronic "structural disequilibria" or "bottlenecks," and argued that the private capital market in such cases was incapable of allocating capital efficiently to the sectors where it was needed.

The idea that growth and wealth-creation depend on capital formation and allocation was, of course, the chief justification for centralized economic planning as practiced almost universally in so-called developing countries for the past four decades or more (with tragically poor results). This is especially true in India, where the Cambridge tradition was most entrenched. Moreover, the central assumption that capital accumulation and allocation to achieve sectoral balance are the critical factors in propelling economic growth had become absolutely untenable after the mid-1950s, when the importance of technological change as a driver of economic growth finally became clear.

Empirical research carried out as early as the 1950s established quite clearly that *per capita* economic growth in the United States and other industrial countries cannot be accounted for primarily in terms of increased inputs [see, for example, Fabricant, 1954; Abramowitz, 1956; Solow, 1957]. In fact, at this time John Kendrick [1956] introduced the related notion of increasing factor productivity as a reflection of technological progress.[26] The relatively poor performance of most centrally planned economic-development programs is probably due in large part to their focus on investment *per se*, to the neglect of technological innovation and parallel adjustments and evolutionary changes in market structure.

6.5. ONTOLOGICAL THEORIES OF DEVELOPMENT

As an alternative to homothetic-growth theories, some economists have advocated "cycle," or "stage," theories based on an analogy with biological growth.

The analogy between social and biological processes goes back at least to British sociologist Herbert Spencer's landmark book, *A Study of Sociology,* of 1872. Among other important insights, Spencer emphasized the functional

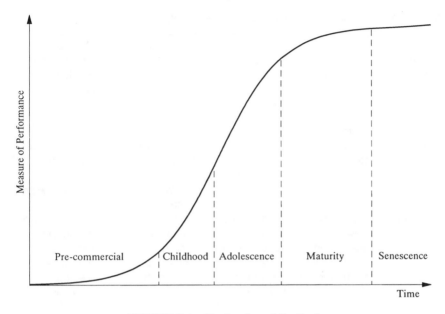

FIGURE 6.1. Technology Life Cycle.

differentiation of social systems into interdependent, specialized parts. He was perhaps the first to formulate the life-cycle model (youth, maturity, decay) of social systems. Another landmark book was German historian Oswald Spengler's *Decline of the West* [1926], which predicted the inevitable decay of Western society. Alfred Toynbee, the British historian, incorporated similar ideas, albeit less deterministically, in his *Study of History* [1934]. Toynbee also gave considerable weight to the challenge-response mechanism borrowed from biology and human behavioral psychology.

The life-cycle model entered U.S. anthropological literature with Julian Steward in the 1940s under the nomenclature **ascendancy, climax,** and **retrogression**. In archaeology it was formalized by Willey and Phillips [1958]. In 1968 it was reformulated by David Clarke, and the stages were renamed as "formative" or "florescent," followed by "coherent" or "classic climax," followed by "post-coherent" or "post-classic."

The idea of an *aging* process in economics goes back at least to the German economist Julius Wolff, whose ideas were taken up later by Simon Kuznets [1930]. "Wolff's law" asserted that the cost of incremental improvement increases as a technology approaches its long-run performance level. A number of other economists of the 1930s explored the phenomenon of industrial succession and displacement, especially in the context of the business cycle [see, for example, Schumpeter, 1939]. Alderfer and Michl [1942] also proposed a law of industrial growth. In summary, it stated that industries mature when techno-

TABLE 6.1. *Summary of Modified Life-Cycle Theory.*

Life-Cycle Stage	Product Technology	Process Technology	Industry Structure (Contestability)	Characteristics Competitive Strategy
Conception	Conception (idea)	n.a.	n.a.	n.a.
Birth	Prototype	n.a.	n.a.	n.a.
Childhood	Diversity of models and designs	Machine-specific skilled labor; general-purpose machines	Low barriers to entry; many early competitors and imitators	Performance maximizing; production near markets
Adolescence	Improved designs, fewer models, reduced rate of change	Product-specific labor skills; special adaptations of machines, e.g., tools, dies, etc.	Entry rate declines; many mergers and dropouts	Market share maximizing; emphasis on marketing, distribution and services; exploitation of scale economies
Maturity	Standardized product, slow evolutionary changes	Semi-skilled labor; large-scale automation	Oligopoly; no new entrants, some mergers	Factor-cost minimizing; worldwide production; investment in fastest-growing markets
Post-maturity (senescence)	Commodity-like product			Disinvestment; selling of technology, turnkey plants, management services, etc.

logical progress slows down, resulting in slower cost reductions and slower market expansion. Later the life-cycle idea was explicitly set forth in something like its present form by Richard Nelson [1962].

In the industrial management and marketing literature, the life cycle emerged as the "product cycle." It was elaborated by Theodore Levitt [1965], among others. Raymond Vernon [1966] used the concept in a classic paper to explain some of the patterns of international trade and investment by multi-national firms. The life-cycle model has been applied extensively to technological development and adoption/diffusion.[27] (See Figure 6.1. and Table 6.1.) The names of the phases of the life cycle change from author to author, but the underlying ontological idea of fixed progression remains the same.

Another specialized relative of the ontological model is the **"long- cycle"** or **Kondratieff-cycle** model, which has established its own niche in the economics literature. The apparent cyclicity of wholesale prices over a (roughly) 50-year period was first noticed by the British economist W. S. Jevons more than a century ago. It was taken up again by the Dutch Marxist economist Van Gelderen [1913] and yet again by the Soviet Russian economist N. D. Kondratieff [1928], whose name has been attached to the phenomenon.[28] That a long wave in economic activity has existed in market economies, at least since the end of the 18th century, is scarcely in doubt, although there is considerable skepticism as to the degree of determinism (i.e., cyclicity) involved.

One other variant on the ontological (life-cycle) theme is the "stage" theory of economic development, advocated particularly by Walt W. Rostow [1960, 1978]. This theory is built in large part on the intellectual foundations laid by Roy Harrod, Evsey Domar, and the other early growth theorists, with their strong emphasis on capital formation. Rostow carried this idea further, suggesting that economic development in any country starts with a critical "takeoff" point, based on the domestic rate of capital formation.

The fact that the life-cycle model in all its variants (including long cycles) is fundamentally ontological and deterministic need hardly be emphasized again. This characteristic has drawn much criticism when such models are applied in various fields (although those who have suggested that the life-cycle model has some explanatory power have often been careful to point out its limitations themselves). The popularity of the life-cycle and its related models is probably due to the lack of a convincing alternative, rather than to its inherent virtues. What is lacking is an equally convenient and easily understood alternative approach to non-deterministic (phylogenic) evolution. Many scientists have realized the need for evolutionary models, and have offered some models for particular cases, but no model has yet combined any degree of realism with transparency and generalizability.

ENDNOTES

1. In particular, the second distinction is somewhat muddy at best (though Faber and Proops insist on it).

2. It should be pointed out that Stephen Jay Gould (and others) dissent, to some extent, from this view. The "punctuated equilibrium" theory emphasizes that relatively sudden changes have occurred at times in the past, especially during the "three great mass extinctions": the Cambrian explosion (570 million years ago, when calcium-based hard parts appeared), the Late Permian (225 million years ago, when 96% of all marine species disappeared), and the Late Cretaceous (65 million years ago, when the dinosaurs disappeared and mammals emerged).

3. In the theory of games, developed initially by von Neumann and Morgenstern [1944],

the class of games easiest to analyze is called zero-sum, meaning that the sum total of gains and losses for all players is zero. Only redistribution is possible. Other possibilities exist, however, including positive-sum games and negative-sum games. Economic growth can be interpreted as an instance of the former, in which "everybody wins" to some degree. Epidemic disease or an arms race can be considered instances of the latter.

4. As noted in Chapter 3, accidental synergies of this sort probably led to the evolution of multi-celled organisms, as specialized cells lost their independence [Margulis, 1970, 1981]. Indeed, the so-called "super organisms," such as the corals and siphonophores (e.g., Portugese man-of-war) appear to be virtually on the edge between colonies of individuals and true organisms.

5. The exact meaning of consciousness and the extent to which animals exhibit consciousness remains highly controversial.

6. The obvious exception is bridge, where the competition is between partners who must cooperate with each other.

7. The lone dissenter, for many years, was Maurice Allais, who illustrated his objection with the so-called Allais Paradox. Allais' suggestion that the so-called independence axiom is the source of the problem now appears increasingly plausible. The Allais paradox, restated by Robin Dawes, is as follows:

> There are two situations, (1) and (2). In both situations, players are invited to choose from an urn containing 1 black ball, 10 blue balls, and 89 red balls. Situation (1) has two payoffs, (a) and (b). In case (a) the payoff is $1 million for all draws. Case (b) has the following payoffs: black = 0, blue = $2.5 million, red = $1 million. The second situation also has two payoffs. Alternative (a) gives black and blue the same payoff, namely $1 million, but pays nothing (0) for red. Alternative (b) pays black nothing, blue $2.5 million, and red nothing. Notice that in both situations (1) and (2) the payoff for a red ball is the same for (a) and (b). Therefore, the choice between (a) and (b) should depend only on the payoffs for blue and black, which happen to be the same in both situations. Comparing the payoffs for (a) and (b) with respect to blue and black draws (ignoring red), it is clear that (b) would always be preferred. (According to the von Neumann-Morgenstern axioms, a rational player should always choose the payoff with the highest expected value). Yet, in situation (1) most people would choose (a) rather than (b), because (a) guarantees $1 million and people prefer to give up a 10% chance of an extra $1.5 million to avoid a 1% chance of receiving nothing. In contrast, in situation (2) people choose (b) because they are willing to increase the chances of getting nothing from 89% to 90% in exchange for a 10% chance to get $2.5 million instead of only $1 million. It is interesting to note that this also illustrates that "utility value" can deviate sharply from "money value."

8. The organization-theory challenge to utility maximization was primarily due to Herbert Simon, and by Richard Cyert and James March. In recent years, moreover, a number of experiments by behavioral psychologists (e.g., and some interesting experiments by economists, among others) have demonstrated beyond serious doubt

that the decision-making rules actually adopted by most people, including managers, in most situations are not rational in the NM sense. (In fact, both Smith's and Sterman's simulations strongly suggest that the most common decision-making rule can be stated very simply as "follow the crowd." Yet the neo-classical mainstream of economics has not surrendered. On the contrary, the so-called rational-expectations theory asserts that impersonal markets (such as the stock market) really do optimize, because they are able to overcome the computability limitations of individuals and take rational account of all information available at a given time [e.g., Sargent and Hanson, 1981]. The argument still rages.

9. Thus a "max-max" strategy is one that maximizes the maximum gain; a "min-max" strategy minimizes the maximum loss, a "minimum regrets" strategy minimizes some function of possible losses, and so on. There are many variations on this theme.

10. The Pareto optimum is a local optimum, not a global utility maximum. It is also known that all Pareto optima correspond to the so-called Walrasian equilibrium, first postulated by the French economist Leon Walras in 1874. This is the static equilibrium in a pure multi-product exchange market such that supply and demand are equal for every product. This is the market-clearing condition.

11. This point is illustrated by an example given by Martinàs [1989]. Suppose Jack and Jill have the same utility function (chosen somewhat arbitrarily to make the point),

$$U = X(20 - X) + Y(20 - Y),$$

where X stands for apples and Y stands for oranges. Suppose Jack starts with 10 apples and Jill starts with 10 oranges. Each has a utility of 100 units. This is not a Pareto-optimal state, because several possible exchanges of apples for oranges will increase both Jack's and Jill's utility. However, there are a number of possible Pareto-optimal "final" states that can be reached. For instance:

Jack trades 3 apples for 7 oranges; $U(\text{Jack}) = 182$, $U(\text{Jill}) = 102$,

Jack trades 4 apples for 6 oranges; $U(\text{Jack}) = 168$, $U(\text{Jill}) = 128$,

and so forth. (There are two other symmetric cases where Jill comes out better off than Jack.) The even trade gives the most equitable result, and also the result that maximizes the combined utilities of the two players together, but the actual result depends on who is the better bargainer. In each of the trades both players gained from their initial positions, and presumably either player would have made any of the trades that improved his/her position if convinced that it was the only available alternative to no trade at all.

Obviously, if each player knew the other's utility function, then both could calculate all possible outcomes and see that the even trade maximized the sum of the two utilities. Even that situation, however, produces no certainty about the actual outcome, unless both players are also strictly rational in the sense defined mathematically by von Neumann and Morgenstern [1944]. Only when all parties have perfect information and behave with perfect rationality is there a unique "best of all worlds" Pareto-optimum final state of a market.

12. The limiting case of "perfect" presbyopia is, of course, "perfect information." However, the notion that farsightedness is relative (and, perhaps measurable) merits further consideration.

13. To simulate this ability on a computer, Holland has constructed parallel message-carrying rule-based systems called classifier systems [Holland, 1988]. Such systems can be thought of as generic programs for playing a game. The rules are the heuristics of the game (or the standard operating procedures of any complex system like an economy). The set of rules is large and overlapping; several different rules may cover a given situation. The classifier activates a given rule (out of the relevant set) based on a higher-level rule that takes into account the results of previous moves in the game. In effect, classifier systems are a form of artificial intelligence. Such systems are still in the early experimental stages, but I would already hazard a conjecture that they will be a powerful tool for exploring some of the crucial evolutionary questions.

In particular, Holland uses a scheme in which each rule, at a given moment, is assigned a specific "strength" that reflects the extent to which it has proved successful (led to desirable outcomes) in the past. Strength is continuously revised by a method he calls the "bucket brigade algorithm." Each rule "competes" for activation whenever it satisfies the relevant conditions. This is done by "bidding" part of its strength. If the rule wins the bid, it pays its upstream "suppliers" (rules that transmitted the message to it in the first place), and posts its message to the next "downstream" node in the network. If the rule is accepted and used by "consumers," it is repaid by their bids. And so on. In effect, the classifiers function internally like an economic system.

14. After hurricane Hugo there was widespread looting on the islands of St. Thomas and St. Croix.

15. Most theoretical attempts to explain the price-equilibrium process in markets, beginning with Leon Walras [1874] and continuing with many others, have assumed either an auctioneer actively soliciting offers to buy and sell, or some sort of mechanism to disseminate price and transaction information, resulting in a single unique price path for the approach to equilibrium. See, for example, Smale [1976].

16. See, for instance, Morishima and Catephores [1988].

17. Assuming N goods (commodities) and M productive resources and/or services, the complete system consists of $2N + 2M - 1$ simultaneous equations, assuming a unique set of inputs (resources, goods, and services) for each output and no multiple or joint products.

18. The existence of a static general equilibrium in such a system was finally proved in the 1930s by Abraham Wald for some special cases. More general proofs were given in 1954 by Kenneth Arrow and Gerard Debreu and by Lionel McKenzie. This achievement has steered a generation of economists into the analysis of highly abstract mathematical models. Actually, the rather tight restrictions of the original Walras model have been significantly loosened. The general equilibrium has even been extended to the dynamic case of exhaustible resources, subject as before to the assumption of perfect futures markets for resources. A great deal of theoretical superstructure has been added to this basic model in recent years.

19. The original mathematical proof of existence of an equilibrium (in the sense above) depends on some restrictive simplifying assumptions. Among them are the following:
    (a) Many suppliers compete for each commodity or service, including labor (there are no monopolies or oligopolies).
    (b) Every commodity or service in the economy is produced from labor or from

other commodities and services produced by the economy and sold in the market. By implication, no resources are taken from the environment and the environment provides no unpriced services (such as waste disposal). In other words, the ecomony is closed, and prices are unaffected by anything outside the market. Moreover, the closed economy produces only goods; there are no "bads" or externalities.

(c) Each purchaser in the market is a perfectly rational optimizer in the sense that he is consciously aware of his own preferences and can instantaneously and consistently decide how he will reallocate his income among all possible goods and services given any change in market prices.

(d) Each purchaser in the market is perfectly informed about the prices and characteristics of all products and services offered for sale at all times. If any change were to occur (e.g., the introduction of a new product), it is assumed that information about it is instantaneous and automatically available to all actors.

The unreality of these four neo-classical assumptions is well known to economists (although some of the implications thereof may not be). Following is a list of commonplace observations about the real economy that standard theory cannot explain (borrowed, more or less, from Niki Kaldor [1985] and quoted by Gerald Silverberg [1987]):

(a) Markets do not always clear. (If they did, inventories would not be necessary and backlogs would not exist.)

(b) Consumers do not always seek the lowest price; they may support local businesses against "foreign" competitors, for instance. (Japanese consumers are notoriously willing to do this, paying, for example, $5 a piece for Japanese apples to support the local farmers.)

(c) Prices do not seem to tend toward uniformity. Businesses mostly price their products on the basis of fixed markups, despite wide differences in costs.

(d) Short-run changes in demand do not affect only marginal firms, as the equilibrium theory would imply; the effects are typically spread more or less uniformly over all producers. (This actually follows from a-c.)

Most economists still seem to think these inadequacies are tolerable. In his introduction to one of the best-known modern texts on economic-growth theory, Nobel laureate Robert Solow wrote:

> "Real economics are not steady states; they are not in any state that can be described in a word, not even in factor—and commodity—market equilibrium. But they do not appear to be very far from, or to be rushing systematically away from steady state conditions. So the steady state may be a fair first approximation. Of course, that is a temporary excuse, not a permanent license."

20. Logically, quasi-state economic contraction should also be a theoretical possibility. There is almost no literature on this. In reality, most economic contractions have been relatively sudden and sharp—the economic analog of Gould's "punctuated equilibrium" theory of biological evolution.

21. Echoing this conventional wisdom, Secretary of the Treasury (and former banker) Andrew Mellon advised President Hoover in 1930 that his policy should be "to liquidate labor, liquidate stocks, liquidate the farmers, liquidate real estate and purge the rottenness from the economy." In Britain, Prime Minister Ramsay MacDonald was similarly advised by the mandarins of his treasury. Both Hoover and MacDonald distrusted orthodox economic theory, though perhaps not sufficiently. Each

attempted somewhat timidly to restore slumping demand by means of small doses of deficit financing and credit. Roosevelt's New Deal was a relatively modest expansion of Hoover's half-hearted interventionism.

22. The influential Harrod-Domar-Mahalanobis single-sector models developed roughly between 1936 and 1956 were built on these foundations; von Neumann [1945], Koopmans [1951], and others explored multi-sector homothetic growth models. The literature is well summarized by Burmeister and Dobel [1970].

23. The classic von Neumann general-equilibrium multi-sector homothetic growth model was first formulated in 1932, though not published in English until 1945. It postulated a closed set of products and a fixed set of processes, such that every product in the set is produced by one or more processes and from other products in the set. Major contributions to this esoteric branch of mathematical theory—later known as "activity analysis"—include [Koopmans, 1951, 1965; McKenzie, 1963; Gale, 1967; Morishima, 1961, 1988; Radner, 1961; Samuelson, 1956, 1965] and others. The literature on multi-sector models of this type is concerned primarily with the derivation of asymptotic "turnpike theorems" and "golden rules" for capital accumulation.

Economic growth models of the optimal control type (wherein the modeler postulates an all-powerful "central planner" whose mission is to maximize some measure of welfare over time) have been especially popular since the publication of Pontryagin's *Mathematical Theory of Optimal Processes* [Pontrjagin, 1962]. Influential precursors using variational methods include Ramsey's *Mathematical Theory of Savings* and Hotelling's *The Economics of Exhaustible Resources*. In the case of an economic aggregate, such as a nation, this objective function is usually taken to be a discounted sum over future aggregate consumption or—equivalently—aggregate production (GNP).

24. An appropriate epitaph on the voluminous neo-classical "growth model" literature is the following:

"Consider, for example, the preoccupation since 1945 of some of the best brains in modern economies with the esoterica of growth theory, when even practitioners of the art admit that modern growth theory is not yet capable of casting any light on actual economies growing over time. The essence of modern growth theory is simply old-style stationary state analysis in which an element of compound growth is introduced by adding factor-augmenting technical change and exogenous increases in labor supply to an otherwise static, one-period, general equilibrium model of the economy. In view of the enormous difficulties of handling anything but steady-state growth (equiproportionate increase in all the relevant economic variables), the literature has been almost solely taken up with arid brain-twisters about 'golden rules' of capital accumulation" [Blaug, 1980, p. 254].

25. Roy Harrod called this balancing process "walking on the razor's edge." It was later shown, however, that the Harrod-Domar model's extreme sensitivity to balancing is an artifact of their particular choice of production function.

26. The development of growth accounting and productivity-growth indexes is discussed in the next chapter.

27. One variant of the life-cycle scheme is the "logistic growth" model, which was initially applied to population dynamics by Raymond Pearl. Together with variants, it has subsequently been applied to numerous other biological and social phenomena by Alfred Lotka, Volterra, and others. It has been widely applied in recent years to

socio-economic phenomena and technological diffusion by numerous authors including. The most recent, and in some ways the most sophisticated, application of the model to technological change, is the product-process technology life-cycle (TLC) model introduced by James Abernathy and James Utterback in 1975.

28. Recent major contributions to this literature include [Mensch, 1975; Mandel, 1980; Freeman et al., 1982; Freeman, 1983; Van Duijn, 1983; Kleinknecht, 1986].

REFERENCES

Abernathy, William J. and James M. Utterback, "A Dynamimc Model of Process and Product Innovation," *Omega* 3(6), 1975.

Abernathy, William J. and James M. Utterback, "Patterns of Industrial Innovation," *Technology Review* 80, June/July 1978 :40-47.

Abramovitz, Moses, "Resources and Output Trends in the United States Since 1870," *American Economic Review* 46, May 1956.

Alderfer, Evan and H. E. Michl, *Economics of American Industry*, McGraw-Hill, New York, 1942.

Arrow, Kenneth J., "Values and Collective Decision-Making," in: Phelps (ed.), *Economic Justice*, Penguin Books, Harmondsworth, England, 1967.

Arrow, Kenneth J. and Gerard Debreu, "Existence of an Equilibrium for a Competitive Economy," *Econometrica* 22(3), 1954.

Arrow, Kenneth J. and F. Hahn, *General Competitive Analysis*, Holden-Day, San Francisco, 1971.

Aubin, Jean-Pierre, "A Dynamical, Pure Exchange Economy with Feedback Pricing," *Journal of Economic Behavior and Organization* 2, 1981 :95-127.

Aubin, Jean-Pierre, *Evolution of Prices Under the Inertia Principle*, Working Paper (WP-89-85), International Institute for Applied Systems Analysis, Laxenburg, Austria, 1989.

Axelrod, Robert, *The Evolution of Cooperation*, Basic Books, New York, 1984.

Axelrod, Robert, "The Evolution of Strategies in the Iterated Prisoner's Dilemma," in: Davis (ed.), *Genetic Algorithms and Simulated Annealing*, Morgan Kaufmann, Los Altos, Calif., 1987.

Axelrod, Robert and D. Dion, "The Further Evolution of Cooperation," *Science* 242, December 9, 1988 :1385-1390.

Bass, Frank M., "A New Product Growth Model for Consumer Durables," *Management Science* 15, January 1969 :215-227.

Blackman, A. Wade, "A Mathematical Model for Trend Forecasts," *Journal of Technological Forecasting and Social Change* 3(3), 1972.

Blackman, A. Wade, "The Market Dynamics of Technological Substitutions," *Journal of Technological Forecasting and Social Change* 6, February 1974 :41-63.

Blaug, Mark, *The Methodology of Economics*, Cambridge University Press, Cambridge, England, 1980.

Burmeister, Edwin and A. Rodney Dobell, *Mathematical Theories of Economic Growth*, MacMillan, New York, 1970.

Cyert, Richard and James March, *A Behavioral Theory of the Firm*, Prentice-Hall, Englewood Cliffs, N.J., 1963.

Dasgupta, Partha and Geoffrey Heal, *Economic Theory and Exhaustible Resources* [Series: Cambridge Economic Handbooks], Cambridge University Press, Cambridge, England, 1979.

Domar, Evsey D., "Capital Expansion, Rate of Growth and Employment," *Econometrica* 14, 1956 :137-147.

Dorfman, Robert, Paul A. Samuelson and Robert M. Solow, *Linear Programming and Economic Analysis*, McGraw-Hill, New York, 1958.

Edgeworth, F. Y., *Mathematical Physics*, Kegan Paul, London, 1881.

Faber, Malte and John L. R. Proops, *Evolution in Biology, Physics and Economics: A Conceptual Analysis*, Discussion Paper (131), University of Heidelberg Department of Economics, Heidelberg, Germany, January 1989. [Mimeo]

Fabricant, Solomon, "Economic Progress and Economic Change," in: *34th Annual Report*, National Bureau of Economic Research, New York, 1954.

Fisher, John C. and Robert H. Pry, "A Simple Substitution Model of Technological Change,"

Journal of Technological Forecasting and Social Change **3**(1), 1971.

Freeman, Christopher, *The Long Wave and the World Economy*, Butterworths, Boston, 1983.

Freeman, Christopher, John Clark, John Soete, and Luc Soete, *Unemployment and Technical Innovation: A Study of Long Waves and Economic Development*, Francis Pinter, London, 1982.

Gale, David, "A Geometric Duality Theorem with Economic Applications," *Review of Economic Studies* **34**, 1967 :19-24.

Goldratt, Eliyahu and Jeff Cox, *The Goal*, North River Press, Croton-on-Hudson, N.Y., 1986.

Gould, Steven Jay, "Another Look at Lamarck," *New Scientist* **84**(1175), 1979 :38-40.

Hamilton, W. D., "The Genetical Evolution of Human Behavior I and II," *Theoretical Biology* **7**(1), 1964 :1-52.

Hardin, Garrett, "The Tragedy of the Commons," *Science* **162**, 1968 :1243-48.

Harrod, Roy F., *The Trade Cycle*, Oxford University Press, London, 1936.

Holland, John H., *Adaptation in Natural and Artificial Systems*, University of Michigan Press, Ann Arbor, Mich., 1975.

Holland, John H., "A Mathematical Framework for Studying Learning in Classifier Systems," *Physica* **22D**, 1986 :307-317.

Holland, John H., "Genetic Algorithms and Classifier Systems: Foundations and Future Directions," in: Grefenstette (ed.), *Genetic Algorithms and Their Applications*, Lawrence Erlbaum, Hillsdale, N.J., 1987.

Holland, John H., "The Global Economy As an Adaptive Process," in: Anderson, Arrow and Pines (eds.), *The Economy as an Evolving Complex Systems* :117-124 [Series: Santa Fe Institute Studies in the Science of Complexity **5**], Addison-Wesley, Redwood City, Calif., 1988.

Hoos, Ida, *Systems Analysis in Public Policy: A Critique*, University of California Press, Berkeley, Calif., 1983. Revised edition.

Hotelling, Harold, "The Economics of Exhaustible Resource," *Journal of Political Economy* **39**, 1931 :137-75.

Johnson, Paul, *Modern Times: The World from the Twenties to the Eighties*, Harper Colophon, New York, 1983.

Kahn, Alfred E., *The Role of Patents*, North Holland, Amsterdam, 1962.

Kaldor, Niko, *Economics Without Equilibrium*, M. E. Sharpe, Armonk, N.Y., 1985.

Kendrick, John W., "Productivity Trends: Capital and Labor," *Review of Economics and Statistics*, August 1956.

Keynes, John Maynard, *General Theory of Employment, Interest and Money*, Harcourt Brace Jovanovich, New York, 1965. Paperback edition.

Kleinknecht, Alfred, *Innovation Pattern in Crisis and Prosperity: Schumpeter's Long Cycle Reconsidered*, MacMillan, London, 1986.

Kondratieff, N. D., "Die Preisdynamik der industriellen und landwirtschaftlichen Waren," *Archiv für Sozialwissenschaft und Sozialpolitik* **60**, 1928 :1-.

Koopmans, Tjalling C. (ed.), *Activity Analysis of Production and Allocation* [Series: Cowles Commission Monograph **13**], John Wiley and Sons, New York, 1951.

Koopmans, Tjalling, C., *On the Concept of Optimal Growth*, Pontificia Academia Scientiarum, Vatican City, 1965. [pp. 225-288]

Kornai, Janos, *Anti-Equilibrium*, North Holland, New York, 1973.

Kuznets, Simon, *Secular Movements in Production and Prices: Their Nature and Bearing on Cyclical Fluctuations*, Houghton Mifflin, Boston, 1930.

Levitt, Theodore, "Exploit the Product Life Cycle," *Harvard Business Review*, November/December 1965.

Lotka, Alfred J., *Elements of Physical Biology*, Dover Publications, New York, 1950. [First published in 1925]

Mandel, E., *Long Waves of Capitalist Development: The Marxist Explanation*, Cambridge University Press, Cambridge, England, 1980.

Mansfield, Edwin, "Technical Change and the Rate of Imitation," *Econometrica* **29**(4), October 1961 :741-766.

Marchetti, Cesare, "Hydrogen and Energy," *Chemical Economy and Engineering Review*, January 1973 :7-25.

Marchetti, Cesare, "Primary Energy Substitution Models: On the Interaction Between Energy and Society," *Journal of Technological Forecasting and Social Change* **10**(4), 1977 :345-356.

Marchetti, Cesare, *Society As a Learning System: Discovery, Invention and Innovation Cycles Revis-*

ited, Research Report (RR-89-29), International Institute for Applied Systems Analysis, Laxenburg, Austria, 1981.

Margulis, Lynn, *Origin of Eukaryotic Cells*, Yale University Press, New Haven, Conn., 1970.

Margulis, Lynn, *Symbiosis in Cell Evolution*, W. H. Freeman and Co., San Francisco, 1981.

Martinàs, Katalin, *About Irreversibility in Micro-economics*, Research Report (AHFT-89-1), Department of Low Temperature Physics, Roland Eotvos University, Budapest, March 1989.

Maynard Smith, John, *Evolution and the Theory of Games*, Cambridge University Press, Cambridge, England, 1982.

McKenzie, Lionel W., "On Equilibrium in Graham's Model of World Trade and Other Competitive Systems," *Econometrica* **22**, 1954 :147-161.

McKenzie, Lionel W., "Turnpike Theorems for a Generalized Leontief Model," *Econometrica* **31**, 1963 :165-180.

Mensch, Gerhard, *Das technologische Patt: Innovation ueberwinden die Depression* [Series: Stalemate in Technology: Innovations Overcome the Depression], Umschau, Frankfurt, 1975. [(English translation: Ballinger, Cambridge, Mass., 1979)].

Morishima, Michio, "Proof of a Turnpike Theorem: The No Joint Production Case," *Review of Economic Studies* **28**, 1961 :89-97.

Morishima, Michio and George Catephores, "Anti-Say's Law Versus Say's Law: A Change in Paradigm," in: Hanusch (ed.), *Evolutionary Economics: Applications of Schumpeter's Ideas*, Chapter 2 :23-53, Cambridge University Press, New York, 1988.

Nelson, Richard R., "Introduction to National Bureau of Economic Research," in: *The Rate and Direction of Inventive Activity*, Princeton University Press, Princeton, N.J., 1962.

Pearl, R., "The Population Problem," *Geographical Review*, October 1922 :638.

Peschel, Manfred; and Werner Mende, *The Predator-Prey Model: Do We Live in a Volterra World?* Akademie-Verlag, Berlin, 1986.

Plott, C. R., "Laboratory Experiments in Economics: The Implications of Posted Price Institutions," *Science* **232**, May 9, 1986 :732-738.

Pontrjagin *et al.*, "The Mathematical Theory of Optimal Processes," Wiley-*Interscience*, 1962. [New York]

Radner, Roy, "Paths of Economic Growth That Are Optimal with Regard Only to Final States," *Review of Economic Studies* **28**, 1961 :98-104.

Ramsey, Frank P., "A Mathematical Theory of Saving," *Economic Journal* **38**(152), December 1928 :543-59.

Rostow, W. W., *The Stages of Economic Growth*, Cambridge University Press, Cambridge, England, 1960.

Rostow, W. W., *The World Economy: History and Prospect*, University of Texas Press, Austin, Tex., 1978.

Samuelson, Paul A., "A Catenary Turnpike Theorem Involving Consumption and the Golden Rule," *American Economic Review* **55**, 1965 :486-496.

Samuelson, Paul A. and R. Solow, "A Complete Capital Model Involving Heterogeneous Capital Goods," *Quarterly Journal of Economics* **70**, 1956 :537-562.

Sargent, T. J. and L. P. Hansen, "Linear Rational Expectations Models of Dynamically Interrelated Variables," in: *Rational Expectations and Econometric Practice*, University of Minnesota Press, Minneapolis, Minn., 1981.

Schelling, T. C., *Micro-motives and Macro-behavior*, Harvard University Press, Cambridge, Mass., 1978.

Schumpeter, Joseph A., *Business Cycles: A Theoretical, Historical and Statistical Analysis of the Capitalist Process*, McGraw-Hill, New York, 1939. [2 volumes]

Seligman, Ben B., *Main Currents in Modern Economics*, Free Press of Glencoe, Glencoe, Il., 1962.

Silverberg, Gerald, "Technical Progress, Capital Accumulation and Effective Demand: A Self-Organization Model," in: Batten *et al.* (eds.), *Economic Evolution and Structural Adjustment*, Springer-Verlag, Berlin, 1987.

Simon, Herbert A., "A Behavioral Model of Rational Choice," *Quarterly Journal of Economics* **69**, 1955 :99-118 [Reprinted in *Models of Man*, John Wiley and Sons, New York, 1957].

Simon, Herbert A., "Theories of Decision-making in Economics," *American Economic Review* **49**, 1959 :253-283.

Simon, Herbert, A., *Administrative Behavior*, Free Press, New York, 1965. 2nd edition.

Simon, Herbert A., "What Computers Mean for Man and Society," *Science* **195**(4283), March 18, 1977.

Simon, Herbert A., "Rational Decision-making in Business Organizations," *American Economic Review* 69, 1979 :493-513.

Simon, Herbert A., *Models of Bounded Rationality*, MIT Press, Cambridge, Mass., 1982.

Simon, Herbert A., "The Behavioral and Rational Foundations of Economic Dynamics," *Journal of Economic Behavior and Organization* 5, 1984 :35-55.

Simon, Herbert A., "A Mechanism for Social Selection and Successful Altruism," *Science* 250, December 1990 :1665-1668.

Smale, Stephen, "Dynamics in General Equilibrium Theory," *American Economic Review* 66, 1976 :288-294.

Smith, V. L., "Experimental Methods in the Political Economy of Exchange," *Science* 234, Oct. 10, 1986 :167-173.

Solow, Robert M., "A Contribution to the Theory of Economic Growth," *Quarterly Journal of Economics* 70, 1956 :65-94.

Solow, Robert M., "Technical Change and the Aggregate Production Function," *Review of Economics and Statistics*, August 1957.

Solow, Robert M., "The Economics of Resources or the Resources of Economics," *American Economic Review* 64, 1974.

Spengler, Oswald, *The Decline of the West*, Alfred J. Knopf, New York, 1926. [2 volumes]

Sterman, John D., "The Long Wave," *Science* 219, March 18, 1983.

Sterman, John D., "An Integrated Theory of the Long Wave," *Futures* 17, April 1985.

Sterman, John D., "Testing Behavioral Simulation Models by Direct Experiment," *Management Science* 33, 1987 :1572-1592.

Sterman, John D., *Deterministic Chaos in an Experimental Economic System*, International Symposium on Evolutionary Dynamics and Non-Linear Economics, University of Texas, Austin, Tex., April 1989.

Taylor, F. W., *Shop Management*, Harper and Row, New York, 1911.

Toynbee, Alfred J., *A Study of History*, Oxford University Press, Oxford, UK, 1934. [10 volumes]

Trivers, Robert, "The Evolution of Reciprocal Altruism," *Quarterly Review of Biology* 46, 1971 :35-39, 45-47.

Tversky, A. and D. Kahneman, "Judgment Under Uncertainty: Heuristics and Biases," *Science* 185, Sept. 27, 1974 :1124-1131.

Tversky, A. and D. Kahneman, "The Framing of Decisions and the Psychology of Choice," *Science* 211, Jan. 30, 1981 :453-458.

Tversky, A. and D. Kahneman, "Rational Choice and the Framing of Decisions," in: Hogarth and Reder (eds.), *Rational Choice: The Contrast Between Economics and Psychology*, University of Chicago Press, Chicago, 1987.

Van Duijn, J. J., *The Long Wave in Economic Life*, George Allen and Unwin, London, 1983.

Van Gelderen, J., "Springvloed: beschouwingen over industriele ontwikkeling en prijsbeweging, de Ni," *de Nieuwe Tijd* 184(4,5,6), 1913.

Vernon, Raymond, "International Investment and International Trade in the Product Cycle," *Quarterly Journal of Economics*, May 1966 :290-307.

Volterra, V., "Principes de Biologie mathématique," *Acta Biotheoretica* 3(1), 1937 :18-30.

von Bertalanffy, Ludwig, "Quantitative Laws in Metabolism and Growth," *Quarterly Review of Biology* 32, 1957 :217-231.

von Neumann, John, "A Model of General Economic Equilibrium," *Review of Economic Studies* 13, 1945 :1-9.

von Neumann, John and Oskar Morgenstern, *Theory of Games and Economic Behavior*, Princeton University Press, Princeton, N.J., 1944.

von Weizsäecker, C. C., "Existence of Optimal Programs of Accumulation for an Infinite Time Horizon," *Review of Economic Studies* 32, 1965 :85-104.

Walras, Leon, *Elements d'Economie politque pure*, Corbaz, Lausanne, 1874.

Wiley, G. R. and P. Philips, "Method and Theory," in: *American Archeology*, 1958.

Wilson, Edward, *Socio-biology: The New Synthesis*, Harvard University Press, Cambridge, Mass., 1975.

Chapter 7

Evolution in Economic Systems

7.1. EVOLUTION AND GROWTH

Before returning to a discussion of economic models, I want to recapitulate the basic idea of evolutionary change: in particular, the distinction between **ontogeny** and **phylogeny**. Development or growth along a predetermined path is ontogeny. Evolutionary change in biological systems at the species level is phylogeny. The economic analogy is imperfect. It is safe to say, however, that radical technological change, with accompanying structural changes in the economy, is essentially phylogenic. Evolution is inherently unpredictable in detail.[1] Nevertheless, one can still seek a general characterization of the *direction* of evolutionary change. The real question is whether, beyond that, one can realistically seek a law, or "covering principle" and a mechanism derivable from that law—that explains the observed direction of evolutionary change and that enables one to predict future change.

The direction of both physical and biological evolution is characterized by increasing *diversity* of forms and structures; increasing *complexity* of forms and structures; and increasing *stability* of complex forms and structures. In biology at least, these generalizations are equivalent to (they imply and are implied by) an accumulation of embodied SU-information. It is reasonable, all things considered, to try to apply this characterization to the economic system. The basic unit of analysis in biology is the *species*. Its economic analog is the *firm*.

At this level of analysis, the evolutionary characterization seems to carry over reasonably well. Firms have grown more complex over time by any reasonable measure. Complexity of organization follows in part from diversity of product (and of production technology). The economy as a whole is also becoming more complex. The standard industrial classification system of today gives a misleading impression of stability, because it uses a fixed number of decimal digits. But industries are subdividing (or being born) every decade.[2]

I think notions like diversity and complexity can be defined without too much difficulty, in the economic context. Thus it should be possible, in principle, to

151

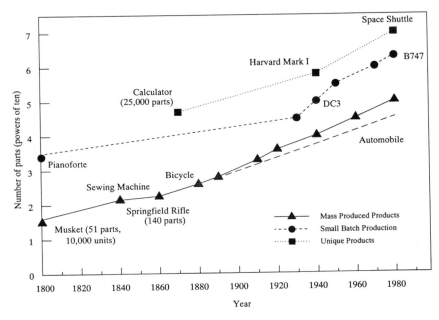

FIGURE 7.1. Trends in Mechanical Complexity.

obtain consistent statistical measures. (Unfortunately, this has not yet been done. See, however, Figure 7.1.) In any case, there is little question that diversity and complexity have been increasing rapidly in economic systems. Whether complex organizations are increasing in *stability* is much less clear, because stability is less easily measured. For instance, there is no straightforward way to assess the relative stability of organizations with a given complexity. My own hunch is that organizational stability is also increasing, other factors remaining constant.

Moreover, among firms, at least, size and complexity seem to be correlated with stability. Certainly, the Fortune 500 list is much less volatile than (for example) a randomly selected list of small firms. The accumulation of useful information, in the semantic sense, is also quite obvious. Only the causal linkages between these facts, if any, are really open to serious debate.

7.2. THE PROBLEM OF ECONOMIC GROWTH REVISITED

As pointed out in Chapter 6, economists have thoroughly explored, and perhaps exhausted, the possibilities of a special class of theoretically interesting but unrealistic models. These are exogenously driven general-equilibrium growth models with the property known as *homotheticity* which means, in effect, that all factors grow in constant proportion to each other. Economists have also dabbled with ontological (deterministic) models of cyclic or staged development that are based on physical or biological analogies. However, these models do not explain most of the important phenomena of evolution, especially radical innovation and structural change.

Homothetic growth can be likened, for instance, to the growth of a coral reef or a stalagmite. In fact, population growth is essentially homothetic. It is not a completely unrealistic idealization to imagine a growing population of herdsman or small farmers expanding into an empty territory and growing in numbers without increasing in individual prosperity. But that is *not* the kind of economic growth that is of interest.

An idealization much more relevant to reality is that of a fixed population occupying a fixed territory, but increasing in per capita productivity and prosperity. As discussed in Chapter 6, neo-classical economists around the beginning of the century assumed that capital formation (savings and investment) was the key to all economic growth. This notion was largely exploded 40 years ago by econometric studies of the U.S. economy that showed convincingly that only a comparatively small fraction of total growth in GNP per capita could be accounted for by investment.[3]

This means that aggregate economic growth *cannot* be explained, as traditionally assumed, in terms of increasing inputs of capital and labor. But something makes the economy grow. What could it be? Economists in the late 1950s quickly labeled the missing ingredient "technological progress." The only problem was that, unlike labor and capital investment, technological progress in the aggregate is not easy to measure directly. Lacking means of direct measurement, empirically-based theories to explain the rate and direction of technological change have been difficult to construct. Even though technological progress has remained a sort of behind-the-scenes *deus ex machina*, typically treated as an exogenous factor (similar to population),[4] mainstream economic theory has finally begun to recognize the need for endogenous theories that give greater recognition to its role in economic growth. A less well recognized need is for theories that are sufficiently evolutionary to be capable of explaining structural change and cooperative behavior, at least "ex post." Finally, a long-term goal would be to develop theories with better predictive capability.

It should be noted, for the sake of historical accuracy, that Karl Marx did propose a sort of evolutionary theory, with the Darwinian competition for survival replaced by a drive for private profit. (He even saw himself as "the Darwin of socio-economic evolution" [Goodwin, 1989].) Marx's most valuable insight may have been his emphasis on cyclical dynamics. Marx's theory was essentially deterministic, however. Moreover, his treatment of economic value was thoroughly unsatisfactory, and his key predictions (of declining rates of profit and real wages, for instance) were even worse.

Technological progress fits the general evolutionary characterization stated at the beginning of this chapter: It meets the test of increasing complexity and diversity. The question of stability must be set aside, pending a more precise definition of the "structures" associated with technological change. However, the essential equivalence of technological progress with an accumulation of SU-information need scarcely be debated. (This point will be expanded on in the next chapter.)

Technological progress is also, in a fundamental sense, unpredictable.[5] Clearly, an invention cannot be predicted in any detailed sense. Yet the *direction* of technological progress is often quite clear long before it is fully realized. Sociologists have argued forcefully that "social need" drives most invention, whence the functionality of major inventions (if not the operational details) can be forecast well in advance [Ogburn and Thomas, 1922; Ogburn, 1937].[6] Economists also now generally accept that demand is the primary engine driving innovation and change. To state it in simpler language, most useful inventions and technological improvements are intended, rather than accidental. They result from deliberate investment in R&D, which in turn is allocated in response to perceived social (or private) needs. A major technological breakthrough rarely occurs before there is any perceived need for it.[7]

An important feature of technology is that one of its components, knowledge, is an intangible. Technology makes its contribution to economic growth indirectly. It must be *embodied,* either in fixed capital, or labor skills, or organizational structure, or "software."[8] Leaving aside labor skills and organizational structure, it is not unreasonable to say that technology is embodied in capital, software being properly regarded as a form of capital. Thus, as economists recognized long ago, the technological contribution to economic growth is largely associated with and "carried by" capital investment. A fast rate of technological change implies a high rate of investment in new capital equipment, especially insofar as the capital equipment is inflexible. (New "flexible" forms of programmable and computer-controlled equipment may reduce the importance of capital investment as the carrier of technology.)

In any case, while capital formation may not be the sole or even the main driver of economic growth, capital accumulation and replacement will almost inevitably accompany technology-driven growth. "The manna of technical progress falls only on the latest machines" [Hahn and Matthews, 1967].[9]

Theories of technological change, like theories of biochemical evolution, fall into two classes. These are sometimes contrasted as *evolutionary* versus *revolutionary*, where the first type depends on a continuous series of many incremental (Usherian) improvements, while the second is driven by a smaller number of radical (Schumpeterian) innovations. The Usherian type of change is gradual, more or less predictable, and essentially ontogenic. Once a trajectory has been established, evolution along the path proceeds almost deterministically. It lacks the essentially unpredictable character of true (phylogenic) evolutionary change.

Until the 1970s, theories of gradual, continuous change were largely either qualitative and descriptive/retrospective or they relied on distinctly ontogenic elements.[10] A more quantitative approach was taken by the so-called epidemic or

S-curve—model of technological diffusion.[11] This simple, deterministic model is closely akin to the life-cycle model discussed in the previous chapter.

7.3. SCHUMPETER'S CONTRIBUTION: RADICAL INNOVATION

In his doctoral dissertation, published as *The Theory of Economic Development*, the Austrian economist Joseph Schumpeter [1912] took economics a step forward. He argued that the main driving force of economic growth is *radical innovation*. (He spoke of "new products, new services, new combinations," without emphasizing technology *per se*). In Schumpeter's view, these innovations are made by entrepreneurs seeking "supernormal" profits. Such extraordinary profits arise from a temporary monopoly position conferred by each innovation until successful imitators are able to enter the market.

Schumpeter called attention, also, to the tumultuous effects of competing innovators when the success of a new product or service often means the demise of an older one, along with the businesses dependent on it. Yet, out of this seemingly anarchic field of conflict, which Schumpeter termed *creative destruction*, arises a higher kind of evolutionary "order out of chaos."

This simple conception caused little stir in the economics profession at first, though it now seems to have resolved all the main problems of neo-classical economics at a single stroke. It provides a qualitative explanation of, and justification for, supernormal profits. It also explains capital accumulation (from profits). It explains the observed fact of technological obsolescence and the corollary fact of technological progress. It even explains, to a degree, the unevenness of technological change. Schumpeter regarded formal R&D as a vital mechanism for enabling corporations (especially large ones) to develop a stream of new products to keep the innovative process going. In so doing, he also challenged the standard prejudice that large corporations, by virtue of being oligopolistic if not monopolistic, were *ipso facto* anti-social. By assigning large corporations a special role as custodians of innovation, he implied the possibility of a "Schumpeterian tradeoff" [see Nelson and Winter, 1978, 1982].

Unfortunately, Schumpeter never developed his theory fully. He occupied himself in later years primarily with questions of business-cycle theory and political economy (and, for a time, government) and did not work out the full range of implications of his important ideas. Schumpeter's simple conceptual model and his later modified version of it [Schumpeter, 1943], were not entirely self-consistent, and they left a number of key questions unanswered [Kamien and Schwartz, 1982]. The most vexing of these has been the extent to which (or the conditions under which) innovative behavior depends on "market structure."[12]

Schumpeter made another important contribution to evolutionary economics beyond his understanding of the central role of radical innovation. This was the clear recognition that static optimization was an inappropriate paradigm for a theory of growth and wealth-creation. In his last and best-known book, *Capi-

talism, Socialism and Democracy, he wrote:

> "A system that at every point in time fully utilizes its possibilities to its best advantage (i.e., achieves static optimization) may yet in the long run be inferior to a system that does so at no given point in time, because the latter's failure to do so may be a condition for the level or speed of long run performance" [Schumpeter, 1943, p. 83].

This statement clearly puts economic growth beyond the realm of Walrasian general-equilibrium theory and its homothetic (quasi-dynamic) variants. As I noted in the opening chapter of this book, neo-classical economists have got into the careless habit of thinking of equilibrium as a state that the economy is *normally* in. In particular, market prices are almost universally assumed to reflect an actual equilibrium (i.e., a market-clearing balance) between supply and demand.

This, of course, contrasts sharply with the situation in thermodynamics, where global equilibrium is an uninteresting, unchanging state toward which all systems tend, but which, for all practical purposes, they never reach. In thermodynamics it is the *approach-to-equilibrium* that matters and that explains all physical processes. Given this perspective (not to mention a fair amount of everyday evidence that I shall forbear to recount in detail), economics clearly needs a non-equilibrium theory, if only because equilibrium is inherently static, while the economic system is extremely dynamic.

An evolutionary disequilibrium theory might begin with a measurement of *distance from equilibrium.* The theory must then move on to explain the driving forces economic analogs of thermodynamic potential that push (or pull) the system toward equilibrium. It may also introduce something analogous to dissipation (non-conservative forces) and even something akin to entropy. With such a conceptual structure, the economic system will probably be seen as strongly non-linear when it is far from equilibrium, possessing multiple feedbacks with bifurcated steady-state solutions amounting to self-organization (see Chapter 1). Exogenous disturbances or fluctuations can flip non-linear dynamical systems of this kind from one trajectory to another. Many kinds of internal evidence show that the economy must be a self-organizing system.

7.4. POST-SCHUMPETERIAN EVOLUTIONARY THEORIES

One can take at least two different approaches to developing an evolutionary disequilibrium theory of economics. The first is essentially *phenomenological*: It starts from the kinds of *behavior* that need to be explained, including cycles (or quasi-cycles), instabilities and discontinuities, long-term growth, cooperative behavior, and so on. It basically seeks to identify mathematical models or model types with the right characteristics (non-linearity, etc.) to produce such behaviors. One may roughly characterize this as the "systems dynamics" school of evolutionary economics.

The second approach starts from first principles, namely, the recognition (admittedly, not yet widely shared by mainstream economists) that Walrasian general equilibrium is not the condition we live in. This follows from the fact that the conditions under which Walrasian equilibrium might be reached such as perfect competition, perfect information and rationality do not and cannot exist. The question to be answered, then, is: Given the conditions that do govern strategic choices in the real world of imperfect markets, namely, limited information and bounded rationality, how can one explain economic growth and the other phenomena noted above?

The "systems dynamics" approach to evolutionary economics begins from dissatisfaction with the ability of neo-classical growth models to generate plausible dynamical behavior without excessive reliance on exogenous shocks and random influences. The pioneer of non-linear economic modeling, Richard Goodwin, recently described the problem thus:

"To be realistic, the conditions I require of the model are: that it must exhibit an unstable equilibrium; that it must be globally stable; that it must endogenously generate morphogenesis [sic] in the form of structural change; that it does so in cyclical form, albeit erratically or aperiodically; that it generates both short and long waves, and finally, that these waves are growth, not stationary, waves" [Goodwin, 1989].

This is a good, albeit incomplete, list of the dynamic behaviors that the real economic system exhibits again and again, but that most economic models do not. The perspective that seems to have emerged, essentially, is that realistic phenomenological behavior requires non-linear models with "chaotic" properties. Many economists have explored non-linear models since Goodwin's first investigations.[13]

Interest in this area of research has multiplied since the discovery of deterministic chaos (the emergence of chaos from order, as one might say).[14] One of the most interesting and important implications of deterministic-chaos theory is the notion that "an evolutionary system may lead to completely different, but self-consistent long-run outcomes, depending on initial conditions and/or small random disturbances" [Silverberg, 1988, p. 549]. This phenomenon is sometimes called **path-dependence** in the context of stochastic models, and **hyperselection** in the case of deterministic models, [e.g., Ebeling and Feistel, 1982, Chap. 7]. Still another synonym for the phenomenon is *lock-in/lock out*.[15]

Whether one calls it path-dependence, hyperselection, or lock-in, it translates roughly into the idea that arbitrarily small exogenous disturbances, or fluctuations, can have major long-run consequences and even lead to different "trajectories." (The "butterfly-wing" phenomenon cited in Chapter 1 illustrates this notion.) To the extent that non-optimal paths can be (or tend to be) locked in, they must be self-determining and self-perpetuating. In fact, still another synonym for path-dependence" is *self-organizing*. These connections are not accidental. Examples of the possible lock-in of suboptimal technologies with varying degrees of plausibility include the QWERTY typewriter keyboard, driving on the left (or right, depending on your viewpoint), the trolley system of urban transport in Europe versus the bus/car system in the U.S., the boiling-water type of nuclear reactor, and so on. These examples tend to show that technological evolution has never found the true (dynamic, long-run) optimum solution.

The second, more axiomatic approach to evolutionary economics begins with a challenge to the elementary criteria for "rationality" in decision-making discussed in the previous chapter.[16] One of the first economists after Schumpeter to recognize that evolutionary mechanisms (variation and selection) may bring about neither optimality nor equilibrium in the long run was Sydney Winter [1964]. According to Winter, predictions based on neo-classical equilibrium and comparative statics need not hold. In other words, Winter took issue with the orthodox view that the real world behaves "as if" neo-classical assumptions were valid.

Winter's important insight opened the door for new approaches on several fronts. Winter himself, together with Richard Nelson, set about developing an evolutionary micro-economic theory of the firm and of a multi-firm economic system, based on *bounded rationality* in decision-making. In effect, the model eschews any notion of global optimization and substitutes a local *search-and-select* strategy to explain the basic patterns of innovation, imitation, industrial structure, and aggregate economic growth discussed by Schumpeter [Nelson and Winter, 1977, 1978, 1982].

The Nelson-Winter simulation model demonstrated that substituting the assumption of bounded rationality for static optimization, at the firm level, yields realistic (and realistically complex) patterns of aggregate growth and changes in industrial structure. It thus comes close to satisfying the "behavioristic" criteria for evolutionary models, noted above, despite the comparative simplicity of its underlying assumptions. The Nelson-Winter model has to be one of the major achievements of evolutionary economics thus far.[17]

Still, while the N-W model seems to include a diffusion mechanism and accounts for an overall increase in the level of technology, it lacks the element of organizational specialization or non-transferable organizational learning. It therefore does not reflect permanent (or at least long-term) advantages resulting from such learning.[18] Nor does the N-W model permit, still less explain, such cooperative behaviors as joint ventures or "strategic alliances." Finally, the N-W model does not appear to give rise to major discontinuities or bifurcations. From the "systems dynamic" point of view, therefore, it is perhaps insufficiently non-linear.

A model that owes much to Nelson and Winter, but that seeks to add more realistic learning mechanisms to the N-W picture, has recently been built by Giovanni Dosi, Luigi Orsenigo, and Gerald Silverberg [Dosi *et al.*, 1986]. It introduces more-complex relationships between firms, sectors, innovation users, and their environment, as reflected by positive and negative feedbacks in the model. With this greater complexity it is able to demonstrate such phenomena as organizational learning, cooperative effects, and bifurcations. The model is still under active development, although results have been reported at conferences [e.g., Silverberg, 1989].[19]

I would like to make one final point, again, with regard to the present stage of evolutionary economic theory: The evolution of cooperative behavior (of which the market itself is one of the most important illustrations) probably cannot be explained in terms of any myopic (or static) evolutionary selection principle. The importance of dynamics is now well recognized by evolutionary theorists. On the other hand, the importance of presbyopia, or long-range vision, in the evolutionary selection process has not yet received enough emphasis.

The ongoing emphasis on the role of organizational learning and of "learning to learn" [e.g., Cohen and Levinthal, 1988, 1989] seems to be pointing in the right direction. It is still far from clear, however, how presbyopic behavior (a prerequisite of cooperation) can evolve in the real world from myopic, short-term maximizing behavior. This is an important area for future research.

7.5. "TECHNOLOGICAL BREAKTHROUGHS" AND NON-LINEAR DYNAMICS

The purpose of the present discussion is to explore the characteristics of technological change from a historical-institutional perspective. In particular, I would like to elucidate the nature of the pattern that finds successive "breakthroughs" followed in turn by rapid improvement, maturation, and an approach to (configuration-dependent) physical limits and then by a new breakthrough.

Clearly, radical Schumpeterian (i.e., breakthrough) innovations are typically followed by an accumulation of Usherian incremental improvements.[20]

Radical innovation is increasingly discussed as "bifurcation," which became familiar with the popularization of René Thom's "catastrophe theory" [Thom, 1972].

The phenomenon of bifurcation was central to the work of Prigogine and his colleagues on the stability of self-organizing, dissipative thermodynamic systems far from equilibrium and in the area of biological evolution, as discussed in Chapter 3.[21] Bifurcation is also central to the mathematical phenomenon of deterministic chaos. Interestingly, chaos, in turn, is increasingly being seen as the source of a fundamentally new kind of order [Prigogine and Stengers, 1984; Anderson et al., 1988].[22]

In the language of systems dynamics, the evolutionary process seems to alternate between periods of turbulence ("chaos") corresponding to a cluster of radical inventions or breakthroughs, and periods of relatively stable behavior corresponding to gradual and incremental improvements. Quite possibly, technological evolution can be characterized (metaphorically, at least) as a progression of a very complex and non-linear system with a number of distinct feedback loops and "strange attractors." As the system evolves in time, shifts in "loop dominance" result in bifurcations; effectively, the system "jumps" from one attractor to another.[23] Each attractor, of course, corresponds to a particular technological configuration.

Definitions of technology abound. For my purpose, technology can be regarded as the knowledge, combined with appropriate means, to *transform* materials, carriers of energy, or types of information from less desirable to more desirable forms. There is a natural hierarchy of technologies. Broad categories like pyrometallurgy or computation constitute the top of the pyramid, and progressively narrower categories constitute the lower levels. In chemical engineering, for example, it is normal practice to identify unit processes (such as the Haber-Bosch process for ammonia synthesis), made up in turn of unit operations (such as pumps, filters, heat exchangers, or absorbers). These generally correspond to standard equipment made up of mechanical, electrical, or electro-mechanical subassemblies (e.g., gear-trains, valves, motors, transformers) and so on down to individual components (e.g., bearings, axles, connectors, transistors, wire). This somewhat tedious litany should nonetheless make the point that the process of transformation (or reaction) is really the objective of technology.

Most of the transformations in question do not occur spontaneously (at least at useful rates) without human intervention. In effect, every transformation initiated by man depends on the deliberate acceleration of some localized natural equilibration process, or the creation of an unstable disequilibrium by artificial means. The oxidation rate of hydrocarbons at ambient temperature and humidity is low, for example. The process can be accelerated by raising the temperature using, for this purpose, the heat of combustion itself to maintain a self-sustaining reaction.

Humans can exploit the high rate of heat output for other useful purposes. Most modern technologies require net energy inputs, and many of them also require other artificial environmental conditions to stimulate reactions (or suppress competing reactions). Such artificial environments are created by means of high or low temperatures, high or low pressures (vacuums), electric fields, magnetic fields, electric current, or neutron fluxes. Technological progress, in a fundamental sense, results from the increasing ability to create these conditions on demand, when and where they are desired.

The two words *on demand* are important, because the ability to produce very high temperatures and pressures (for example, by exploding a bomb in a confined space) is much less valuable than the ability to produce high temperatures or pressures independently of each other, and continuously, without blowing up the apparatus. It is the difference between nuclear bombs and nuclear power. One of the most important trends permeating technology today is miniaturization. The localization of a desired capability in an ever-smaller volume, miniaturization is often crucial to making technology available on demand.

In the past, physical limits were considered primarily in the context of selecting measures and assessing methodologies for technological forecasting [Ayres, 1968, 1969; van Wyk, 1985]. Physical limits may be thought of as ultimate limits imposed by nature on *any* technology. The speed of light is an example. No object with finite mass may move faster than the speed of light; indeed, as any object approaches the speed of light its mass and energy approach infinity! Other examples of physical limits are the "absolute zero" of temperature or pressure and absolute purity of materials.

Between an absolute physical limit and a practical engineering limit lies a range of intermediate cases. (The subject would be far easier to discuss if this were not so.) Perhaps the best example of an intermediate case is the famous **Carnot limit** for the efficiency of reversible-heat engines operating between hot and cold reservoirs. It is an absolute limit for one generic category of energy-conversion devices. On the other hand, it is not an absolute upper limit on the efficiency of energy-conversion devices in general.

It is evident that there is really a hierarchy of performance limits. The most general physical limits are independent of any situational conditions or constraints. The more that "real-world" conditions are imposed on the means of achieving the performance goal, the tighter the limits are likely to be.[24] Seen this way, technological limits are essentially tradeoffs. At the more abstract levels of description, the tradeoffs involve only generalized physical variables, such as temperature, pressure, density, voltage, current, magnetic field strength, and so on. But at the lower, more concrete levels of description, technologies are defined in terms of specific arrangements or configurations of devices and specific materials. Limits, in turn, are determined by these characteristics.

It should be clear that the evolutionary path of technology is influenced much less by "ultimate" physical limits than by "configuration-dependent" limits, which reflect the specific properties of materials in particular configurations. The relevance of configuration-dependent limits to the problem of technological innovation and economic growth has been emphasized in the *barrier-breakthrough* model of technological innovation [Ayres, 1988a].

The basic hypothesis of the barrier-breakthrough model is that the rate of technological progress tends to be higher when a technology is "far" from its limit and tends to slow down as the technology approaches its limit. (In this respect, the model is ontogenic rather than phylogenic.) Thus, progress is most difficult and slowest and R&D is most expensive and least rewarding as one approaches a technological "barrier," or configuration-dependent limit. Conversely, progress is easiest and R&D likely to be lowest in cost and highest in returns just after such a barrier has been overcome. The latter event, when it occurs, is what we mean by a *breakthrough*.

That Usherian improvements will likely follow Schumpeterian breakthroughs is not a particularly new idea to economists. What may be less familiar is the converse notion, namely, that each technological combination has its intrinsic limits and that Usherian improvements become harder and harder to make as the limits are approached.

It is an important fact of life that some technologies are already close to their theoretical limits, while others apparently still remain far from any limit. Steam turbines and carbothermic iron-smelting, for example, cannot in principle progress much further — they are demonstrably close to their theoretical limits. Indeed, only by careful consideration of the function(s) of a given technology and its historical path of development can a measure of **technological distance** be selected with any confidence.

The search for a new Schumpeterian combination (breakthrough) may simply be conducted by trial and error. It is random only where there is no basis for a targeted strategy, this being most likely when the physical principles underlying the desired phenomenon are not well understood. For instance, the early searches for chemical agents with bacteriocidal, insecticidal, or carcinogenic properties were essentially random. The same was true for the early phases of the search for an anti-knock additive for gasoline (which resulted in the discovery of tetra-ethyl lead). It is still partially true of the search for drugs effective against cancer and AIDS.

The random phase of the search process seldom lasts long. As observations are accumulated and correlated with one another, patterns emerge and scientists are able to screen out unlikely possibilities and focus attention on more-promising ones with increasing success. The more reliable the pattern-based search becomes, the more the patterns can be regarded as representative of a general phenomenon. When all the observations can be fitted to some quantifiable relationship, one has achieved a successful empirical model, which can then be used for prediction. When, ultimately, the model can be derived from generally accepted physical principles, one has achieved an analytical theory.[25] Thus science advances in parallel with technological advancement and, in the later stages at least, increasingly contributes to it.

The selection of a technology is based partly on competitive performance and partly on other considerations (e.g., cost, appropriability, and comparative advantage). In any case, it is not strictly based on technological considerations or entirely on economic ones. Technological advancement, i.e., the rate of progress, is inherently uneven. To oversimplify somewhat, the process may be regarded as a sequence of substitutions of successively better combinations or breakthroughs. Periods of rapid progress follow each breakthrough. As the limits of the new combination are approached, progress slows down and costs rise. Eventually the high costs and slow progress signal that conditions favor a new breakthrough.

It should not be surprising that some breakthroughs are much further away, in some sense, than others. Despite a very high rate of progress in semiconductor and computer technologies over the past four decades, the ultimate limits for solid-state electronic devices still appear to be far enough away as to permit this rate of progress to continue for at least another decade or two, if not more. It can be argued, in fact, that in many cases the configuration-dependent limits themselves are not likely to be well understood until they are closely approached. The limits of standard types of heat engines for energy conversion are now quite well understood, for example, whereas the limits of a somewhat ill-defined technology like manufacturing are poorly understood perhaps because they are so far away.

Not every technological barrier can be overcome. The electric car is a case in point. The advantages of using electricity for vehicular propulsion have long been obvious, but the means of accomplishing this are extremely elusive. The conditions of low cost, long life, reliability, safety, and high energy density seem to be impossible to combine in a secondary battery using known electrochemical combinations. A breakthrough obviously cannot be ruled out, but none has been forthcoming despite decades of intensive search and a number of apparently promising prototypes. The highly touted fuel cell has proven similarly difficult to reduce in engineering practice to small packages for ordinary applications.

I have not yet given the notion of *technological distance* any formal definition, although I hope that the term makes sense in the context in which it has been used. On the other hand, to develop the relationships presented here in more quantitative form, a more rigorous definition is needed. What we seek is a generalizable way of expressing the *distinguishability* of a reference system from its environment. The question has been addressed most systematically in thermodynamics, where density, pressure, volume and temperature are familiar state measures (internal energy, heat content or enthalpy, work content or Helmholtz free energy, and entropy being others). Technological distance from ambient or average conditions corresponds to a high degree of distinguishability.

In a later chapter I shall argue that the most general measure of distinguishability, hence of technological distance, is *information* in the technical (Shannonian) sense [Ayres, 1987b,c]. For thermodynamic systems, this measure of information is exactly proportional to the most general thermodynamic potential, sometimes called *available useful work* [Evans, 1969; Tribus and McIrvine, 1971]. To illustrate: The higher the temperature or pressure used in an industrial process, the greater the technological distance from ambient conditions which can be roughly characterized as the technological zero point. This notion should apply to other situations where technological distance is measured in terms of figures of merit like power, voltage, speed, acceleration, accuracy, and so on. It remains for future research to specify techniques for computing the information equivalent of non-thermodynamic figures of merit, such as the power-to-weight ratio of prime movers, or the complexity of electronic circuits.

7.6. PROGRESS FUNCTIONS (WITH K. MARTINÀS)

Another aspect of evolutionary change deserves comment at this point. Many biologists and physicists from Darwin and Dollo to the present day have emphasized the (apparent) irreversibility of evolution. This has prompted a variety of attempts to "explain" biological evolutionary processes in terms of the second law of thermodynamics.[26] The notion of borrowing irreversibility arguments from biology or physics and applying them to economics, based on a rather shaky analogy, meets immediate objections. Neither population growth, nor economic growth (in the adiabatic or homothetic sense), nor capital accumulation seem to be necessarily irreversible. Most economic variables are oscillatory, if not cyclic. Ups are notoriously followed by downs.

Yet there are also good reasons to suspect that the economic system does evolve irreversibly, in some sense, at the macro-level. Not only is this notion consistent with the second law of thermodynamics [e.g., Georgescu-Roegen, 1971] and with biological evolution [Faber and Proops, 1986; Ayres, 1988b], it also makes strong intuitive sense that economic progress should parallel the irreversible increase in human knowledge, especially technology (see also Chapter 8, section 6).

These notions of macro-irreversibility, however, have limited relevance to micro-economics. There does exist *micro-level* irreversibility in economics, namely, the irreversibility of pairwise exchanges [Martinàs, 1989]. This follows from the condition—sometimes called *incremental Pareto efficiency*[27] —that no economic agent will voluntarily undertake an economic activity leaving him or her less well off. In terms of transactions in the market domain, the argument is simple: If A has more apples than he wants, and fewer oranges, while B has too few apples and a surplus of oranges, A may be willing to trade some of his surplus apples for B's surplus oranges. But A will not be willing to trade in the reverse direction (oranges for apples). He does not trade for the sake of trading. The trade will not take place unless both A and B are better off after the trade in their own terms. The same rule applies when a medium of exchange (i.e., money) is present, the relative preferences (for oranges and apples) being expressed as internal exchange rates or "reservation prices." The rule also holds for production decisions.

Transactional irreversibility, in this sense, is implicit in Walras's law and the *tâtonnement* process. Strangely, the 19th century pioneers of utility theory, such as Jevons, Walras, Pareto, Fisher, and Edgeworth never explored the implications of transactional irreversibility.[28] It was first formulated explicitly only 40 years ago as *Ville's axiom:*

"No [price] path exists which moves always in the preferred direction but ends at its starting point" [Ville, 1951].

Ville's axiom was originally set forth as a necessary condition for the existence of a differentiable total utility function, depending only on quantities of exchangeable goods. Ville's axiom is applicable at the level of an economic system (i.e., a market) for which a unique price is defined for each commodity. The corresponding axiom in the foregoing case can be stated:

No voluntary exchange transaction between two firms (economic decision-makers) results in negative (or zero) surplus values for either party. Similarly, no voluntary production decision results in a negative (or zero) surplus value for the firm.

I refer to this as the "no loss" decision-making rule, to contrast it with the more familiar "profit maximization" rule. In contrast with Walras, Ville, and virtually all the post-Walrasian literature on equilibrium, the axiom above makes no assumption about the existence of price paths in the usual sense. I return to this important difference later.

For Jevons, Walras, Pareto, Edgeworth, and the other neo-classicists, the central problem of mathematical economics (or *psychics*) was to determine the conditions for maximization of pleasure or utility. In particular, neo-classical economics since Walras has focused intensively on the conditions for a general equilibrium, rather than on the properties of non-equilibrium states and the approach to equilibrium. In this respect, the neo-classical program was very different from the one under discussion. Our key results do not rely on the principles of profit maximization and utility maximization, on which most of neo-classical economics was built.

We will argue hereafter that irreversibility as applied to voluntary exchanges and production decisions by satisficing economic agents, together with other straightforward assumptions including the existence of a universal medium of exchange is sufficient to prove the existence of a non-decreasing function of the quantities of money and goods possessed by each agent, in the absence of depreciation. It will be argued that this function can be interpreted without inconsistency as *progress*. The economic agents need not be profit-maximizers, nor do they need perfect information. As it turns out, the exercise yields an unexpected reward in terms of suggesting new micro-economic models and new interpretations of existing data.

7.6a. Micro-Foundations

Assume the existence of a number of **economic units** (EUs) and an aggregation of many units that together constitute an **economic system** (ES). Within an ES, assume for convenience a set of rules governing exchange transactions and a systemwide medium of exchange. It is called **money**. An EU is defined for our purposes as the smallest entity possessing either an implicit or explicit decision-making rule with the property that no voluntary economic transaction occurs that leaves the EU *less well off* than before the transaction. One of the EUs is assumed to consist of the labor force cum consumers. Worker-consumers, taken as an aggregate, can also be regarded as economic decision-makers: They have (aggregate) preferences (demand) and voluntarily exchange labor for money and money for goods. The consumer aggregate can accumulate goods, but do not sell them. It can accumulate money or go into debt. (It would also be reasonable to postulate another EU that produces no goods, but can borrow money from either producers or consumers and lend it to anyone else. However this complexity adds nothing of vital importance to the theory at this stage.)

The total amount of money in an economic system is assumed to be conserved over transactions. The theory allows for the possibility of involuntary money transfers (essentially **taxes**) from firms to a common pool (the ES), from which it may be redistributed as subsidies. It also allows for the possibility of purely financial transactions, such as loans in exchange for interest payments or dividends. (These possibilities are not pursued further in this summary.) An economic system might, in principle, also possess goods owned in common. This model makes no fundamental distinction between non-exchangeable property that cannot be privately owned, such as the air or the biosphere, and exchangeable property that could be owned privately but is not, as in a socialist state. For now, however, We do not consider the problem of public goods further.

By assumption, an EU may be capable of either of two types of voluntary actions: **production** or **exchange**. An EU in this model would normally be a **firm**.[29] Let us assume that EUs may interact directly with one another only in binary fashion, via exchanges of goods[30] or money. We will *not* assume the existence of a unique posted market price (known, therefore, to all EUs) for pairwise economic exchanges between EUs. In each individual exchange there is an actual or—in the case of a barter exchange, a virtual money transfer corresponding to the goods transfer. Both material goods and money are conserved in exchange transactions.

Prices, in this theory, are defined only for specific pairwise transactions. In a static pure-exchange model, pairwise exchange prices converge to a unique equilibrium price.) Nevertheless, market prices for an ES can be defined only in the (for the moment, conjectural) Pareto-equilibrium limit, i.e., after all possible transactional gains have been exhausted. In the non-equilibrium case, individual voluntary exchange transaction prices can, and do, vary among economic actors and over time. The fact that exchange prices are not posted precludes the existence of a *price path* in the strict sense of Walras, Ville, *et al.*

Production, in our theory, is a deliberate conversion (by firms) of material **goods** into other goods, using **labor** and **capital**. Labor is assumed to be available at a given money price (wages) and the aggregate of all wage payments is equated to **aggregate demand** for consumer goods. Firms receive money when they sell consumer goods. The aggregate amount of labor in the economic system can be assumed to be conserved (or it could be assumed to grow at a constant rate).

Capital can be considered, for our purposes, as a type of good that can be either produced (by some firms) or purchased for money. In our theory, capital is needed for production and is accumulated for that purpose. It is neither made obsolete nor worn out. The related problems of consumption by individuals and individual decisions (by workers) to sell labor for wages and (by consumers) to buy goods for money, are not considered here. Nor do we discuss the problem of dissipative uses of some intermediate goods, and depreciation or obsolescence of capital.

It is assumed that no voluntary transaction of any type occurs in the absence of an explicit decision to act, based on the "no loss" decision-making rule. The criterion for a positive decision is that the EU be left *no worse off* than it was before. This can be restated in more precise terms. We assume that *well-offness* (welfare) is a function of the **economic state** of the EU. The latter is determined by a set of **observables**. Examples of observables include stocks of money and goods (including capital goods). The observables are, of course, basic variables of the economic system.

The assumptions above make it possible to formally introduce the concept of **reservation price** or, equivalently, internal worth. This quantity is definable for each EU. For convenience, the shorter term *worth* sometimes appears hereafter as a synonym. The internal worth of the ith good is V_i

$$V_i = V_i(X_1, X_2, .., M),\qquad(7.1)$$

where X_i stands for the quantity of the ith good, and M for the money. It is important to distinguish **worth** from both market price (**exchange value**) and **utility**. The internal worth of a good is defined for (and by) each EU and may be known only to the EU. Assuming the "no loss" decision-making rule postulated earlier, an EU agrees to sell a good if, and only if, its internal worth is less than the price offered by another EU; conversely, an EU agrees to buy the good if and only if the internal worth of the good being offered equals or exceeds the price offered. Similar logic can be applied for the production decision.

In the special case of an exchange in which the EU does not care whether the transaction takes place, the internal worth of the good to the would-be seller must be exactly equal to the money price that would be received. Similarly, for a marginal production decision, the internal worth of the product must be exactly equal to the internal worth of the resources that would be utilized. (Thus, only marginal transactions are reversible.) The foregoing definition of internal worth (reservation price) differs from that used conventionally for **value** in micro-economics [Debreu, 1959], in three specifics:

1. It presumes only a weak form of rationality akin to common sense (the "no loss" rule) and limited availability of information; it does not presume perfect rationality or perfect information. (Obviously, however, it does not exclude those cases.) In the standard case, producers maximize profit. In our case, the weak decision-making rule is applicable to all production decisions.

2. It depends only on the internal state (quantities of goods and money) and the preferences of an economic unit (EU); it is independent of the economic system (ES) to which the EU belongs, except insofar as the economic system creates money. In the case of general equilibrium, by contrast, value is system-determined.

3. Internal worth does not determine the actual *path* of the approach to equilibrium, or of economic development. It only determines whether an exchange or production decision is possible in each specific case. The rate at which an economic process takes place, involves additional factors, including technological capabilities and constraints, and individual characteristics. The real process cannot be described without specifying these factors and constraints.

Any production or exchange transaction will result in a change in the economic state of the EU through a change in the quantities of goods and money. For instance, a production decision will involve a conversion of raw materials and purchased labor into finished goods for sale. An exchange transaction is a sale of goods for money at an agreed-upon price. It is important to note that, because each EU may have a different preference and decision-making rule, two EUs would be likely to assign a different internal worth for each good, even if both were in the same economic state.

It is convenient, for what follows, to distinguish between *extensive* variables, which, in some sense, measure the *size* of the system, and *intensive* variables, ratios that measure characteristics independent of size. Stocks of goods and money are examples of extensive variables. Intensive variables can be ratios of extensive variables. Worth (defined above) is another example.

The time rate of change of extensive variables can always be expressed in terms of flows:[31]

$$\frac{\Delta X_i^\alpha}{\Delta t} = J_i^\alpha + S_i^\alpha, \tag{7.2}$$

where J_i^α is net imports of the ith commodity (imports minus exports) and S_i^α is the net production (production minus intermediate use) of the ith commodity within the EU. By similar logic, one can write

$$\frac{\Delta M^\alpha}{\Delta t} = - \sum_i P_i^\alpha J_i^\alpha + I^\alpha, \tag{7.3}$$

where P_i^α is the money cost (market price) of one unit of the ith good or commodity and I^α is the net financial inflow, i.e., the difference between credits, subsidies, interest or dividends received, loans or investments from outside the EU (e.g., dividends or capital gains) and debits (interest or dividends paid, taxes paid, losses on external investments, etc.).

It is convenient to introduce a new notation $J_i^{\alpha\beta}$, which is interpreted as the flow of commodity i from EU β to EU α, where

$$J_i^{\alpha\beta} = - J_i^{\beta\alpha}. \tag{7.4}$$

That is, every commodity flow to the EU can be identified by origin. Let $I^{\alpha\beta}$ be the (non-trade) financial flows from EU β to EU α. Let $P_i^{\alpha\beta}$ be the price of the ith good in the exchange between the αth and βth EU. Then the money flow can be written as

$$I^\alpha = - \sum_{\beta i} P_i^{\alpha\beta} J_i^{\alpha\beta} + \sum_\beta I^{\alpha\beta}. \tag{7.5}$$

The assumptions above imply that $S_i^\alpha = S_i^\alpha(X_1^\alpha, X_2^\alpha \cdots)$, $J_i^{\alpha\beta} = J_i^{\alpha\beta}(X_1^{\alpha\beta}, X_2^{\alpha\beta} \cdots)$, and $P_i^{\alpha\beta} = P_i^{\alpha\beta}(X_1^{\alpha\beta}, X_2^{\alpha\beta} \cdots)$. These relations characterize the EU, so they can be *determined* experimentally, at least in a *Gedanken* sense.

With one exception, the assumptions above are basically familiar to economists. The exception is important, however. In contrast to the usual case, it is possible to consider an exchange process without a *market price* in the usual sense. Even the weak form of (bounded) economic rationality (the "no loss" rule) with no posted prices and limited to pairwise transactions is sufficient to create the equivalent of an "invisible hand" that assures convergence to a Pareto-optimal state indistinguishable from Walrasian equilibrium. Perfect rationality and perfect information are not necessary conditions for an efficient market.

7.6b. A Progress-Function for Firms

As already noted, a productive EU (i.e., a firm) can possess and accumulate two kinds of wealth: material goods and monetary assets. The former consists of capital goods, raw materials, inventories of works in-progress, and unsold final goods. The latter comprises investments, loan portfolios, bank accounts, and cash.

The change in wealth ΔW^α of the αth EU during a time period Δt can now be expressed by the accounting balance

$$\Delta W^\alpha = \sum_i V_i^\alpha \Delta X_i^\alpha + \Delta M_i^\alpha, \tag{7.6}$$

where V_i is the internal worth of the ith good expressed in monetary units of the ith material good or commodity, X_i is the stock of the ith material good (or commodity in the EU) and M^α is the quantity of monetary assets held by the αth EU. It would be convenient if ΔW^α, ΔX_i^α, and ΔM^α could simply be converted into perfect differentials, resulting in an integrable expression. In its present form, however, this is not possible.

One way to formulate the problem is to note that the change of wealth $\int \Delta W$ is not a well-behaved, differentiable function of X_i, M alone, but also depends on other factors (variable parameters). In other words, the change in level of wealth after a finite time is *path-dependent;* it depends on the particular sequence of transactions that is followed in X_i, M space. In mathematical language, an *integrating factor* is needed. But for the expression (7.6) as it stands, such a factor cannot even be proved to exist. While $\Delta W \geqslant 0$ for all voluntary processes (a version of Walras's law),[32] there exists the possibility of non-market economic processes (e.g., taxes) such that $\Delta W < 0$ is possible.

The next step, therefore, is to seek a mathematical transformation of (7.6) into a form such that the existence of an integrating factor is provable. This means manipulating the expressions into a form that explicitly reflects some additional information about the nature of economic transactions. To be specific, we seek an expression that explicitly reflects the *irreversibility* of exchanges and production decisions under the assumed "no loss" decision-making rule.

To accomplish the desired transformation, one can, for convenience set $\Delta t = 1$ and insert (7.2) and (7.3) into (7.6). This yields, after combining terms,

$$\Delta W^\alpha = \sum_i (V_i^\alpha - P_i^\alpha) J_i^\alpha + \sum_i V_i^\alpha S_i^\alpha + I^\alpha. \tag{7.7}$$

In the special case of the labor/consumer sector, no goods are produced ($S = 0$) and financial income includes the sum of all labor payments by producers, less taxes (if any), plus any interest, dividends, etc. Assuming both trade and production decisions are governed by the "no loss" rule, it follows that the first two summations on the right-hand side of (7.7) must be non-negative. In other words, the flows are *uni-directional*, reflecting the irreversibility of economic actions. Specifically

$$\sum_i (V_i^\alpha - P_i^\alpha) J_i^\alpha \geqslant 0 \tag{7.8}$$

is equivalent to asserting that exchange transactions occur only when the EU perceives an economic benefit (increased internal worth). Similarly,

$$\sum_i V_i^\alpha S_i^\alpha \geqslant 0 \tag{7.9}$$

is equivalent to asserting that a unit of output is produced (net) only if the internal worth to the EU is non-negative. It follows from (7.8) and (7.9) that, for the αth EU

$$\Delta W^\alpha \geqslant I^\alpha. \tag{7.10}$$

We seek an integrating function $T(X_i, M)$ and a new variable Z such that, for each EU independently,

$$\lim_{\Delta t \to 0} \frac{\Delta W}{\Delta t} = T \frac{dZ}{dt}. \tag{7.11}$$

Carathéodory [1909] proved a number of years ago that such an integrating function T exists for an irreversible process. The rigorous proof depends on characterizing the irreversibility as follows: An economic process (in our case) is irreversible if, in the near neighborhood of *every* point in the state-space there is another point arbitrarily nearby that *cannot* be reached by any reversible process as long as there is no net financial inflow ($I^\alpha = 0$). [A reversible process in our case is one such that $\Sigma_i(V_i - P_i) J_i = 0$ and $\Sigma_i V_i S_i = 0$]. The full proof of the theorem is too complicated to repeat here. Its applicability to the economic case, as described above, was first shown by Bródy, Martinàs and Sajó [Bródy *et al.*, 1985]. Actually, a similar proof of integrability was given by Hurwicz and Richter [1979], based on Ville's axiom. The Ville formulation is essentially equivalent to the Carathéodory irreversibility condition.[33]

In short, the necessary conditions for (7.11) are satisfied. It follows that, substituting back into (7.7) and dropping the superscript α for convenience

$$TdZ = \sum_i (V_i - P_i)J_i + \sum_i V_i S_i + I \qquad (7.12)$$

and from (7.6) we have

$$TdZ = \sum_i V_i dX_i + dM. \qquad (7.13)$$

For future reference, note that for voluntary processes ($I \geqslant 0$) equations (7.8) and (7.9) imply $dZ \geqslant 0$. That is to say, Z is absolutely non-decreasing in this case. (This no longer holds true when depreciation of durable goods and consumption processes are introduced.)

The next step is to characterize and select a function $Z(X_i, M)$. Equation (7.13) implies that

$$\frac{\partial Z}{\partial X_i} = \frac{V_i}{T} \qquad (7.14)$$

and

$$\frac{\partial Z}{\partial M} = \frac{1}{T}. \qquad (7.15)$$

The progress function Z contains essentially the same information as the internal-worth function V_i together with the integrating function T. Equation (7.13) defines only the TdZ product. To define them individually, some arbitrary choices need to be fixed. First, we want Z to increase in a spontaneous economic process. It follows that T should be positive.

A further requirement on T is that it should be *homogeneous of zeroth order* (to ensure that Z is a first-order homogeneous function.) Homogeneity to zero*th* order for T means, in effect, that we want it to depend only on *intensive* variables, i.e., on ratios of extensive variables. On the other hand, we want Z to be homogeneous to the first order, meaning that its dependence on *size* (extensive variables) is essentially linear. (It can be shown that additivity of the Z function is consistent only with first-order homogeneity. The first-order homogeneity condition is

$$\lambda Z = Z(\lambda X_i, \lambda M). \qquad (7.16)$$

Differentiating (7.16) with respect to λ yields, after straightforward manipulation and setting $\lambda = 1$,

$$Z = \sum_i \frac{V_i X_i}{T} + \frac{M}{T}. \qquad (7.17)$$

An implication of differentiability is that

$$dZ = \sum_i \frac{V_i}{T} dX_i + \frac{1}{T} dM + \sum_i X_i d\left(\frac{V_i}{T}\right) + M d\left(\frac{1}{T}\right). \qquad (7.18)$$

But (7.13) also holds, whence by matching terms we can derive another equation for either T or V_i:

$$0 = \sum_i X_i d\left(\frac{V_i}{T}\right) + M d\left(\frac{1}{T}\right). \qquad (7.19)$$

When T and V_i are specified, Z is determined by (7.17)

There are an infinite number of possible functional forms for T, Z. One of the simplest expressions among them for T satisfying all the required conditions is

$$T = \frac{M}{\Sigma g_i X_i}, \qquad (7.20)$$

where the g_i are coefficients yet to be specified. Substituting (7.20) into (7.19), differentiating, and collecting terms, one obtains an integrable equation for V_i/T. Integrating (and multiplying by T) yields for the internal worth (or reservation price):

$$V_i = \frac{g_i M}{\Sigma g_i X_i}\left[\ln\left(\frac{M}{X_i}\right) + c_i\right], \qquad (7.21)$$

where c_i is a constant of integration. Substituting (7.21) back into (7.17) completely defines the *form* of Z:

$$Z = \sum_i g_i X_i\left[\ln\left(\frac{M}{X_i}\right) + c_i\right] = \sum_i g_i X_i \ln\left(\frac{M}{k_i X_i}\right), \qquad (7.22)$$

(since the g_i have not yet been defined). It remains only to find a consistent interpretation of the coefficients g_i and k_i and the quantities X_i. The physical interpretation of these terms is deferred until the next section.

An important caveat that must be emphasized is that (7.22) is only one possible form of Z. In fact, there is no guarantee that this particular form is the correct one in any given case. Nor is it necessarily true (or even likely) that all EUs in the real world will be characterized by a T-function (or the corresponding V- and Z-functions) having the same form. The actual form would have to be determined by experiment or observation on a case-by-case basis. However, for idealized models involving transactions among *indistinguishable* (i.e., interchangeable) EUs, it is clear that the mathematical forms of T and Z should also be indistinguishable, hence identical for our purposes.

Finally, given that Z has the useful property to be proved later that its maximum value corresponds to a static equilibrium (and Pareto optimum), then the standard second-order condition (declining marginal internal worth, or reservation price) holds; namely,

$$\frac{\partial V_i}{\partial X_i} \leqslant 0. \tag{7.23}$$

7.6c. Interpretation

It is now appropriate to seek reasonable economic interpretations of the expressions for T, V, and Z. The integrating factor T was defined by (7.20) as the ratio of money assets to a weighted sum of goods assets in the EU. This ratio has an obvious interpretation as **liquidity**. For a producer, this makes the expression (7.21) for the worth (reservation price) V_i easy to interpret. The internal worth of the ith good to a producer is directly proportional to the liquidity of the EU, and directly proportional to the weight of that good in its inventory. This much is entirely in accord with intuition.

The logarithmic term is less obvious. It says that, for fixed liquidity T, the internal worth of any good to a producer decreases logarithmically the greater the quantity of that good is on hand, whereas the internal worth of a given stock of money increases logarithmically. Yet, on reflection, few corporate chief executives would find this rule counterintuitive. In general, money in the bank is preferable to inventory, always provided there are goods available on the market to buy. For the labor/consumer sector, taken as a whole, there is a similar tradeoff between liquidity and physical possessions.

It would be a natural mistake, in view of (7.11) to interpret the product TZ as "wealth." But it must be recalled that the integral $\int \Delta W$ is path-dependent. It was precisely for this reason that we had to find an integrating factor. On the other hand, Z is not wealth, either; among other problems, it cannot have units of money (TZ does). How, then, shall we interpret Z?

It is easy to see that Z is at a local maximum when the EU reaches a condition such that (under the "no loss" rule) it cannot improve its economic state by engaging in further economic activities. This final state of non-activity, in fact, can be interpreted as an equilibrium for the EU. Thus all economic activity can be interpreted as an approach to (Pareto) equilibrium. The local maximum would not, in general, constitute a global maximum in the absence of further specific constraints.

It is important to emphasize here that Z is *not* a classical utility function U, although the utility function (when it exists) is also maximized at the equilibrium point. The utility function for an EU is better interpreted as the *difference* between the progress function before a transaction or production decision, and after it. The relationship between Z and U is discussed in more detail later.

We believe that it is natural to interpret Z as a *stock of economically useful information*, because every good can be expressed in informational terms [Ayres,

1987*a,b*]. In fact, one can assert that each manufactured good is characterized by a distance from thermodynamic equilibrium, hence a certain quantity of *embodied* information:

$$X_i = \frac{H_i}{b_i},$$

(7.24)

where H_i is measured in *bits* and the coefficient b_i has dimensions of information/ quantity.[34] Substituting (7.24) in (7.17) we get

$$Z = \sum_i \left(\frac{V_i H_i}{b_i}\right) \frac{1}{T} + \frac{M}{T}.$$

(7.25)

Defining

$$w_i = \frac{V_i}{b_i}$$

(7.26)

we can rewrite Z

$$Z = \sum_i \frac{w_i H_i}{T} + \frac{M}{T}.$$

(7.27)

There is no *a priori* restriction on the dimensionality of Z. We know that TZ has the dimension of money. If we choose to express Z in *bits* of information, then T has the dimensions of money/information and w also has dimensions of money/information. Here T may be interpreted as the unit worth (reservation price) of a standard type of information, while w_i is the relative unit worth of the ith type. Thus w_i/T is the relative worth of the ith type of information. One of the w_i can be set equal to unity, for convenience.[35]

The index i over *commodities* or *goods* might equally well be considered as an index over types of information embodied in materials, structures, organizations, etc. These include thermodynamic information, morphological information, symbolic information, organizational information, and so on. In this context, money M and the worth of expected future labor (and life), L can also be viewed as special kinds of information.

7.6d. Concluding Remarks

Three concluding comments suggest themselves. In the first place, it is very tempting to try to define a *progress function* for the economic system ES as a whole. As noted already, however, the "no loss" rule cannot be applied to the economic system as a whole, whence it is impossible to derive either an integrating factor or a progress function for an ES. It is worth emphasizing yet again that, whereas most axioms and theorems of neo-classical economics deal with an ES, we have restricted ourselves initially to individual economic units (EUs) and pairwise transactions between EUs.

Our exclusive concern with individual EUs and pairwise transactions explains why we do not need to use utility maximization, and why the weaker "no loss" rule suffices. On the other hand, it seems plausible that for an EU, where it is definable, one could develop a variational principle for optimization purposes. The basic idea of constrained maximization has been often applied in economics, with a variety of objective functions often arbitrarily chosen. We have already noted that Z is a maximum when the EU is in a Pareto-optimum state, i.e., a state such that further economic activity will not improve its condition. It would seem, therefore, that Z should be maximized. This being so, we can also consider maximizing the sum over all EUs in the ES:

$$Z = \sum_{\alpha} Z^{\alpha}. \qquad (7.28)$$

Absent a decision-making rule for the system as a whole, one cannot prove that Z for the system as a whole actually tends toward such a maximum. However, it is not implausible that a government might reasonably adopt the objective of maximizing Z. It is also quite plausible that the individual EUs within the ES would agree (if consulted) to such a maximization policy for the ES as a whole.

The third and final comment is that there is, indeed, a close analogy between economics and thermodynamics. In fact, the foregoing derivation proceeds in detail along the same lines as Carathéodory's axiomatic development of thermodynamics. The basis of the analogy is that irreversibility plays a key role in each case. It is also true that function T in our derivation is *like* the temperature; the non-decreasing function Z in our derivation is *like* entropy, the product TZ is *like* enthalpy, wealth W is *like* heat, and so on [Martinàs, 1989]. We freely admit having referred to the analogous arguments in thermodynamics. But the analogy was only a guide. The economic derivations presented above are rigorous; they stand on their own.

Actually, there may be an even deeper connection between thermodynamics and economics. The fact that Z plays the same role in our theory as entropy does in classical thermodynamics is not coincidental. Irreversibility is the key in both cases, as noted earlier.

It is interesting that a number of physicists from Leo Szilard and Leon Brillouin on have argued that information *is* negative entropy (or *negentropy*, in Brillouin's language.)[36] On the other hand, we have argued that the function that plays the same role as entropy in this formulation of economics *is* information. The logical circle appears to be closed.

ENDNOTES

1. One is tempted to emphasize the role of random processes (such as mutation) in evolution. There are indications, however, that some processes that appear to be random may be instances of deterministic chaos rather than true randomness.

2. In 1900, for example, petroleum was used mainly as a source of "illuminating oil." That product is no longer even available in the industrialized countries; today petroleum is refined to make fuel for cars, trucks, and aircraft; heating oil; and petrochemicals. None of these products existed in the 19th century. In 1900 aluminum was a curiosity metal with no significant use. Today it is the basis for an important industry because aircraft could not fly without it. In 1950 there were half a dozen computers in the world—elaborate calculators, really—and they were almost exclusively used for scientific purposes. Today the computer industry is a giant. The semi-conductor manufacturing industry was created from nothing in the past thirty years. The "software" industry is now in the process of establishing its own separate identity. And so on.

3. The most important studies were those of Solomon Fabricant [1954], Moses Abramowitz [1956], and Robert Solow [1956, 1957]. Solow's work also contributed to the theory of growth and resulted in his being awarded the Nobel Prize for economics.

4. At the macro-level, technological change typically appears in models only indirectly through a productivity index of some sort. The aggregated productivity indexes used today in macro-models date back mainly to the work of John Kendrick.

5. The orthodox view of economists on this question is nicely summarized as follows: "Technical knowledge, being the product of a production process in which scarce resources are allocated, can be produced. We do not know exactly what will be produced, but we are certain we will know more after an unknown period" [Heertje, 1983].

6. The sociologist S. Colum Gilfillan [1937] noted, for example, that the rapid increase in demand for civil airline travel in the 1930s created a need for a technology permitting aircraft to land and takeoff safely at night. Gilfillan listed 20 possible approaches to the problem and concluded there was no reason to suppose that a feasible solution could not be found within a few years. He was quickly proven right, though the final answer (radar) was not on his original list of possibilities.

7. But such events are not unknown. The laser, for instance, seems to have been invented more or less by accident, and long before anybody knew of a good use for it.

8. "Software" comprises more than instructions for computers. It includes chemical formulae, blueprints, scientific data, and so on. Although software is growing in importance, the share of investment in this form of capital remains relatively minor in comparison with others.

9. The recognition that technological improvement makes an impact on the economy primarily insofar as it is embodied in capital goods has spawned a large neo-classical literature on capital "vintage" models. This trend began with Kenneth Arrow's [1962] reformulation of the theory of economic progress, in which he emphasized the

importance of "learning by doing" in the capital-goods industry and the embodiment of technological progress in capital. This idea implies that technological progress diffuses exactly in proportion to the rate of adoption of new equipment. Of course, this is simplistic. New machines do not necessarily incorporate the latest technology and old ones can sometimes be adapted to do so. Moreover, technology is not exclusively associated with machines. Nevertheless, the link between capital-goods replacement and technology diffusion is important and real.

10. The works of technological and economic historians illustrate the ontogenic approach; it has some explanatory power but claims little or no predictive power, e.g., Albert Payson Usher [1929], Nathan Rosenberg [1969, 1972, 1976, 1982] and Paul David [1975, 1985, 1988].

11. The S-curve approach was introduced in this context by Zvi Griliches [1961] and Edwin Mansfield [1961]. Alternative models in this tradition include Bass [1969]; Blackman [1972, 1974], Ayres and Shapanka [1976], Sharif and Kabir [1976], Sahal [1981], Easingwood et al. [1983], Skiadas [1985], Peschel and Mende [1986]. Good reviews include Mahajan and Peterson [1985], Thirtle and Ruttan [1986]. The mechanical, quasi-static nature of these adoption/diffusion models has been challenged, in particular, by J. S. Metcalfe [1981], Giovanni Dosi [1982], Christopher Freeman [1982], Carlotta Perez [1983], and Paul Stoneman [1983].

12. Schumpeter's basic qualitative model of innovation implies that, if innovation is a successful strategy for survival in an initially competitive market, then innovative behavior itself should be imitated. In other words, firms should attempt to retain their market share by continuous innovation, or "success breeds success". Successful innovation also implies a departure from conditions of pure competition, so there is certainly some feedback between innovation and market structure.

In addition, while Schumpeter emphasized the importance of radical innovation in economic life, the so-called Schumpeterian hypothesis (re market structure) does not clearly distinguish between radical and incremental innovations. Indeed, virtually all the econometric analysis and modeling that has been done to test the hypothesis has used data on small innovations that fit the Usherian (gradualist-incrementalist) tradition. This is the case because most innovations are small, after all, and there is no clean way to segregate radical innovations from incremental ones.

13. See, for instance, Goodwin [1951, 1967, 1982, 1987, 1989]. Another prolific writer on non-linear dynamics in economic systems has been Richard Day [1981, 1982]. As far back as 1967, Day also investigated growth models, both micro and macro, with nonlinear properties. These models displayed stable stationary states followed by 2-period, 4-period, and higher-order cycles and "chaotic" behavior. In recent years a large number of other more or less simple non-linear models with "chaotic" properties have been investigated, e.g., Torre [1977], Benhabib and Nishimura [1979, 1985], Stutzer [1980], Simonovits [1982], Bhaduri and Harris [1987], Deneckere and Pelikan [1986], Medio [1987], Radzicki [1989]. Empirical research on chaos in economics is now voluminous, e.g., Baumol and Quandt [1985], Day and Shafer [1985, 1987], Barnett et al. [1986a,b], Brock [1986, 1988], Baumol & Benhabib [1987], Boldrin [1988]. The last two are survey papers.

14. See May p1975], Hofstadter [1981], Crutchfield et al. [1986].

15. *Lock-in* was popularized by Brian Arthur [1983, 1988] and Paul David [1988*a,b*]. Arthur and others have focused attention both on simple mathematical models of the lock-in phenomenon and on economic mechanisms—such as "returns to adoption"—to account for it. David has explored the topic of path-dependence mainly from the standpoint of economic history [David, 1989]; Cowan [1989], and Ayres and Ezekoye [1991].

16. This may be said to have begun with the path-breaking work of Herbert Simon.

17. Nevertheless, from the theoretical perspective, the N-W model lacks some important features (most likely due, it must be said, to its deliberate simplicity). The model assumes that RandD is essentially equivalent to a lottery: You spend your money to buy tickets; the expected payoff is simply proportional to the number of tickets bought. The payoff in RandD is a short-term market advantage. Competitors can imitate at modest cost after a short lapse of time.

18. A critical and largely unconsidered aspect of organizational learning is "learning to learn" (see Cohen and Levinthal [1988]). Almost certainly the Japanese economic success can be attributed in large part to the Japanese commitment to find and adopt the best available technology, regardless of source. Many observers have contrasted this with the tendency of established U.S. firms to reject ideas "not invented here."

19. Other dynamic evolutionary models incorporating various combinations of the features noted above have been reported recently. Of considerable interest is work on industrial/technological evolution by Halina Kwasnicka and Witold Kwasnicki and a model reported by John Conlisk [1989]. Also relevant is the work of Peter Allen [Allen, 1988; Allen *et al.*, 1984; Allen and Sanglier, 1981], Michelle Sanglier, and others on regional/ecological self-organizational models.

20. The "fast-slow" phenomenon as reflected in an apparent clustering of major techno-logical innovations during periods of slow economic growth has also been suggested as a possible partial explanation of the so-called Kondratieff cycle, or "long wave" [Freeman *et al.*, 1982; Freeman, 1983]. Mensch has suggested that innovations them-selves can be usefully distinguished between new products and new processes, with the latter predominating during periods of less radical innovation and faster economic growth. Unfortunately, both hypotheses are difficult to test empirically, due to data gathering problems.

21. See, for instance, [Glansdorff and Prigogine, 1971; Nicolis and Prigogine, 1977; also Prigogine *et al.*, 1972].

22. Among other applications in economics, bifurcation in chaotic systems has been proposed as an explanatory model of the long (Kondratief) wave [Rasmussen *et al.* 1985].

23. The language here is paraphrased from [Radzicki, 1989], who attributes it to [Rich-ardson, 1984].

24. As an illustration, consider the heat pump—a refrigerator operating in reverse as a means of "pumping" heat from a cold reservoir (5 °C) to a warm reservoir, the house to be heated (+ 20 °C). An idealized heat pump using the Carnot cycle and oper-ating between these two temperatures would have a coefficient of performance

(COP) of 19.5, meaning that 19.5 times as much energy is delivered as heat at the high temperature end as is required to operate the pump. However, "real-world" considerations reduce this performance dramatically. In the first place, heat can be delivered only through a heat-exchanger of some sort. The flow of heat through a heat-exchanger depends on the existence of a temperature gradient. The smaller and more compact the heat-exchanger, the larger the gradient must be. (A heat-exchanger with a large surface area is also more costly). For a heat-exchanger small enough to get through the door of a house, and cheap enough to afford, a fairly big gradient is necessary. In effect, the introduction of "real" heat-exchangers means increasing the effective temperature at which the heat is delivered from 15 °C to 45 °C. The COP drops from 19.5 to 6.4. The next real-world constraint is that no practical configuration of components closely approximates a Carnot cycle. A practical alternative is the Rankine cycle, because there are many well-engineered small Rankine-cycle compressor-expander units to choose from. But the Rankine cycle has unavoidable irreversibilities in the cooling and expansion stages that reduce its theoretical efficiency below that of the Carnot cycle. The COP drops from 6.4 to 5.1. Allowing for realistic compressor losses brings the COP down to 4; other miscellaneous losses cut it further to 3, which is barely 15% of the ideal case. To be sure, better materials and component designs can (and will) reduce some of these losses in the future, but intermediate constraints, in principle, will always exist. See Heap [1983].

25. This is not to suggest that such a theory is necessarily "true," in the sense of being complete. Newton's theory accurately explained all astronomical observations that could be made in his time with one exception: the precession of the perihelion of the orbit of the planet Mercury. Einstein's special theory of relativity later replaced Newton's theory. It successfully explained all the observations that Newton could explain, plus the anomaly. It also made completely new predictions, including the important relationship $E = mc^2$.

26. The best-known recent effort along these lines is *Evolution as Entropy: Towards a Unified Theory of Biology*, by Daniel Brooks and E. O. Wiley [1986].

27. Incremental Pareto efficiency is closely related to bounded rationality (BR)—sometimes called satisficing—as defined by Herbert Simon [Simon, 1955, 1959, 1982].

28. Mirowski [1984], has pointed out that the analogy between physical science and moral science was very clear to the early neo-classicists and cites a variety of evidence supporting this assertion; Stanley Jevons (1905) for instance, stated, "The notion of value is to our science what that of energy is to mechanics", although Mirowski contends that Jevons misunderstood the physics. Walras wrote in 1862 of his intention to try to create "a science of economic forces analogous to the science of astronomical forces...the analogy is complete and striking." Later he wrote an article entitled "Économique det Mécanique" full of analogies (some erroneous) between mechanics and economics [ibid]. Fisher included a table showing the concordance between physics and economic variables in his 1926 book [see Mirowski, 1989].

29. In the real world an EU might also be an individual family, government agency, co-op, foundation, commune, criminal organization, church, or some other entity.

30. In general, services can also be exchanged for goods (or other services). At this stage, however, transactions are restricted to those involving only tangible goods or labor.

The extension to other services will be considered later.

31. The general bookkeeping equation for any conserved quantity X (such as a physical commodity) is

$$\frac{dX}{dt} = F + G,$$

where F is a generalized current (inflow) that crosses the boundary of the economic unit, and G is a generalized source (or, with a negative sign, a sink). By assumption, X can be any commodity that can be bought, sold, produced, or consumed, including money or shares of stock.

32. The usual statement of Walras's law is that the vector product of market prices P and excess demand E is always equal to zero in a pure exchange economy, even when equilibrium has not been established.

33. Hurwicz and Richter showed that Ville's axiom suffices to prove the integrability of an expression corresponding to ΔW for an economic system (ES), rather than an individual economic unit (EU), provided a unique price vector $p(x)$ exists for each bundle of goods x. However, such a price vector need not exist. For this reason, the total utility function for an ES is undefined.

34. Note that information H can be defined directly in terms of entropy and interpreted as "distance from thermodynamic equilibrium"

$$H_i = S_{i_0} - S_i,$$

where S_{i_0} is the entropy in the equilibrium state and S_i is the entropy in the actual state. (The symbol S is normally used for entropy in thermodynamics, and must not be confused with the usage in equation (7.2). The natural unit for H is the "bit."

35. It is interesting to note that in thermodynamics a similar situation exists. There the temperature scale is arbitrary, and one is free to choose two points on the scale for convenience. (In the case of the Celsius scale, the zero point is set by the freezing point of water and the 100 point is set by the boiling point of water).

36. See, for instance Szilard [1929]; Brillouin [1953]; Jaynes [1957a,b].

REFERENCES

Abramovitz, Moses, "Resources and Output Trends in the United States Since 1870," *American Economic Review* **46**, May 1956.

Allen, P. M., "Dynamic Models of Evolving Systems," *System Dynamics Review* **4**, Summer 1988 :109-130.

Allen, P. M., G. Engelen and M. Sanglier, "Self-Organizing Dynamic Models of Human Systems," in: Frehland (ed.), *Synergetics: From Microscopic to Macroscopic Order*, Springer-Verlag, New York, 1984.

Allen, P. M. and M. Sanglier, "Urban Evolution, Self-Organization, and Decision-Making," *Environment and Planning* **13**, 1981 :167-183.

Anderson, Philip W., Kenneth J. Arrow and David Pines (eds.), *The Economy As an Evolving Complex System* [Series: Santa Fe Institute Studies in the Sciences of Complexity **5**], Addison-Wesley, New York, 1988.

Arrow, Kenneth J., "The Economic Implications of Learning by Doing," *Review of Economic Studies* **29**, 1962 :155-173.

Arthur, W. Brian, *Competing Technologies and Lock-In by Historical Small Events: The Dynamics of*

Allocation Under Increasing Returns, Research Paper (43), Committee for Economic Policy Research, Stanford University, Palo Alto, Calif., 1983.

Arthur, N. Brian, "Self-Reinforcing Mechanisms in Economics," in: Anderson, Arrow and Pines (eds.), *The Economy As an Evolving Complex System* :9-32 [Series: Santa Fe Institute Studies in the of Complexity 5], Addison-Wesley, Redwood City, Calif., 1988.

Arthur, W. Brian *et al.* "Path-dependent Processes and the Emergence of Macro-structure," *European Journal of Operations Research* **30**, 1987 :294-303.

Arthur, W. Brian, Yu. M. Ermoliev and Yu M. Kaniovski, *Non-Linear Urn Processes: Asymptotic Behavior and Applications*, Working Paper (WP-87-85), International Institute for Applied Systems Analysis, Laxenburg, Austria, September 1987b.

Ayres, Robert U., "Forecasting by the Envelope Curve Technique," in: J. Bright (ed.), *Technological Forecasting for Business and Government*, Prentice-Hall, New York, 1968.

Ayres, Robert U., "Technological Forecasting and Long-Range Planning, McGraw-Hill, New York, 1969.

Ayres, Robert U., *Manufacturing and Human Labor As Information Processes*, Research Report (RR-87-19), International Institute for Applied System Analysis, Laxenburg, Austria, July 1987a.

Ayres, Robert U., *Optimal Growth Paths with Exhaustible Resources: An Information-Based Model*, Research Report (RR-87-11), International Institute for Applied Systems Analysis, Laxenburg, Austria, July 1987b.

Ayres, Robert U., *The Industry-Technology Life Cycle: An Integrating Meta-Model?* Research Report (RR-87-3), International Institute for Applied Systems Analysis, Laxenburg, Austria, March 1987c.

Ayres, Robert U., "Complexity, Reliability and Design: Manufacturing Implications," *Manufacturing Review* **1**(1), March 1988a :26-35.

Ayres, Robert U., "Self Organization in Biology and Economics," *International Journal on the Unity of the Sciences* **1**(3), Fall 1988b. [Also IIASA Research Report #RR-88-1, 1988].

Ayres, Robert U. and Ike Ezekoye, "Competition and Complementarity in Diffusion: The Case of Octane," *Journal of Technological Forecasting and Social Change* **39**, 1991 :145-158.

Ayres, Robert U. and Adele Shapanka, "The Use of Explicit Technological Forecasts in Long-Range Input-Output Models," *Journal of Technological Forecasting and Social Change*, 1976 [Also published in H. S. Linstone and D. Sahal (eds.) Technological Substitution: Forecasting Techniques and Applications, Elsevier, New York, 1976.]

Barnett, William A., J. Geweke, and K Shell (eds.), *Economic Complexity: Chaos, Sunspots, Bubbles and Non-Linearity*, Cambridge University Press, Cambridge, England, 1986.

Barnett, William A., Ernst R. Berndt and Halbert White (eds.), *Dynamic Econometric Modeling: Proceeding of the Third International Symposium in Economic Theory and Econometrics*, Cambridge University Press, New York, 1986.

Bass, Frank M., "A New Product Growth Model for Consumer Durables," *Management Science* **15**, January 1969 :215-227.

Baumol, W. and J. Benhabib, *Chaos: Significance, Mechanism and Economic Applications*, RR (87-16), C. V. Starr Center for Applied Economics, New York University, City, N.Y., 1987.

Baumol, W. and R. Quandt, "Chaos Models and Their Implications for Forecasting," *Eastern Economic Journal* **11**, 1985 :3-15.

Benhabib, J. and K. Nishimura, "The Hopf Bifurcation and the Existence and Stability of Closed Orbits in Multi-sector Models of Optimal Economic Growth," *Journal of Economic Theory* **21**, 1979 :421-444.

Benhabib, J. and K. Nishimura, "Competitive Equilibrium Cycles," *Journal of Economic Theory* **35**, 1985 :284-306.

Bhaduri, A. and D. J. Harris, "The Complex Dynamics of the Simple Ricardian System," *Quarterly Journal of Economics* **102**, 1987 :893-902.

Blackman, A. Wade, "A Mathematical Model for Trend Forecasts," *Journal of Technological Forecasting and Social Change* **3**(3), 1972.

Blackman, A. Wade, "The Market Dynamics of Technological Substitutions," *Journal of Technological Forecasting and Social Change* **6**, February 1974 :41-63.

Boldrin, Michele, "Persistent Oscillations and Chaos in Dynamic Economic Models: Notes for a Survey," in: Anderson, Arrow and Pines (eds.), *The Economy As an Evolving Complex System*, Chapter 2 :49-75 [Series: Santa Fe Institute Studies in the Sciences of Complexity 5], Addison-Wesley, Redwood City, Calif., 1988.

Brillouin, Leon, "Negentropy Principle of Information," *Journal of Applied Physics* **24**(9), 1953 :1152-1163.

Brock, William A., "Distinguished Random and Deterministic Systems," *Journal of Economic Theory* **40**(1), 1986 :168-195.

Brock, William A., "Non-Linearity and Complex Dynamics in Economics and Finance," in: Anderson, Arrow and Pines (eds.), *The Economy As an Evolving Complex System* :77-98 [Series: Santa Fe Institute Studies in the Sciences of Complexity **5**], Addison-Wesley, Redwood City, Calif., 1988.

Bródy, Andras, Katalin Martinàs and Konstantin Sajó, "Essay on Macro-economics," *Acta Oec.* **36**, 1985 :305.

Brooks, Daniel R. and E. O. Wiley, *Evolution As Entropy: Towards a Unified Theory of Biology*, University of Chicago Press, Chicago, 1986.

Carathéodory, Constantin, "Untersuchungen über die Grundlagen der Thermodynamik," *Mathematical Annals* **67**, 1909 :355-386.

Cohen, Wesley M. and Daniel A. Levinthal, *Learning to Learn: An Economic Model of Firm Investment in the Capacity to Learn*, mimeo, Carnegie-Mellon University, Pittsburgh, Penn., September 1988.

Cohen, Wesley M. and Daniel A. Levinthal, "Innovation and Learning: The Two Faces of RandD," *Economics Journal* **99**, September 1989 :569-596.

Conlisk, John, *Optimization, Adaptation and Random Innovations in a Model of Economic Growth*, International Symposium on Evolutionary Dynamics and Non-Linear Economics, University of Texas, Austin, Tex., April 16–19, 1989. [Proceedings in press]

Cowan, Robin, *Nuclear Power Reactors: A Study in Technological Lock In*, Evolutionary Dynamics and Non-Linear Economics, Massachusetts Institute of Technology, Cambridge, Mass., April 16–18, 1989.

Crutchfield, J. *et al.* "Chaos," *Scientific American* **255**, December 1986 :46-57.

Cyert, Richard and James March, *A Behavioral Theory of the Firm*, Prentice-Hall, Englewood Cliffs, N.J., 1963.

Dasgupta, Partha and Joseph Stiglitz, "Uncertainty, Industrial Structure and the Speed of RandD," *Bell Journal of Economics* **11**, Spring 1980 :1-28.

David, Paul A., *Technical Choice, Innovation and Economic Growth*, Cambridge University Press, London, 1975.

David, Paul A., "Clio and the Economics of QWERTY," *American Economic Review (Papers and Proceedings)* **75**, 1985 :332-337.

David, Paul A., *Path-dependence: Putting the Past into the Future of Economics*, Technical Report (533), Institute for Mathematical Studies in the Social Sciences, Stanford, Calif., November 1988*b*.

David, Paul A., *The Future of Path-dependent Equilibrium Economics*, Technical Paper (155), Center for Economic Policy Research, Stanford, Calif., August 1988*a*.

Day, Richard H., "Dynamic Systems and Epochal Change," in: Sablov (ed.), *Simulations in Archeology*, University of New Mexico Press, City, N. Mex., 1981.

Day, Richard H., "Irregular Growth Cycles," *American Economic Review* **72**(3), 1982 :406-414.

Day, Richard H., "The Emergence of Chaos from Classical Economic Growth," *Quarterly Journal of Economics*, May 1983 :210-213.

Day, Richard H., "Disequilibrium Economic Dynamics," in: Day and Eliasson (eds.), *The Dynamics of Market Economies*, North Holland, Amsterdam, 1986.

Day, Richard H. and Wayne Shafer, "Keynesian Chaos," *Journal of Macro-economics* **7**, 1985 :277-295.

Day, Richard H. and Wayne Shafer, "Ergodic Fluctuations in Deterministic Economic Models," *International Journal on Behavior and Organization* **8**, 1987 :333-361.

Debreu, Gerard, *Theory of Value*, John Wiley and Sons, New York, 1959.

Deneckere, R. and S. Pelikan, "Competitive Chaos," *Journal of Economic Theory* **40**, 1986 :13-25.

Dosi, Giovanni, "Technological Paradigms and Technological Trajectories," *Research Policy* **11**, 1982 :147ff.

Dosi, Giovanni, Luigi Orsenigo and Gerald Silverberg, *Innovation, Diversity and Diffusion: A Self-Organization Model*, SPRU, University of Sussex, Sussex, England, June 1986. [Second draft]

Easingwood, C. J., Vijay Mahajan and E. Muller, "A Non-symmetric Responding Logistic Model for Technological Substitution," *Journal of Technological Forecasting and Social Change* **20**, 1983 :199-213.

Ebeling, Werner and Rainer Feistel, *Physik der Selbstorganization und Evolution*, Akademie-Verlag, Berlin, 1982.

Evans, Robert B., *A Proof That Essergy Is the Only Consistent Measure of Potential Work for Chemical Systems*. Ph.D. Thesis [Dartmouth College, Hanover, N.H.], 1969.

Faber, Malte and John L. R. Proops, "Time Irreversibilities in Economics," in: Faber (ed.), *Studies in Austrian Capital Theory, Investment and Time: Lecture Notes in Economics and Mathematical Systems*, Springer-Verlag, Berlin, 1986.

Fabricant, Solomon, "Economic Progress and Economic Change," in: *34th Annual Report*, National Bureau of Economic Research, New York, 1954.

Freeman, Christopher, *The Economics of Industrial Innovation*, MIT Press, Cambridge, Mass., 1982. 2nd edition.

Freeman, Christopher, *The Long Wave and the World Economy*, Butterworths, Boston, 1983.

Freeman, Christopher, John Clark, John Soete and Luc Soete, *Unemployment and Technical Innovation: A Study of Long Waves and Economic Development*, Francis Pinter, London, 1982.

Georgescu-Roegen, Nicholas, *The Entropy Law and the Economic Process*, Harvard University Press, Cambridge, Mass., 1971.

Gilfillan, S. Colum, "The Prediction of Inventions," in: Ogburn *et al.* (eds.), *Technological Trends and National Policy, Including the Social Implications of New Inventions*, National Research Council/National Academy of Sciences National Resources Committee, Washington, D.C., 1937.

Glansdorff, P. and Ilya Prigogine, *Thermodynamic Theory of Structure, Stability and Fluctuations*, Wiley-Interscience, New York, 1971.

Goodwin, Richard M., "The Non-Linear Accelerator and the Persistence of Business Cycles," *Econometrica* **19**, 1951 :1-17.

Goodwin, Richard H., "A Growth Cycle," in: Feinstein (ed.), *Socialism, Capitalism and Economic Growth*, Cambridge University Press, Cambridge, England, 1967.

Goodwin, Richard M., *Essays in Economic Dynamics*, Macmillan Press Ltd., London, 1982.

Goodwin, Richard M., *A Growth Cycle, Socialism, Capitalism and Economic Growth*, Cambridge University Press, London, 1987.

Goodwin, Richard M., *Economic Evolution and the Evolution of Economics*, International Symposium on Evolutionary Dynamics and Non-Linear Economics, University of Texas, Austin, Tex., April 1989.

Griliches, Zvi, "Hybrid Corn: An Exploration of the Economics of Technological Change," *Econometrica* **25**(4), October 1961.

Hahn, F. and R. C. O. Matthews, "The Theory of Economic Growth," *Surveys of Economic Theory* **2**, 1967 :1-124.

Heap, R. D., *Heat Pumps*, Publ., City, State, 1983.

Heertje, A., "Can We Explain Technical Change," in: Lamberton *et al.* (eds.), *The Trouble with Technology: Explorations in the Process of Technological Change*, St. Martin's Press, New York, 1983.

Hofstadter, D. R., "Strange Attractors: Mathematical Patterns Delicately Poised Between Order and Chaos," *Scientific American*, November 1981 :22-43.

Hurwicz, Leonid "An Integrability Condition with Applicatins to Utility Theory and Thermodynamics," *Journal of Mathematical Economics* **6**, 1979 :7-14.

Jaynes, Edwin T., "Information Theory and Statistical Mechanics I," *Physical Review* **106**, 1957*a* :620.

Jaynes, Edwin T., "Information Theory and Statistical Mechanics II," *Physical Review* **108**, 1957*b* :171.

Kamien, Morton and Nancy Schwartz, *Market Structure and Innovation* [Series: Cambridge Surveys of Economic Literature], Cambridge University Press, Cambridge, England, 1982.

Kendrick, John W., "Productivity Trends: Capital and Labor," *Review of Economics and Statistics*, August 1956.

Kendrick, John W., *Productivity Trends in the United States*, Technical Report, National Bureau of Economic Research, Princeton University Press, Princeton, N.J., 1961.

Kwasnicka, Halina and Witold Kwasnicki, "Diversity and Development: Tempo and Mode of Evolutionary Processes," *Journal of Technological Forecasting and Social Change* **30**, 1986 :223-243.

Kwasnicka, Halina and Witold Kwasnicki, *Multi-Unit Firms in an Evolutionary Model of a Single Industry Development*, Evolutionary Dynamics and Non-Linear Economics, Massachusetts Institute of Technology, Cambridge, Mass., April 16–18, 1989.

Levin, Richard C., "Technical Change, Barriers to Entry and Market Structure," *Econometrica* **45**, 1978 :347-361.

Mahajan, Vijay and R. A. Peterson, *Models for Innovation Diffusion*, Sage Publications, Beverly Hills, Calif., 1985.

Mansfield, Edwin, "Technical Change and the Rate of Imitation," *Econometrica* **29**(4), October 1961 :741-766.

March, J. G. and Herbert A. Simon, *Organizations*, John Wiley and Sons, New York, 1958.

Martinàs, Katalin, *About Irreversibility in Micro-economics*, Research Report (AHFT-89-1), Department of Low Temperature Physics, Roland Eotvos University, Budapest, March 1989.

May, R. M., "Simple Mathematical Models with Very Complicated Dynamics," *Nature* **26**, June 1975 :459-67.

Medio, A., "Oscillations in Optimal Growth Models," *Journal of Economic Dynamics and Control* **11** 1987 :201-206.

Mensch, Gerhard, *Das technologische Patt: Innovation ueberwinden die Depression* [Series: Stalemate in Technology: Innovations Overcome the Depression], Umschau, Frankfurt, 1975. [English Translation: Ballinger, Cambridge, Mass., 1979]

Metcalfe, J. S., "Impulse and Diffusion in the Study of Technical Change," *Futures* **13**(5), 1981 :347-359.

Mirowski, Philip, *Physics and the Marginalist Revolution*, Publ, City, State, 1984.

Mirowski, Philip, "On Hollander's 'Substantive Indentity' of Classical and Neo-Classical Economics: A Reply," *Cambridge Journal of Economics* **13**, 1989 :471-477.

Nelson, Richard R. and Sydney G. Winter, "In Search of a Useful Theory of Innovation," *Research Policy* **6**(1), January 1977.

Nelson, Richard R. and Sydney G. Winter, "Forces Generating and Limiting Concentration Under Schumpeterian Competition," *Bell Journal of Economics* **9**, 1978 :524-548.

Nelson, Richard R. and Sydney G. Winter, *An Evolutionary Theory of Economic Change*, Harvard University Press, Cambridge, Mass., 1982.

Nicolis, Gregoire and Ilya Prigogine, *Self-Organization in Non-Equilibrium Systems*, Wiley-Interscience, New York, 1977.

Ogburn, William F. (chairman), *Technological Trends and National Policy*, Report, National Research Council, Washington, D.C., 1937.

Ogburn, William F. and Dorothy Thomas, "Are Inventions Inevitable? A Note on Social Evolution," *Political Science Quarterly* **37**, March 1922 :83-98.

Perez-Perez, Carlotta, "Towards a Comprehensive Theory of Long Waves," in: Bianchi *et al.* (eds.), *Long Waves, Depression and Innovation*, 1983. [Proceedings of Siena Conference]

Peschel, Manfred and Werner Mende, *The Predator-Prey Model: Do We Live in a Volterra World?* Akademie-Verlag, Berlin, 1986.

Prigogine, Ilya, Gregoire Nicolis and A. Babloyantz, "Thermodynamics of Evolution," *Physics Today* **23**(11/12), November/December 1972 :23-28(Nov.) and 38-44(Dec.).

Prigogine, Ilya and I. Stengers, *Order Out of Chaos: Man's New Dialogue with Nature*, Bantam Books, London, 1984.

Radzicki, Michael J. *Institutional Dynamics, Deterministic Chaos and Self-Organizing Systems*, International Symposium on Evolutionary Dynamics and Non-Linear Economics, University of Texas, Austin, Tex., April 16–19, 1989. [Proceedings in press]

Rasmussen, S., Erik Mosekilde and J. D. Sterman, "Bifurcations and Chaotic Behavior in a Simple Model of the Economic Long Wave," *System Dynamics Review* **1**, Summer 1985 :92-110.

Richardson, G. P., "Loop Dominance, Loop Polarity and the Concept of Dominant Polarity," in: *Proceedings of the 1984 International Conference of the System Dynamics Society* :156-174, Oslo, Norway, 1984.

Rosenberg, Nathan, "The Direction of Technological Change: Inducement Mechanisms and Focusing Devices," *Economic Development and Cultural Change* **18**, 1969 :1-24.

Rosenberg, Nathan, *Technology and American Economic Growth*, Harper and Row, New York, 1972.

Rosenberg, Nathan, "Science, Invention and Economic Growth," *Economic Journal* **84**, 1974 :90-108.

Rosenberg, Nathan, *Perspectives in Technology*, Cambridge University Press, New York, 1976.

Rosenberg, Nathan, *Inside the Black Box: Technology and Economics*, Cambridge University Press, New York, 1982.

Sahal, Devendra, *Patterns of Technological Innovation*, Addison-Wesley, Reading, Mass., 1981.

Schumpeter, Joseph A., *Theorie der wirtschaftlichen Entwicklungen*, Duncker and Humboldt, Leipzig, 1912.

Schumpeter, Joseph A., *Capitalism, Socialism and Democracy*, Harper and Row, New York, 1943. [Reprinted Harper Colophon Edition (1975)]

Sharif, M. N. and C. Kabir, "A Generalized Model for Forecasting Technological Substition," *Journal of Technological Forecasting and Social Change* 8, 1976 :353-364.

Silverberg, Gerald, "Modelling Economic Dynamics and Technical Change: Mathematical Approaches to Self-Organization and Evolution," in: Dosi *et al.* (eds.), *Technical Change and Economic Theory* :531-557, Pinter Publishers, London, 1988.

Silverberg, Gerald, *Selection and Diffusion of Technologies As an Evolutionary Process: Theory and Historical Reality*, International Symposium on Evolutionary Dynamics and Non-Linear Economics, University of Texas, Austin, Tex., April 1989.

Simon, Herbert A., "A Behavioral Model of Rational Choice," *Quarterly Journal of Economics* 69, 1955 :99-118. [Reprinted in *Models of Man*, John Wiley and Sons, New York, 1957]

Simon, Herbert A., "Theories of Decision-Making in Economics," *American Economic Review* 49, 1959 :253-283.

Simon, Herbert A., *Administrative Behavior*, Free Press, New York, 1965. 2nd edition.

Simon, Herbert A., *Models of Bounded Rationality*, MIT Press, Cambridge, Mass., 1982.

Simonovits, A., "Buffer Stocks and Naive Expectations in a Non-Walrasian Dynamic Macro-model: Stability, Cyclicity and Chaos," *Scandinavian Journal of Economics* 84(4), 1982 :571-581.

Skiadas, Christos, "Two Generalized Rational Models for Forecasting Innovation Diffusion," *Journal of Technological Forecasting and Social Change* 27, 1985 :39-61.

Solow, Robert M., "A Contribution to the Theory of Economic Growth," *Quarterly Journal of Economics* 70, 1956 :65-94.

Solow, Robert M., "Technical Change and the Aggregate Production Function," *Review of Economics and Statistics*, August 1957.

Stoneman, Paul, *The Economic Analysis of Technological Change*, Oxford University Press, London, 1983.

Stutzer, M., "Chaotic Dynamics and Bifurcation in a Macro-Model," *Journal of Economic Dynamics and Control* 1, 1980 :377-393.

Szilard, Leo, "Über die Entropieverminderung in einem thermodynamischen System bei Eingriffen intelligenter Wesen," *Zeitschrift für Physik* 53, 1929 :840-856.

Thirtle, Colin G. and Vernon W. Ruttan, *The Role of Demand and Supply in the Generation and Diffusion of Technical Change*, Publ., City, State, 1986.

Thom, R., *Structural Stability and Morphogenesis: General Theory of Models*, Publ., City, State, 1972.

Torre, V., "Existence of Limit Cycles and Control in Complete Keynesian Systems by Theory of Bifurcations," *Econometrica* 45, 1977 :1457-1466.

Tribus, Myron and Edward C. McIrvine, "Energy and Information," *Scientific American* 225(3), September 1971 :179-188.

Usher, Albert Payson, *A History of Mechanical Inventions*, Harvard University Press, Cambridge, Mass., 1929.

Van Wyk, R. J., "The Notion of Technological Limits," *Futures*, June 1985.

Ville, Jean, "The Existence Conditions of a Total Utility Function," *Review of Economic Studies* 19, 1951 :123-128.

Winter, Sydney G., "Economic Natural Selection and the Theory of the Firm," *Yale Economic Essays* 4, 1964 :225-272.

Chapter 8

The Economy as a Self-Organizing Information-Processing System[1]

8.1. THE ANALOGY WITH LIVING SYSTEMS

An economic system is, in three important respects, a kind of living system. First, an economic system is similar to biological systems (and some non-living physical and chemical systems) in exhibiting *self-organization* [Ayres, 1987]. In simple terms, a self-organizing system has a characteristic internal pattern, or structure, that tends to preserve or reproduce itself over time in spite of external perturbations or disturbances. In biology such patterns reveal themselves both in *form* (morphology) and in *function.* The two are obviously related: Form limits function and is determined by it. This is true at the cellular level, the organism level, and the community level. For an economic system, the analogs of cells are individuals. The analogs of specialized organs are firms or industries; and the products of industry are analogous to the minerals, amino acids, sugars, and lipids, as well as the more specialized vitamins that circulate through the organism and are metabolized for energy and/or incorporated in biomass. Raw materials correspond to food for the living organism, and so forth.

The second point of similarity between an economic system and a living system is the need for continuous inputs of useful energy available for metabolic purposes. Actually, "energy" *per se* is conserved in every physical, biological, or industrial (economic) process: It is neither gained nor lost. (This is the meaning of energy conservation.) However, the useful or available component of energy is not reusable.[1] It is used up (or dissipated) by living systems and converted to entropy. An economic system is evidently dissipative. Resource inputs to an economic system include fuels, food, and fiber crops, all of which embody free energy or available work.[2] Other inputs include metal ores, building materials and the like, which embody a kind of order left by nature and often destroyed by human use. Outputs of the economic system, on the other hand, are final goods. They are ultimately discarded or, in rare cases, recycled. Wastes that are not recycled also disappear from the economic system, although they do not disappear in reality and may continue to cause harm, either directly to humans or indirectly by disturbing the ecosystem [Ayres and Kneese, 1969]. Such harm generates costs, of course, but those whose activities caused the harm do not bear the resulting costs. To rectify this, environmentalists have set forth the so-called *polluter-pays principle,* although this principle is enunciated in the abstract far more often than it is obeyed in the concrete.

Free energy is expended and lost at every stage—extraction, refining, manufacturing, construction, and even final consumption [Ayres, 1978]. Though total energy is always conserved, free energy is not. Fuels are rich in free energy, while outputs of industrial processes or energy-conversion devices mostly take the form of low-temperature waste heat, oxidation products, or degraded materials. Thus, an economic system, in reality, is absolutely dependent on a continuing flow of free energy from the environment. In preindustrial times, the sun provided almost all free energy in the form of wood, food crops, animals, water power, or wind power. Today, the major source of energy, by far, is fossil fuels: petroleum, natural gas, and coal from the earth's crust. These exhaustible resources will of course eventually have to be replaced.

The third point of strong similarity between economic systems and living systems is their ability to grow and evolve, both in magnitude and functionality. As noted in Chapters 1 and 5, biological organisms exhibit two distinct types of developmental change. Darwinian evolution (of species) is a non-deterministic process of qualitative change driven by random mutations (differentiations) and governed by both intra- and interspecies competition. The technical term for this process is *phylogeny.* But during the early stages of fetal development and growth of each individual, the evolutionary process repeats, as it were, like a tape-recorder set to fast forward. This is *ontogeny.* As we learned in high school biology: "Ontogeny recapitulates phylogeny." Fetal development is essentially deterministic. It has few random elements, being governed mainly by genetic information already stored in the DNA of the fertilized egg.

As was pointed out in Chapters 6 and 7, an economic system exhibits both types of change. *Ontogenic* change is exemplified by the life cycle of a product, or its industrial counterpart. While not as deterministically pre-programmed as fetal development in biology, most new products do seem to go through a sequence of developmental stages closely analogous to infancy, adolescence, maturity and senescence. Occasionally, however, the system undergoes discontinuous and unpredictable change normally associated with radical (Schumpeterian) innovation. This sort of technological (hence, economic) evolution is essentially *phylogenic.*

One can see right away that, in an economy, competition between firms and products (or processing technologies) closely resembles biological competition among individuals and species.[4] It is interesting to examine the analogy more closely in terms of other key biological phenomena: *1.* metabolism and homeostasis, *2.* basic growth (ontogeny), *3.* reproduction and population differentiation, and *4.* species differentiation.

For economists, the standard unit of analysis is *the firm*: a hypothetical organization devoted strictly to production and profit-making. Here the biological notions of **metabolism** and **homeostasis** are not particularly relevant to standard economic theory, which is rather abstract. In the real world, each firm engaged in production must purchase material and energy, as well as labor, and these inputs are roughly analogous to the food, water, and air ingested by a living organism.

The biological analogy is less useful with regard to output, however. The biological organism uses inputs of material and energy for three distinct purposes:

Metabolism in steady state (including excreta).

Growth (the addition of biomass).

Production of scents, flowers, fruit, nuts and/or seeds to attract biological partners and ensure reproduction (reproduction).

In contrast, the firm uses materials and energy for only two purposes:

To provide internal services to operate facilities (metabolism).

To provide external products and/or services that can be sold profitably (production).

The operation of facilities to provide internal (overhead) services closely resembles steady-state biological metabolism. On the other hand, a real firm produces neither physical copies of itself nor increments to its biomass. Here the analogy breaks down. It holds only if the firm is viewed in ultra-simplified monetary terms: its biomass corresponding to *capital*, its metabolic requirements corresponding to *fixed costs*, its inputs corresponding to *revenues*, and biomass growth being the result of accumulated *profits*. Looking at it another way the firm, unlike an organism, does not (normally) produce its own capital goods (biomass).

In an abstract firm that consists only of capital, costs, revenues, and profits, homeostasis is a meaningless concept. In real organizations run by humans, of course, homeostasis is an important phenomenon. For instance, a firm can be viewed as a culture in miniature, with a variety of mechanisms for protecting its identity and integrity. (The language of culture is gradually spreading in formal management literature: Terms like *company culture* are now as commonplace as terms like *company man* once were. The latter obviously presupposes the former.)

The economic analog of asexual reproduction is the *spinoff*, and the analog of sexual reproduction is the *joint venture* by two or more established firms. A spinoff is a startup by a person (or a group that breaks away intact) from an established firm.[5] Of course many, if not most, new starts have no simple or well-defined ancestry, so again the analogy with biological reproduction is weak. In economics and, especially, technology—unlike biology—one can make a strong case for spontaneous creation. Because firms do not reproduce themselves in the biological sense, one cannot distinguish species differentiation from population differentiation. Differentiation does occur, however, both through the adaptation of existing firms and the creation of new ones by any one of several means.

From the perspective of this book, the evolutionary process for both biological systems and the economy means the accumulation of useful information in reproducible form. Here the word *useful* is intended to distinguish certain kinds of genetic information stored in DNA, or in the brain cells of a higher animal, from *information* in the more general (Shannonian) sense of reduced uncertainty, or inherent distinguishability. Useful information in the biological context is information that contributes to an organism's ability to survive and reproduce itself. Useful information for a firm is information that enables it to survive and grow. The two objectives are related but not identical.

It is tempting to equate selectively useful (SU) information with knowledge. For humans the two concepts are closely related. Unfortunately, the word *knowledge* has an unavoidable association with human mental processes and self-aware learning. The learning process, as such, should be distinguished from the knowledge that it conveys, however. In fact, SU-information can be accumulated just as well by unconscious processes (Darwinian evolution, for example) as by conscious ones. It can be embodied in macro-molecules of DNA or in unconscious behavior-patterns, as well as in gear shapes, blueprints, microprocessor chips, or chemical formulas. All knowledge is evidently information, but not all information is knowledge. Any sequence of a thousand distinguishable symbols say, letters of the alphabet, has a definite and computable information content (related to its probability of occurrence) regardless of whether the particular sequence conveys any useful knowledge.[6]

A major difference between biological and technological evolution is that the former is myopic and unconscious, while the latter is increasingly presbyopic and conscious. Mutations at the molecular level are essentially random. Inventions, in an economic-technological system, are anything but random. They are, in general, a conscious response to a perceived need or demand, and *ipso facto* presbyopic.

In evolution, the accumulation of useful information or knowledge (using the latter word in a generalized, non-anthropocentric sense), is distinct from physical growth. Yet quantitative growth and territorial expansion is not unimportant, since the expansion of the domain of living organisms throughout the oceans, and over virtually all land surfaces, has required a series of successful adaptations to different environments.

Similarly, the evolutionary growth of the human economic system can be characterized as increased output combined with an accumulation of technological knowledge (or simply technology). Again, qualitative and quantitative growth are related, although qualitative growth is much more important. Growth of output, as measured by GNP, has accompanied a series of successful technological innovations. Such innovations were first necessary to replace exhausted stocks of resources (such as coal in place of wood for charcoal, or petroleum in place of whale oil) and then to satisfy new, or unsuspected, human desires as primary human needs were met. Without a doubt, technological change, including both improvements and major innovations, is an essential prerequisite of long-term evolutionary economic growth.

Evidently, except for its non-local character, the real economic system looks very much like a self-organizing dissipative structure in Prigogine's sense (see Chapters 1 and 5): It is dependent on a continuous flow of free energy (the sun or fossil fuels), and it exhibits coherent, orderly behavior. Moreover, like living organisms, it embodies structural information as morphological differentiation and functional specialization. In economic terms, specialization and differentiation of form and function were first recognized as the "division of labor." Production itself is now highly differentiated into sectors, products, and services. Similarly, labor skills are increasingly subdivided into occupational classifications.

I would like to recapitulate a few of the main points of this book so far:

1. The economy is an evolving, self-organizing, dissipative system. Economic evolution resembles biological evolution (phylogeny) in that it is driven by radical innovations analogous to mutations.
2. Economic evolution differs sharply from biological evolution, insofar as innovations do not occur at random, nor are they exogenous. There is an element of guiding (albeit bounded) rationality.
3. Dynamic economic growth is driven by technological advances, which are, in turn, stimulated by economic forces. Such growth also results in continuous structural change in an economic system. For instance, the so-called Leontief input-output coefficients ratios between inputs and outputs for each sector do not (and cannot) remain constant in a growing economy driven by innovation or one dependent on an exhaustible resource.
4. The energy and materials feeding an economic system have shifted over the past two centuries from renewable to non-renewable sources.
5. It follows that a long-term sustainable survival path must sooner or later reverse this shift. This will be feasible only if human technological capabilities rise to levels much higher than current ones. Consistent with this need, the role of human knowledge-generating activity is rapidly growing in importance relative to other activities.

8.2. THE ECONOMY AS AN INFORMATION-TRANSFORMATION SYSTEM

An economy is a human construct that depends in a fundamental sense on the existence of external (extra-somatic) means of information storage and transfer.

An anthill, a hive of bees, or a colony of baboons exhibits social behavior but nothing even closely analogous to an economic system. Information, as defined earlier, is an attribute of any physical system *relative* to a defined reference system. In some cases the reference state is the environment. In other cases the reference state is randomness or chance. In any case, information is not an *absolute* attribute of a system in quite the same way as pressure, temperature, or chemical potential. Nevertheless, once the reference system has been defined, the information content of a subsystem exists independently of the presence or absence of any observer. It is but a short step to regard thermodynamic information as being embodied in material objects relative to the natural environment.[7]

The extension of information concepts beyond the realms of communication and electronic data processing and/or thermodynamics and statistical mechanics (i.e., the realm of phenomena that can be described in terms of large numbers of a few varieties of symbols or particles) requires a leap of faith. It is only a minor leap, perhaps, because information theory has already become deeply intertwined with computation theory and computer science. It is quite straightforward, at least in principle, to extend the notion of embodied information to computer programs and, thence, to physical designs. The concept of distinguishability is clearly generalizable to the realm of forms, shapes, and structures (see Chapter 2). Shapes, in turn, are constructed from geometric surfaces. Patterns carrying information can be embodied in surfaces or printed on them. Hereafter we term this **morphological information**.

A page of print (such as this one) carries morphological information embodied in the patterns of type. The usual confusion between *information* in the technical sense of distinguishability, and its more familiar definition as the carrier of meaning, can now be addressed by invoking the famous image of the room full of typewriting monkeys. The page typed by a monkey, full of random letters and punctuation, carries roughly as much information in the Shannonian sense as a page of this text or a list of telephone numbers. Information as such is not necessarily meaningful or useful.

Knowledge, however, which I equate here with useful information, can be conveyed only by words or symbols. The meaning attached to the symbols must be supplied exogenously by the reader of the page, or the receiver of a message. A given set of symbols (information) may or may not carry meaning to a particular observer. If one does not know a language, it is difficult to determine whether a sequence of sounds or graphic fragments constitutes a sequence of symbols with meaning, or whether it is simply *noise*. In practice, this can be done only with computer-aided statistical tests. Any real written or spoken language is characterized by a non-random distribution of symbols (and of symbolic elements, or codons) unique to that language. This general fact about languages permits the following test: If two independent strings of reasonable length exhibit the same non-uniform distribution symbol, one is dealing with a language. Otherwise, they are probably just noise.[8]

It is clear, following this line of reasoning, that a *smart* observer can extract more meaning or knowledge from a given amount of information (a string of symbols or codons) than a less-intelligent observer can. Thus the knowledge conveyed to an observer by a message (a sequence of codons) is not necessarily proportional to the information-content of the message itself. It may be much smaller, or even zero. Unlike information, knowledge exists only for particular (human) observers. On the other hand, once knowledge exists, it can be transmitted only in the form of coded symbolic information. Thus useful information, in the sense of information that facilitates economic activities and production, is a limited subset of all symbolic information.

There is, on reflection, still another kind of information to be considered, namely, *control information*. Its archetype is the sequences of signals sent from the brain, via the nervous system, to operate the muscles and thus perform actions.[10] But in this case the information needed to control a printing device, for instance, is proportional to the information embodied in what is printed. Control information does not, in essence, differ from morphological information. Of course, the actions of workers and the decisions of managers are based directly on control information processed by human senses and human brains (with help from computers). This is the kind of information-processing familiar to ergonomists, psychologists, and computer scientists.

To summarize, one can tentatively distinguish three kinds of information with different but related economic roles:

1. Thermodynamic information (associated with chemical composition).
2. Morphological information (associated with shape, form, and structure) and control information needed to guide motions or make decisions.
3. Symbolic information (associated with control processes and knowledge).

These three kinds of information can all be measured in terms of a common unit (**bits**). In general, however, they are not necessarily physically interchangeable or substitutable. Indeed, thermodynamic and morphological information pertain to different domains. The former refers to distinguishability in the atomic and molecular domains, while the latter measures it in the human domain. Moreover, as discussed previously (see Chapter 2, Table 2.3), the implied reference systems in the various cases differ.

As stated above, neither symbolic nor control information differs in any fundamental sense from morphological information. A kind of unity becomes evident when one reflects that shapes can be (and increasingly are) described either in abstract symbolic terms, or in terms of the information required to generate them. Thus, morphological information, control information, and symbolic information appear to be effectively equivalent.

8.3. INFORMATION TRANSFORMATION AND VALUE ADDED

Value added to products by manufacturing depends on functional capability added to crude materials. Functional capability arises, in turn, from physical properties associated with a specified physical and chemical composition, shape, and finish (an aspect of shape). Each of these must be held within given limits or tolerances. The required precision of compositional and dimensional specifications defines the minimum amount of information embodied in each component. Information is added when parts are assembled into subassemblies, machines, processing equipment, structures, and systems.

It is suggested, hereafter, that the so-called factor services of capital and labor comprise the information input to the production process, while the output or product is analogous to the information output, or *message*. The information ultimately embodied by labor and capital in the output (message) is far less than the information embodied in the inputs, including labor. Most of the input information is, in fact, wasted. An efficient production process or an efficient economic system must surely be one that wastes as little information as possible. A good design, on the other hand, should require as little information as possible to fulfill a given functional purpose, other factors remaining equal. The idea elaborated below is that the factor services concentrate information (or reduce entropy) in certain specified materials or products, even though the total amount of information in the environment as a whole is actually reduced (i.e., global entropy increases).

Georgescu-Roegen's characterization of the economic process as a "transformation of states of low entropy to states of high entropy"[10] is, so far as it goes, merely an implication of the second law of thermodynamics. As such, it is neither more nor less true of an economic system than of other natural processes, although Georgescu-Roegen does well to remind economists that an economic system is inherently dissipative. Based on the perspective outlined above, an economic system can be conceptualized as a transformation of materials with three major stages (Figure 8.1). In the first stage, raw materials are extracted from their environmental matrix, beneficiated, reduced from ores, refined and purified, and finally alloyed or chemically recombined into final materials of desired composition and physical microstructure. At each step within this process the physical composition is altered and thermodynamic information (negentropy) is added to the material (Figure 8.2).

In the second stage, finished materials are converted into products and structures. There is no further change to composition or microstructure in general, but materials are shaped and formed. Morphological information is now added to finished materials, step by step, as materials are formed into crude shapes (such as castings, extrusions, bars, rods, sheets, or pipe), then further shaped by cutting, drilling, milling, stamping, grinding, or forging. Finally, the components are assembled into recognizable subsystems, such as motors, window frames, or

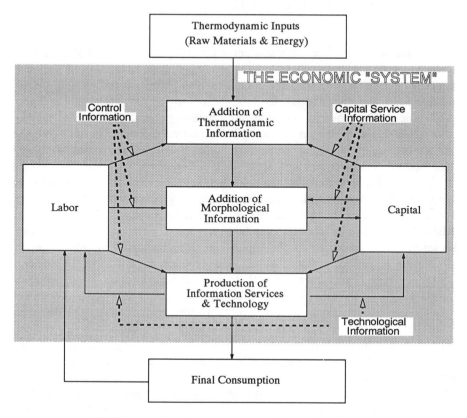

FIGURE 8.1. The Economy as an Information Processor.

TV tuners. Subsystems, in turn, are assembled into final products, structures, or networks, such as the highway system, the water system, the sewerage system, the electric power system, or the telephone system.

In the third stage of the transformational process, material products and systems are used to provide services, of several different kinds, to consumers. Fuels provide energy to both producers and final consumers, via prime movers. Capital goods serve producers. Consumer durables and non-durables serve final consumers. As services are delivered, wear and tear depreciation dissipates some of the thermodynamic information (in the case of food and fuel) and/or of the morphological information embodied in the service-providing object. At the same time, the service itself can be interpreted as an information input to the consumer. Machlup [1962, 1979, 1980] has characterized the sectors education, entertainment, finance, insurance, law, government regulation, etc. as the *knowledge industry*. The equivalence of these services to information transfer, in some form, is obvious. I would argue that such an equivalence also holds for other,

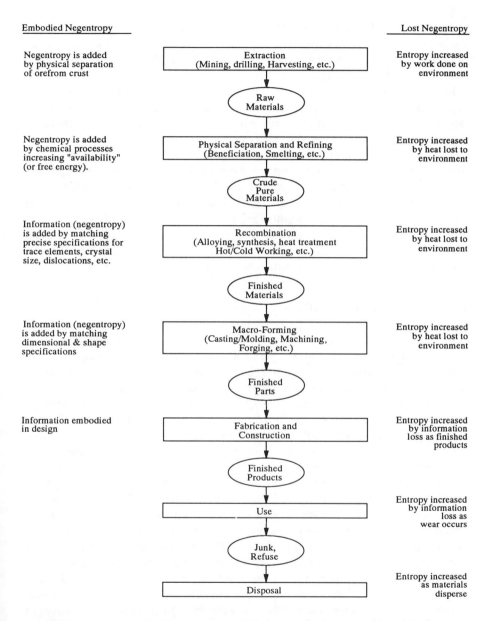

FIGURE 8.2. Representation of the Economic System as Materials-Process-Product Flow.

more physical services, such as distribution, housing, health, or public safety. All these services contribute, in the end, to facilitating the survival and propagation of our corporeal selves. But food intake, health, and safety are, in turn, reflected in feelings of well-being. These feelings are surely another form of embodied information. This follows from the fact that external stimuli (information) can cause people to forget hunger, pain, tiredness, or fear. Equally, external stimuli can arouse and magnify these feelings.

In addition to the three stages of the transformation of thermodynamic information extracted from the environment, an economic system has three feedback loops. These can be identified very easily as labor, capital, and technology (knowledge). I discuss these feedbacks below.

Theories attempting to relate the economic value of a product or service to a single factor (such as labor or energy) have a long and somewhat disreputable history in economics. It must be emphasized that no such notion is contemplated here. To be sure, I do argue that labor skills, capital, available energy, and technology are all more or less embodied or *condensed* forms of information. It does *not* follow that the market price of a given product or service is (or should be) simply or directly related to its quantitative information content. Thermodynamic and morphological information are not interchangeable.

8.4. THE ROLE OF LABOR

In the simplified model of the economy as an information-processing system, presented in Figures 8.1 and 8.2, the roles of human labor and capital were obscured to emphasize the importance of thermodynamic inputs. Actually, human labor, in the most primitive economic systems, was a significant source of thermodynamic input (mechanical work). However, that was long ago. By the 18th century, on the eve of the Industrial Revolution, human labor in Europe accounted for an insignificant fraction of thermodynamic inputs (power to do work or stationary heat-energy sources) available for economic purposes.[11]

The primary contribution of human labor in the 18th century was, as it still is, to control the *means of production*—animals, tools, machines, and equipment that actually did most of the work [Ayres, 1987]. Of course, the control function of human labor is not always easy to separate from the work function. These functions have separated slowly and incrementally over the centuries. An early step was the substitution of simple sails for galley slaves, and camels or mules for human porters. It was a significant step forward to use yoked oxen, waterwheels, or windmills to turn grindstones for making flour or to pump water by means of bucket chains. One of the great breakthroughs of the Industrial Revolution was to substitute water power (later steam power) for human arm and leg strength in operating cotton-spinning machines. Power-driven machine tools and textile looms followed only a few years later. Nevertheless, some jobs have been very difficult to mechanize effectively, and physical strength has been a requirement in many occupations until very recently. This has probably obscured the fact that, for the economic system as a whole, humans now provide a negligible share of the power and energy consumed.

An interesting question arises. As human workers have been required to use less and less physical strength, has the average worker's net output of *control information* increased correspondingly? Here it must be borne in mind that, for a manual worker, especially one who works from a standing position, the information output of the brain is needed mainly to exercise conscious control over the motions of his limbs, avoid obstructions, maintain balance, and so on.[12] Little excess capacity is available to control the motions that contribute directly to the accomplishment of the task. At the other extreme, a comfortably seated assembler or typist using an electric typewriter needs very little information for controlling body motions (except for the fingers) and can therefore concentrate efficiently on producing PC boards, manuscripts, or whatever.

Ergonomists since Frederick W. Taylor in the early years of this century have made substantial progress in finding the simplest ways to do various jobs, eliminating extraneous motions and distractions, and introducing training programs to increase worker output per unit of time. Thus, although there is no obvious way to quantify it, the net useful control-information output of human labor has undoubtedly risen significantly over the past two centuries. This is partly due to the increased education and training of workers. It is due in much greater degree to the extreme division of labor that characterizes modern manufacturing. The division of labor has made it possible to mechanically assist almost all the workers' motions reducing the information needed to control the workers' own actions. These changes have, together, permitted a Tayloristic rationalization of factory jobs to optimize procedures and eliminate unnecessary motions.

8.5. THE ROLE OF CAPITAL

As implied above, physical capital comprises the means of production itself. At the time of the French Revolution in 1789, the role of inanimate capital goods was overshadowed by that of animals. Today the reverse is true. But until relatively recently, the exclusive function of capital goods[13] was to convert thermodynamic information inputs—crude materials and fuels—into more-useful forms, to embody morphological information therein, and to deliver the products to consumers. Direct hands-on human labor was the only viable source of control information for the production process.

Historically, capital equipment began as extensions of, or substitutes for, the active parts of the human body.[15] Knives, chisels, and arrow points were extensions of teeth and fingernails; clubs, stones, and hammers extended the fist; the cooking fire was an extension of the digestive system;[15] fur coverings, clothing, floors, walls, and roofs are all extensions of the human skin. The horse and cartwheel were both extensions of the human leg; later, sailing ship, steamship, railroad, automobile, and airplane served the same basic purpose.

Domesticated animals were also capital,[16] notwithstanding the fact that many of them were used as sources of energy or power. In this sense, they were simply extensions of human muscles; this function has since been taken over largely by mechanical "prime movers," beginning with the waterwheel and windmill. Their modern replacements are the hydraulic or steam turbine, the internal combustion engine, and the electric motor. Mechanization simply applies mechanical power to the various extensions of the human body.

Capital also (and increasingly) includes extensions of the human senses, at least insofar as they are needed for purposes of production. The eyeglass, the telescope, and the microscope are obvious examples, but the entire telecommunications system is really a vast extension of human eyes and ears.[17] Early calculating machines, conceived by Pascal, Leibnitz, and especially Babbage, were finally implemented on a commercial scale by the end of the 19th century along with the punch-card sorting and tabulating machines developed by Herman Hollerith for the 1890 U.S. Census. These devices process symbolic information and, to that degree, can be regarded as extensions of the human brain.

With the advent of electronic computers since WWII, and especially since the 1960s, machines have begun processing information on a much larger scale. In recent years, indeed, it has been increasingly feasible for electronic computers in conjunction with instruments and sensory devices to generate information outputs suitable for machine control. The extent to which machines are able to control other machines will increase enormously as and when the control information can be modified to adapt to variable conditions using feedback from sensors in the tools.[18]

At the same time, electronic computers vastly increase the ability to interpret sensory inputs, whether from humans or from other machines. This combination of new capabilities permits electronic devices to take over many more machine supervisory functions. The revolution in electronic data processing has, on the one hand, produced an enormous increase in the amount of potentially useful information that can be delivered to human supervisors. On the other hand, it has already eliminated some lower-level data-processing jobs, like those of payroll clerks and actuaries.

To summarize: The economic role of physical capital is changing. On an aggregate level, it is probably fair to say that, from prehistoric times up to the present, physical capital amplified the physical capabilities of human workers, thus increasing the amount of productive work that could be accomplished by a single worker. It has taken over the basic metabolic function of converting thermodynamic information from the environment into useful work. In factories, the worker is now essentially a machine-controller who provides control information only. The conventional production-function model of neo-classical economics is fairly consistent with this conceptualization of the role of capital as a labor-amplifier, although it seems equally consistent with an alternative conceptualization of capital as an independent factor providing capital services.

On the other hand, it is no longer appropriate (if it ever was) to conceptualize capital as a simple labor substitute or multiplier. Since the invention of the wheel, capital equipment has also provided new kinds of final services. This tendency has sharply accelerated in the present century, with the development of the major infrastructure systems of our society—from water and sewerage systems to roads, electric power, and telecommunications. Computers can substantially increase the productivity of other forms of physical capital. But they have also vastly increased the effectiveness of the generation of new knowledge—especially, perhaps, technology. They have also reduced the cost and hence increased the accessibility of certain kinds of information to end-users. This function may be more significant than economists have as yet recognized.

Disregarding the increasing importance of capital as a direct provider of services to end-users, which has its own implications for the measurement of capital productivity,[19] in the production process itself, capital may be regarded either as a labor-amplifier or as a provider of capital services. From the first point of view, the output of capital is effectively a multiplier of the labor inputs (control information) to that process. The magnitude of the multiplier is a function of technology. From the second point of view, the output of capital is a flow of information defined as **capital services**.

The multiplier concept is anything but straightforward, inasmuch as most goods produced today could not be produced at all by unaided human workers except by first reproducing the basic tools and other capital goods required. This would require a recapitulation of the entire evolution of technology, at least in some abbreviated version.[20] On the other hand, to measure aggregated capital services in information terms seems scarcely less problematic. Economists have always had trouble measuring capital, stemming in part from fundamental conceptual problems.

One further comment seems appropriate at this point. It is important to distinguish the *effect* of capital from its *quantity*. The fact that economic growth over the past two centuries has largely coincided with a rapid buildup in the apparent quantity of invested capital as compared to labor (i.e., an increased capital-intensiveness) has induced a habit of confusing physical quantity with effect. Yet, there is growing evidence that the two are at best unrelated and may even be related inversely. Mini-mills are replacing integrated steel plants because recycling is more efficient than starting from ore. In the case of computers, smaller is inherently faster, more convenient, and ultimately cheaper. Micro-computers currently offer more than ten times the output per dollar of investment than mainframes. Bio-technology offers a similar perspective. Yesterday's iron lung is replaced by today's polio vaccine. Tomorrow's fertilizer factory may be a couple of genes implanted in the chromosomes of crop plants to give them the ability to fix their own nitrogen.

Physical capital is not only an intermediate in the economic sense, but may also be intermediate in the technological sense.[21] A truly advanced technology may require relatively little physical capital, at least as we recognize it.[22]

Technology can be defined (for our purposes) as knowledge applied to production. It can be assumed that the stock of technological knowledge grows as a function of formal R&D investment, on the one hand, and hands-on experience, or learning by doing (not to mention other kinds of learning), on the other. The effectiveness of the production activities at each stage of the value-added transformation increases if and when the new knowledge is applied. This occurs mainly in two ways: (1) by being embodied in labor (human capital) as training, skills, organizational improvements, and more-efficient procedures and (2) by being embodied in new equipment. More often than not, these two go together, inasmuch as new machines require new methods, and new methods usually require new skills and new organization.[23] This link is the third, and arguably the most important, of the feedback loops accounting for economic growth in terms of the general model (more accurately, a meta-model) of the economic system displayed in Figure 8.1.

8.6. THE ROLE OF MONEY

Money plays a central role in modern economic theory. It is the "universal medium of exchange" required for any market to operate efficiently. But its nature has changed radically since the days when money was merely a tangible commodity that had a "store of intrinsic value," as well as the properties of being compact, portable, easily recognizable, and hard to counterfeit. Precious metals have been in use for this purpose for at least three millennia.[24] Many other commodities, such as copper, salt, tobacco, furs, cotton cloth, amber, coral, elephant tusks, and even cowrie shells, have played the same role in other societies. All these forms of money were, of course, tangible.

The use of gold and silver coins for exchange was gradually undermined by the variety and unreliability of coins, which came about partly because of progress in metallurgy. Kings or counterfeiters could easily substitute cheaper metals (copper, tin, lead, zinc) for silver and gold, thus adulterating the metal content. In the 13th and 14th centuries, European merchants replaced coins in trade with *bullion*, a specified weight of metal. The adulteration persisted, however. To overcome it, European merchants in the 16th century adopted the Chinese system of deposit banking, wherein the metal itself was left in a bank. The first such bank in Europe seems to have been the Rialto bank in Venice. Others followed shortly in Amsterdam, Hamburg, Nürnberg, Augsburg, and other cities. Actual transactions were based on documents similar to checks or letters of credit.

Next, the letters of credit evolved into general-purpose, exchangeable certificates, entitling the bearer to a named equivalent in silver or gold. These certificates, of course, had to be issued by some trustworthy banking institution or by a government. Just as coinage became a government monopoly, so eventually did paper money. The first use of paper money required a major leap of faith that is difficult to imagine today. In effect, paper certificates had to be accepted *in lieu* of coins. People would not accept paper in exchange for goods unless they believed that others would accept it too. It was apparently the American colonists who invented fractional reserve banking and first issued paper money on a large scale. They did so originally because trade was stifled by lack of liquidity, i.e., a shortage of coins. The colonists lacked sufficient coinage because they had been forbidden by the British to mint their own. They could, however, print money based on coins or precious metals held in a central reserve. This innovation was quickly followed by abuse; under pressure of wartime the various colonies gave in to the temptation to print money not backed up by gold or silver. To cure the problems of multiple competing currencies and runaway inflation, the newly united colonies adopted Alexander Hamilton's proposal of central (i.e., state) banking.[25]

Notice that paper money can be introduced only in a literate society. Thus, paper money "embodies" more information, and requires significantly greater knowledge from its users, than did gold or silver coins. When paper money is competing directly with coinage, it is partly the information printed on the paper (and the quality of the printing) and partly general information about the society, the economy, the government, or other relevant institutions, that enables the prospective holder of paper money to decide what value to assign to it. Is the physical appearance of the note credible? Is the named issuer trustworthy? (If a bank, is it solvent)? Do others accept the certificates? Is the paper money backed by "real" money? And so on.

The Western world went off the so-called fractional gold-exchange standard in 1932, for reasons that some economists have never accepted. To oversimplify a little, the gold standard was perceived to be an obstacle to economic recovery at a time when inflation was not a problem; the main problem then, according to Keynes and his followers, was inadequate demand (or credit). Keynes's ideas have had enormous influence over the past five decades, but the cycle may be ending. Deficit financing as a macro-economic tool may have reached its limits. The current (1991-92) recession, unlike earlier ones, is not responding to Keynesian "pump-priming" to increase consumer demand.

In an outgrowth of Keynes's school, the monetarists argue that the best tool for macro-economic management is the money supply as controlled by the central bank. Milton Friedman is the best-known proponent of this theory. Monetarists believe that almost all macro-economic ills, from depression to inflation, can be traced to government mismanagement of the money supply, usually for political reasons.[26] Even today, there are voices calling for a return to the gold standard, in a purer form than before, as a kind of automatic restraint on excessively inflationary government policies [e.g., Skousen, 1990]. Even the proponents of this idea, however, admit that it will probably not happen, unless the world financial system collapses.

Meanwhile, the nature of money is still changing rapidly. In fact, the real argument against returning to a gold standard is that the central bank, and the commercial banks, have lost control over the monetary-transfer mechanism and, hence, the money supply. The banking crisis of this recession is due, above all, to the fact that banks—hence, the central bank—no longer have an effective monopoly over the creation of credit. The issuing of corporate bonds by large firms as a means to avoid bank debt began the process. The growth of the enormous market in government bonds to finance government debt was a bigger step in this direction. The creation of a market for junk bonds, by Michael Milken and Drexel-Lambert, carried the process further. Thus banks had lost their role as primary supplier of capital for domestic business and industry by the end of the 1980s. They sought other avenues for profit and shifted to lending to large-scale real estate developers (where they compete with savings banks and insurance companies) and foreign governments, especially in the Third World. By the end of the decade many of these loans were going sour. Banks also suffered competition for consumer loans—and small time-deposits—from savings banks, credit unions, and "money funds." And, while banks have fought back hard to retain consumer credit business (Visa and MasterCard are issued by banks), the banks are under severe competitive pressure from non-bank credit-card issuers, especially American Express.

All forms of credit are effectively forms of money. The official measures of money supply used by the Federal Reserve, however, include only (1) coin and currency in circulation, excluding currency held by banks; (2) demand deposits at commercial banks, excluding those held by other banks and the U.S. government, less reserves on checks being processed for collection ("the float"); and (3) time deposits at commercial banks, other than those held by other banks or the U.S. government. There is a separate series on time deposits held by the U.S. government in commercial banks. At one time, the omissions, such as time deposits in savings banks, were not important enough to distort the statistics. But nowadays, most corporate finance and much consumer finance escapes the banking system altogether. Corporations sell stocks or bonds through brokers or investment bankers, mostly to other financial institutions, such as pension funds and insurance companies. Many individuals and small firms use credit cards in place of checks for day-to-day purchases and pay the credit card bills from money market accounts. There is a growing tendency in Japan and Europe, soon to be followed in the U.S., for utilities and merchants such as supermarkets and service stations to substitute prepaid debit cards for cash, checks or credit cards. This will further erode the central role of the banks by eliminating the float, which has been one of their major sources of profit.

The processing of all these transactions involving money and "para-money" is information processing in the purest sense. Electronic information flows have already largely superseded paper flows. The paper check-clearing process has become so burdensome that people readily seek ways to avoid it and speed up transactions. International currency exchange transactions are almost exclusively electronic. What is actually exchanged is information (data). The long-term implications of this trend toward more and more abstract forms of money can hardly be guessed at, today, except that it is surely a precursor of trends in other sectors. Increasingly, economic activity is a self-referential information-exchange process in which the physical "hardware" component of production processes and products is becoming less and less significant as compared to the information component (service or "software").

8.7. THE ROLE OF TECHNOLOGY

The previous chapter mentioned self-modification as an evolutionary mechanism. Learning is obviously the major form of self-modification. But learning is not heritable. Thus, extra-somatic means of information storage and processing offer major gains in efficiency. In an economic system, of course, learning and information transfer between individuals is a central element. For simplicity I refer to this as **technology**.

Humans were not the first organisms to transmit non-genetic information from generation to generation. Most mammals and birds teach their young to some degree. Humans are, however, the first species to systematically store non-genetic information *as such* in external repositories, such as libraries, that are maintained by the society as a whole. The development of spoken and, later, written languages was obviously critical to this storage. Indeed, although humans may have existed as a distinct species for several hundred thousand years, extra-somatic information storage and accumulation began only about 6,000 years ago (Table 8.1). Moreover, in recent centuries, the accumulation and transfer of cultural information, recently enhanced by the use of computers, has undoubtedly approached, and possibly surpassed, the genetic transfer of information, as suggested in Figure 8.3.

For example, the systematic use of extra-genetic information to improve food gathering and production began in prehistoric times. Knowledge about plant reproduction, animal behavior, weather, climate, geography, and even astronomy vastly enhanced the ability of a human population to support itself and grow in numbers. At first, much of this knowledge was acquired and retained by a process not dissimilar to natural selection. (As this chapter emphasizes, manufacturing can be regarded as the implanting of "useful" morphological information in materials.) More recently, the application of knowledge to increase the efficiency of production has become increasingly intentional, systematic, and analytic.

The processes of learning about nature, and disseminating and applying that knowledge for the benefit of humankind, are themselves now institutionalized. Scientific knowledge has accumulated rapidly in the past few centuries, as Figures 8.4 and 8.5 show. (Figure 8.5 shows the annual rate of publication, and Figure 8.4 the cumulative total.)

What, then is knowledge? One plausible definition might be: the minimum amount of information (in bits) needed to express the essential message being transmitted. In other words, *knowledge is the minimum amount of information embodied in a decoder or reproducing device required to copy (or reproduce) an object.* The object being reproduced may be any human-manufactured item. It need not be tangible: It may be a behavior pattern, a picture, a design, a piece of music, or a set of data points. Or, of course, it may be an organism or an organization. The intuitive definition of knowledge as an understanding of "how things work" is roughly valid, if limited. Theoretical knowledge contributes little

TABLE 8.1. *Evolution of External Means of Transmitting Information.*

Ideographs	400 B.C.	Sumeria
Pictograms	3500 B.C.	Sumeria
Government (state)/law	3500 B.C.	Sumeria
Writing (cuneiform)	3000 B.C.	Sumeria
Phonograms	3000-2800 B.C.	Sumeria
Hieroglyphics	3000-2800 B.C.	Egypt
Dictionaries	2800 B.C.	Sumeria/tablets
Book (papyrus roll)	2800 B.C.	Sumeria/Egypt
Archives	2500 B.C.	Sumeria
University	2500 B.C.	Sumeria
Library	2500 B.C.	Sumeria
Multiplication tables	2500 B.C.	Sumeria
Code of law	2200 B.C.	Sumeria
Mail (letters)	2100 B.C.	Sumeria
Fractions	2000 B.C.	Babylon
Algebra	1800 B.C.	Mesopotamia
Alphabet	1500-1400 B.C.	Palestine (Phoenicians)
Coinage	700 B.C.	Lydia
Zero	300 B.C.	Babylon
Printed books	1500	Germany
Encyclopædia	1800	France
Public education	1800	New England
Telegraphy, Postal Service	1840's	U.S., U.K., Germany
Telephony	1880's	U.S.
Radio-telegraphy	1890's	U.K.
Radio-broadcasting	1920's	U.S., U.K.
TV broadcasting	1930's	U.S., U.K.
FAX	1980's	Japan

or nothing to productivity unless it is incorporated in machine capabilities, organizations, labor skills, or product designs.

It is not too misleading to think of knowledge manifest in labor skills as a set of programs for a biological computer (the brain) operating a general-purpose, self-aware machine (the body). General education starting with infancy provides the comprehensive internal world-model and data base; eyes, ears, and hands constitute the external monitoring system, while job-related training provides the specific operating programs for controlling processing equipment, handling tools, driving vehicles, or carrying out other functions. Managerial skills, including allocation of resources, scheduling, motivating employees, marketing, design, and so on, are also forms of knowledge. The increasing replication of knowledge of this kind in so-called expert systems—crude as they are— testifies to the close analogy between knowledge stored in human brains and knowledge in computer programs. Productivity of labor is partly attributable to

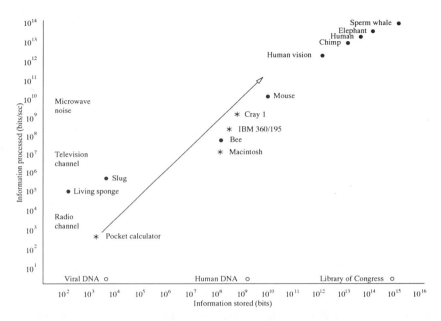

FIGURE 8.3. Computers VS. Living Systems. Source: Reproduced with Permission from Moravec, 1988.

invested capital and partly to all these skills and forms of knowledge.

Evidently, the productivity of capital—or, roughly speaking, its quality—is essentially related to the knowledge embedded in it. Successive generations of capital equipment are typically more and more productive because they embody an accumulation of knowledge, based on research, development, and practical experience. Over successive generations of capital, skills initially learned by the workers are gradually shifted to the machine permitting the machines to be operated by less-skilled workers or even without direct supervision. Some clerical and managerial functions, too, can be shifted from humans to machines (e.g., computers). These points will be discussed further in the next chapter.

The notion of actively using stored information (knowledge) to reduce the need for resource inputs follows immediately from the notion of using knowledge either to increase the efficiency of capturing and accumulating "natural" information (free energy) from the environment, or to increase the efficiency with which it is subsequently used. In effect, resources are created by human intelligence. Hence looking at it another way, intelligence is an effective *substitute* for resources. Solar energy is, of course, captured by photosynthesis and stored as fossil fuels, but to find and use these resources requires knowledge. The essential equivalence of knowledge and resources is neatly summarized by the proverb "Give a man a fish, and he can feed his family for a day. Teach him how to fish, and he can feed his family forever." Long ago, humans learned how to make useful objects from copper, silver, iron, and other metals from nearly pure

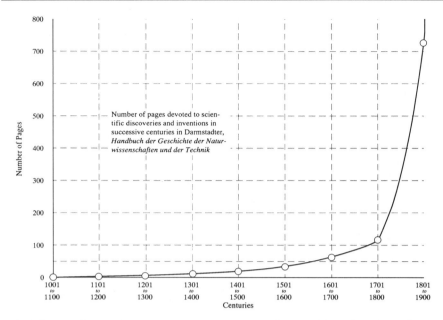

FIGURE 8.4. Growth Curve of Human Knowledge. Source: Repoduced with Permission from Lotka, 1939.

nuggets. As one resource is exhausted, typically, another of lower grade is exploited. Thus, charcoal gave way to coke for iron-smelting, and whale oil was replaced by kerosene for illumination.

Knowledge also permits more efficient utilization of any resource. As the demand for motor gasoline grew rapidly in the early 20th century, means of increasing the gasoline and octane output from each barrel of crude oil were actively sought. Using sophisticated modern refinery technology, and anti-knock additives, the gasoline-recovery fraction has risen from 15 percent to as much as 60 percent of each barrel, while average octane levels have almost doubled from about 50 (for natural gasoline) to more than 90 in the late 1960s. This, in turn, permits higher engine-compression ratios and correspondingly higher levels of thermal efficiency. Through technological knowledge, the amount of useful transportation work that can be extracted from crude oil has increased enormously [Enos, 1962]. It is not unreasonable to attribute this enhancement to technological knowledge created by the techno-economic system. This system comprises both embodied knowledge (in capital equipment) and disembodied knowledge in various portable forms, such as books and computer software.

A further observation regarding resources: The vast storehouse of fossil fuels in the earth's crust that humankind is currently exploiting is an accumulation of natural negentropy left from incompletely decayed living organisms in earlier geological periods. It is stored (from a molecular perspective) as chemical structure—energy-rich hydrocarbon molecules that are quite stable at ambient

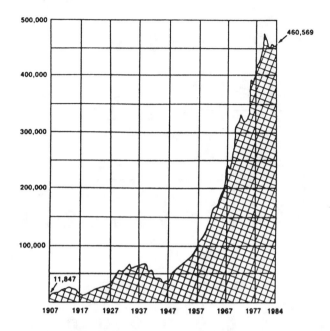

FIGURE 8.5. Annual Growth of Number of *CA* Abstracts Published (1907-1984). Source: Reprinted with Permission from *J. Chem. Inf. Comput. Sci. 1985*, **25**, 186–191. Copyright 1985, American Chemical Society.

temperatures, but that combine exothermically with oxygen above ignition temperature. Obviously, humans will have to find other energy sources within a century or so. It is already clear that several alternative possibilities exist, including fission, fusion, and photovoltaic cells on the earth or in space. All of them, however, will require large investments of capital (i.e., stored information), not to mention technologies more advanced than are currently available. In any case, some surplus negentropy in the form of capital will have to be set aside from the existing fossil-fuel store to "finance" the eventual changeover.

To summarize this section, the extra-somatic accumulation of knowledge is simply an extension of learning. While some types of learning can be achieved only by direct experience and practice, other types of knowledge, especially in the sciences, can be accumulated and passed on in "reduced" form. It is no longer necessary for a young scientist to repeat all the detailed arguments and review the evidence that led Galileo, Newton, and others to their key conclusions. It is perfectly true that each generation builds on the platform of knowledge accumulated painfully by its predecessors.

8.8. THE ROLE OF STRUCTURE AND ORGANIZATION

I have pointed out that physical structures embody information. This is no less true of economic structures. The structure of an economic system is usually

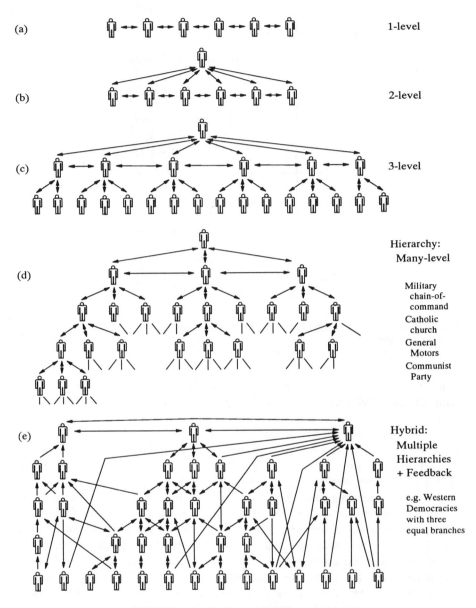

(a) 1-level

(b) 2-level

(c) 3-level

Hierarchy:
Many-level

(d) Military
chain-of-
command
Catholic
church
General
Motors
Communist
Party

(e) Hybrid:
Multiple
Hierarchies
+ Feedback

e.g. Western
Democracies
with three
equal branches

FIGURE 8.6. Sociopolitical Structures.

defined in terms of a standard industrial (sectoral) classification and a set of input-output relationships. The structure of an economy is fully defined, in these terms, by the specification of a final demand vector or a gross output vector, a capital investment vector, and a matrix of input coefficients or output coefficients.[28] The mathematics of such models, and their use in economics, was pioneered by Wassily Leontief [1986] and need not be recapitulated here.

More to the point, perhaps, is the obvious and close analogy between an economic system and the food web of an ecological system or community. In the latter case one can consider each species to be a node. The energy (biomass) inputs to any node are derived from other species from a lower trophic level, except in the case of the lowest level (vegetation), which derives its energy from photosynthesis. The node and flow structure can be defined in diagrammatic form, as illustrated in Figure 2.1 and its associated text.

Not surprisingly, essentially the same diagram applies to the economic case. Moreover, the corresponding mathematics is almost identical if one conceptually substitutes *money flow* for *energy flow*. There are significant structural differences, of course. An ecosystem has a single level of primary photosynthesizers on which all other levels depend. In the economic case, interdependence is far more dispersed; nothing quite like a trophic level exists. Also, of course, money is not a strictly conserved quantity (in some countries, at least) except in the very short run.

Some justification of this interpretation can be derived from Figure 8.6. This rather abstract set of diagrams might represent a food chain or a set of increasingly complex socio-political hierarchies. The arrows could represent flows of material/energy/money or of information.[29] Generally speaking, the flatter the pyramid, the more efficient it is in terms of information flow, information processing, and system control.

The first case, *a,* represents a number of independent non-interacting unitsthat do not really constitute a system at all. The second case, *b,* resembles a simple tribal structure, an orchestra, or a herd of deer grazing on grass. The third case, *c,* is something like a university (faculty, deans, president), a small firm, or a predatory wolf or lynx added to the deer/grass system. The fourth case, *d,* could represent a military hierarchy, the Catholic Church, the Communist Party, a large company like GM, or a multi-level food chain. In a simple hierarchical structure (*c, d*), the flow of information is impeded by the intervening levels. The more such levels there are, the greater the information *impedance* (a term from electrical engineering) and the *noisier* the message transmitted.

One of the principal problems with hierarchies is that they are better suited for the propagation of *commands* from the top (strategic objectives, papal edicts, five-year plans) than for arriving at feasible and acceptable solutions incorporating all relevant factors. Their great weakness is that information passes inefficiently in the reverse (upward) direction. As a result, the top leaders of hierarchies are often badly informed about the situation confronting those at the lower levels. To compensate, one could introduce special channels to gather intelligence from the rank and file, as illustrated in Figure 8.6e.[30] Analysis along ecosystemic lines has proved to be productive in such fields as historical anthropology and management science.

ENDNOTES

1. Material in this chapter has previously appeared in R. U. Ayres "Self-organization in Biology and Economics," *IIASA* RR-88-1, January 1988.

2. The usable component of energy has been variously called "thermodynamic potential," "free energy," "available useful work" or (in Europe) "exergy." The term *essergy*—derived from essence of energy—has also been suggested. However, its unfamiliarity argues against adoption here. I sometimes use the simpler term *free energy* despite its misleading economic connotations (not to mention the confusing alternative special cases named after Gibbs and Helmholtz, respectively).

3. In the abstract neo-classical (Walrasian) model of the economic system, resources are implicitly assumed to be generated by labor and capital. Thus, the neo classical system is, in effect, a perpetual motion machine. This fact was emphatically pointed out by the Nobel prize-winning chemist Frederick Soddy in 1922. It was also stressed by the biophysicist Alfred Lotka, among others [Lotka, 1950]

4. Indeed, equations used originally to describe population dynamics of two interacting species (known as the Lotka-Volterra equations) have also been applied with some success to the dynamics of technological competition. This would seem to support the assertion of fundamental similarity between biological and economic competition although the analogy is weakened by the deterministic nature of the Lotka-Volterra model.

5. A well-known example is the founding of Intel Corp. by a group from Fairchild Camera and Instrument Co. (itself founded by a scientist from Bell Labs).

6. A species is defined, in effect, by the inability of its members to produce fertile offspring by sexual intercourse with members of another species.

7. Available useful work (denoted, in this book by the symbol B, is sometimes called "the essence of energy," or, occasionally, *essergy*. By the same token, one can, perhaps, argue that knowledge is the essence of information. However, the analogy is imperfect and unnecessary.

8. My view, as expressed in this chapter, was largely anticipated by Jean Thoma, although I did not discover Thoma's unpublished paper [Thoma, 1977] until this book was nearly ready for the printer.

9. What would constitute a "reasonable" length is debatable. In this regard, deliberate encryption might make such a test impracticable. Obviously an encrypted message is not "noise," even if it cannot be deciphered.

10. The realization that controls, in biological and engineering systems alike, often involve sensory feedback loops, led to the development of cybernetics as a mathematical subspecialty of general systems theory. The seminal work was that of Norbert Wiener [Weiner, 1948]. The basic ideas have also found applications in biology, ecology, psychology, and organization theory. In view of the growing academic interest in the matter, it is worthwhile pointing out that the phenomenon of self-organization necessarily involves feedback loops [Ayres, 1988a]. In this context,

controls are simply points of leverage where a conscious human intervention can have an amplified impact.

11. This quotation attributed to Georgescu-Roegen was taken from an unpublished background paper, "Application of Entropy Concepts to National Energy Problems," by J. C. Allred, prepared for a symposium on Entropy and Economics held in October 1977 in Houston, Texas.

12. The following estimates are taken from Braudel [Braudel, 1981]. The population of Europe in 1800 was about 187 million, of which roughly 50 million were active workers. There were also about 14 million working horses, 24 million oxen, at least 600 thousand water-powered flour mills, and perhaps 150 thousand windmills. The animals, working 10 hours per day, generated something like 90 million horse-power hours per day, while the mills—which could operate longer hours, at least in principle—may have generated anywhere from 24 to 50 million hp-hr. By contrast, only about half of the human workers were employed exclusively in unskilled occupations (porters, miners, agricultural laborers), and most of the rest were concerned with controlling animals or machines (such as looms or spinning wheels) than on driving them.

Braudel estimated the continuous power output of the human workforce at 0.9 million hp, assuming the power output of a man to be 0.03–0.04 hp and making some adjustment for the fact that more than half of all European workers, even then, were not utilized for their physical strength. This implies an output of something like 9 million hp-hr day.

In short, no more than 7% of all mechanical work was done by humans in 18th century Europe. The direct human contribution to total energy use was even smaller. According to Braudel, Europe in 1789 burned about 200 million tons of wood—which Braudel misleadingly equates with 16 million hp—plus a significant amount of coal, for space-heating, cooking, washing, and industrial processes. Altogether, no more than 3–4% of all thermodynamic inputs in late 18th century Europe were provided by human labor.

13. A useful summary of the relevant ergonomic literature can be found in [Miller 1978; Yaglom and Yaglom, 1983].

14. With a very few and minor exceptions such as the punch-card controls for Jacquard looms, and cam-controls for automatic machine tools, such as Spencer's "automat."

15. This is not the place for a discussion of the rather special role of liquid, or "working," capital.

16. As is the chemical industry (and, for that matter, the metallurgical industry).

17. Animals were a particularly useful form of capital, insofar as they could provide for their own metabolic needs by grazing *ad hoc*. In modern times, of course, most animal feed is produced deliberately for that purpose.

18. The German word for television is *Fernsehen*: literally, "far-seeing."

19. Two early studies at the Harvard Business School, both inspired by Wiener's pioneering work on cybernetics, were *Making the Automatic Factory a Reality* by John Diebold and others [1951] and *Automation and Management* by James Bright

[1958]. Both saw the computer taking over as machine-controller in a relatively short time. Compare these with the more recent and far less optimistic perspective offered by Wright and Bourne [1988].

20. To be strictly consistent with the framework adopted here, capital goods that provide final services should be treated in the same way as long-lived consumer durables.

21. It would also require a great deal of time, since many of the steps would have to be carried out in strict sequence. One cannot produce a hard steelcutting edge, needed for any machining operation, without a means of making steel from iron; one cannot reduce iron ore to pig iron without first building a blast furnace, which requires high-quality bricks and bellows (driven by a steam engine or water wheel), and so on. The task of a modern Robinson Crusoe would require many decades, even with the benefit of a complete library of science and engineering and an army of multi-skilled workers.

22. An intermediate is a product consumed in the production of some other product. To be sure, capital goods are long lived and are not physically incorporated in the products made with their help. Early economists effectively defined capital in this way. However, numerous other intermediates (such as solvents, lubricants, detergents, flotation agents, etc.) are used up without being embodied in final products. Why, except for long life, should capital goods be regarded any differently?

23. This notion has been expressed very well by Lewis Thomas in *The Lives of a Cell* [1974] and other writings.

24. Occasionally is is possible to achieve major gains by organizational change (supplemented by training) as illustrated by Toyota's *kan-ban* system for "just-in-time" delivery of parts to the assembly line. The use of statistical quality-control methods offers another illustration.

25. Bars of silver or gold were stamped by merchants for exchange purposes in Babylonia, India, and China. The shekel was named for a merchant family that had a reputation for accurate weights. The first standardized coins were minted in the 7th century B.C. in Lydia. The king of Lydia (Croesus) was the first to make coinage a royal monopoly. The Romans were the first to standardize the coinage system. Julius Caesar established the gold *aureus* as the equivalent of 100 silver *sesterces*. The British pound sterling was originally a Saxon pound (weight) of standard silver. The word *dollar* came from *Thaler*, which originated as the Joachimsthaler, a silver coin produced in northern Bohemia in the 16th century. Gold coins were introduced in the 19th century when higher-value coins were needed (due to economic growth and inflation) and significant quantities of gold became available from new mines in the United States and South Africa.

26. Central regulation at the federal level was not imposed until 1913, with the creation of the Federal Reserve system.

27. For instance, monetarists argue that it was not the stock market crash of 1929 that caused the Great Depression, but a sharp contraction of the money supply, due mainly to commercial bank failures. (Half of all the banks in the country failed, which decreased the money supply by 33% from 1929 to 1933). One of the major contributing causes was an increase in the discount rate imposed by the Federal

Reserve Bank in 1931 to counter fears of a gold drain.

28. Astronomy evolved historically as means of predicting dates of annual spring runoff in the Nile Valley, where seasonal changes are very slight.

29. The mathematics of such models, and their use in economics, was pioneered by Wassily Leontief [e.g. 1986] and need not be recapitulated here.

30. The importance of information (perfect or otherwise) for markets was discussed in Chapter 6. Evidently, the implication also works in reverse. That is to say, monetary flows imply market price (and hence supply/demand) information. See, for instance, Henri Theil [1967].

31. A successful hybrid system with these features is the U.S. government. The executive branch is a typical hierarchy, with a great many levels. Both the legislative branch and the judicial branch have far fewer levels and can be regarded as specialized feedback systems designed to stabilize the system. The legislative branch, in practice, has three to five levels [public→public interest group→lobbyist→legislative staff→legislator]. The judicial branch is comparable [public→lower court→higher court→etc.].

REFERENCES

Allred, J. C., *Application of Entropy Concepts to National Energy Problems*, Technical Report, Entropy Economics Symposium, Houston, Tex., Nov. 2, 1977. [Unpublished]

Ayres, Robert U., *Manufacturing and Human Labor As Information Processes*, Research Report (RR-87-19), International Institute for Applied Systems Analysis, Laxenburg, Austria, July 1987.

Ayres, Robert U., "Self Organization in Biology and Economics," *International Journal on the Unity of the Sciences* **1**(3), Fall 1988*b*. [Also IIASA Research Report #RR-88-1, 1988]

Braudel, Fernand, *Civilization and Capitalism (15th-18th Century)*, Harper and Row, New York, 1981. English edition. [3 volumes]

Bright, James R., *Automation and Management*, Harvard Business School Press, Boston, 1958.

Butzer, Karl W., *Early Hydraulic Civilization in Egypt*, University of Chicago Press, Chicago, 1976.

Butzer, Karl W., "Civilizations: Organisms or Systems?" *American Scientist* **68**, September-October 1980 :517-523.

Daly, Herman E., "The Economic Thought of Frederick Soddy," in: *History of Political Economy* :469-488 **12**(4), Duke University Press, Chapel Hill, N.C., 1980.

Diebold, John *et al.*, *Making the Automatic Factory a Reality*, Harvard Business School, Boston, 1951.

Enos, J. L., *Petroleum Progress ;and Profits: A History of Process Innovation*, MIT Press, Cambridge, Mass., 1962.

Friedman, Milton and Anna Schwartz, *A Monetary History of the United States, 1867-1960*, Princeton University Press, Princeton, N.J., 1963.

Leontief, Wassily, *Input-Output Economics*, Oxford University Press, London, 1986. 2nd edition.

Lotka, Alfred J. (ed.), *Proceedings of the American Philosophical Society*, American Philosophical Society, Philadelphia, Pennsylvania, 1939.

Lotka, Alfred J., *Elements of Physical Biology*, Dover Publications, New York, 1950. [First published in 1925]

Machlup, Fritz, *The Production and Distribution of Knowledge in the U.S.*, Princeton University Press, Princeton, N.J., 1962.

Machlup, Fritz, "Uses, Value and Benefits of Knowledge," *Knowledge* **1**(1), September 1979.

Machlup, Fritz, *Knowledge and Knowledge Production* **1**, Princeton University Press, Princeton, N.J., 1980.

Marchetti, Cesare, *Society As a Learning System: Discovery, Invention and Innovation Cycles Revisited*, Research Report (RR-89-29), International Institute for Applied Systems Analysis, Laxenburg, Austria, 1981.

Miller, James G., *Living Systems* 5, McGraw-Hill, New York, 1978.

Moravec, Hans, *Mind Children: The Future of Robot and Human Intelligence*, Harvard University Press, Cambridge, Mass., 1988. [Work performed at the Robotics Institute, Carnegie-Mellon University]

Peschel, Manfred and Werner Mende, *The Predator-Prey Model: Do We Live in a Volterra World?* Akademie-Verlag, Berlin, 1986.

Skousen, Mark, *Economics of a Pure Gold Standard*, Mises Institute, Auburn, Ala., 1990. 3rd edition.

Soddy, Frederick, *Cartesian Economics*, Hendersons, London, 1922.

Theil, Henri, *Economics and Information Theory*, North Holland, Amsterdam, 1967.

Thoma, Jean, *Energy, Entropy and Information*, Research Memoranda, International Institute for Applied Systems Analysis, Laxenburg, Austria, June 1977.

Thomas, Lewis, *The Lives of a Cell: Notes of a Biology Watcher*, Viking Press, New York, 1974.

Wiener, Norbert, *Cybernetics: Control and Communications in the Animal and the Machine*, John Wiley and Sons, New York, 1948.

Wright, Paul K. and David A. Bourne, *Manufacturing Intelligence*, Addison-Wesley, New York, 1988.

Yaglom, A. M. and I. M. Yaglom, *Probability and Information*, D. Reidel, Boston, 1983. 3rd edition. [English translation from Russian]

Chapter 9

Information Added by
Materials Processing[1]

9.1. INTRODUCTION

It was argued in Chapter 8 that man-made things (artifacts) can be regarded as materials that have been imprinted with useful information. The term *useful* in this case explicitly implies economic value. It must be emphasized at the outset that, in general, information need not have economic value and that economic value is not simply proportional to information content. These points will be reiterated later in context. Throughout, the term *information* is used in the strict technical (Shannonian) sense, as a measure of the inherent likelihood or distinguishability of an object (or message) with respect to a reference system or environment.

A key point here is that a materials subsystem gains or loses information (with respect to its environment) with any change in its thermodynamic state. In fact, I will argue that the information embodied in such a subsystem is proportional to the *available useful work* in that subsystem. (I use the more familiar term *free energy* more or less synonymously, hereafter, for convenience. The two are not strictly identical, but the differences tend to be small.) Of course, free energy can be concentrated in one materials subsystem—for instance, a metal—only at the expense of others, such as fuels. When subsystems are combined, there will always be a reduction in the total amount of free energy. This loss is the thermodynamic cost of extracting, separating, and refining materials. The information on negentropy added to or concentrated in the refined materials is therefore also roughly proportional to the overall expenditure of free energy (mainly from fossil fuels) to drive the metabolic functions of the economic system. This "used up" free energy is converted to entropy. Some economic implications of this are discussed later.

Value added by manufacturing depends on the functional capability added to crude materials. Functional capability arises, in turn, from physical properties associated with specified physical and chemical compositionss, shape, and finish (an aspect of shape). Each of these must be held within given limits or toler-

215

ances. The required precision of compositional and dimensional specifications defines the minimum amount of information embodied in each component part. Further information is added when parts are assembled into subassemblies, machines, processing equipment, structures, and systems.

It was suggested, in the last chapter, that the factor services of capital and labor comprise the information inputs to the production process, while the output or product is analogous to the information output, or *message*. Normally, the information ultimately embodied in the output (message) by labor and capital is far less than the information embodied in the inputs, including labor. Most of the input information is, in fact, wasted. An efficient production process or an efficient economic system must surely be one that wastes as little information as possible. A good design, by the same token, should require as little information as possible to achieve a given functional purpose, other factors remaining equal. The idea elaborated below is that the function of the factor services is to *concentrate* information (or reduce entropy) in certain specified materials or products, even though the total amount of information in the environment as a whole is reduced (i.e., global entropy always increases).

It will be recalled that *thermodynamic information* (information associated with composition, and information associated with temperature, pressure, and chemical potential) was discussed in some detail in Chapter 2, section 4. The reader is referred to that section.

9.2. ENERGY CONVERSION

Steam engines (Rankine cycle) and internal combustion (Otto or diesel) engines are the *prime movers* in our industrial system. They convert energy from chemical fuels to mechanical power and/or to electrical power. The energy-conversion process is usually analyzed in terms of a thermodynamic cycle (or sequence) of heat transfers and expansions and compressions of an ideal gas. The most familiar is the theoretical (and practically unachievable) **Carnot cycle**. All such energy-conversion cycles are considered for convenience to operate between a high temperature reservoir at temperature T_h and a low temperature reservoir at temperature T_0. Also, for convenience, most simple cycles are assumed to consist of four stages, each of which is desirable in terms of changes in one *extensive variable* (such as U or V) or one *intensive variable* (such as P or T), holding the others constant. Many idealized cycles are constructed from sequences of steps such as

Isobaric (constant pressure) heating or cooling ($dP = 0$);

Isochoric (constant volume) heating or cooling ($dV = 0$);

Isothermal compression or expansion ($dT = 0$);

Isentropic (**adiabatic**) compression or expansion ($dS = dQ = 0$).

TABLE 9.1. *Theoretical Limits of Performance for "Air-standard" Engines, Using Non-Condensing Working Fluid.*

Cycle	1-2 Compress	2-3 Add Heat	3-4 Expand	4-1 Cool	Maximum Efficiency
Carnot	Adiabatic $S_1=S_2$	Isothermal $T_2=T_3$	Adiabatic $S_3=S_4$	Isothermal $T_4=T_1$	$1-\dfrac{T_1}{T_4}$
Stirling	Isothermal $T_1=T_2$	Isochoric $V_2=V_3$	Isothermal $T_3=T_4$	Isochoric $V_4=V_1$	$1-\dfrac{T_1}{T_4}$
Erickson	Isothermal $T_1=T_2$	Isobaric $P_2=P_3$	Isothermal $T_3=T_4$	Isobaric $P_4=P_1$	$1-\dfrac{T_1}{T_4}$
Otto	Adiabatic $S_1=S_2$	Isochoric $V_2=V_3$	Adiabatic $S_3=S_4$	Isochoric $V_4=V_1$	$1-\dfrac{T_4-T_1}{T_3-T_2}=1-\dfrac{1}{r^{k-1}}$
Diesel	Adiabatic $S_1=S_2$	Isobaric $P_2=P_3$	Adiabatic $S_3=S_4$	Isochoric $P_4=P_1$	$1-\dfrac{1}{k}\left(\dfrac{T_4-T_1}{T_3-T_2}\right)$
Brayton	Adiabatic $S_1=S_2$	Isobaric $P_2=P_3$	Adiabatic $S_3=S_4$	Isobaric $P_4=P_1$	$1-\dfrac{T_4-T_1}{T_3-T_2}=1-\dfrac{1}{r^{k-1}}$

For adiabatic processes: $r =$ compression, $k = 1.41$ for air.

$$\frac{T_2}{T_1}=\left(\frac{V_1}{V_2}\right)^{k-1} \quad \text{and} \quad \frac{T_3}{T_4}=\left(\frac{V_4}{V_3}\right)^{k-1}$$

Most of the standard idealized energy-conversion cycles are shown in Table 9.1. The theoretical efficiency of each idealized cycle is also shown. (The idealized Rankine cycle is not shown because it applies not to an ideal gas but to a condensing fluid. It is therefore somewhat more complex, in principle, with several sub-cases depending on the characteristics of the so-called "vapor dome," which is the locus of all points where liquid and vapor are in equilibrium.)

In this conceptualization, the maximum available work (or essergy), B, is a simple function of the temperature of the hot reservoir, based on the maximum (Carnot) efficiency with which heat can be converted into useful mechanical work. This is shown in Figure 9.1. Evidently, the available work obtainable in practice via one of the other energy-conversion cycles is lower. Figure 9.2 shows the available work recoverable from an idealized Otto cycle engine, as a function of compression ratio r. (It shows, incidentally, the benefit of using higher fuel octanes.)

To avoid confusion at this stage, it is important to emphasize that energy-conversion processes *per se* do not result in the embodiment of useful information in final materials.[2] When a fuel is burned, combustion products (CO_2, H_2O) are created, which then dissipated in the atmosphere. In this process, taken

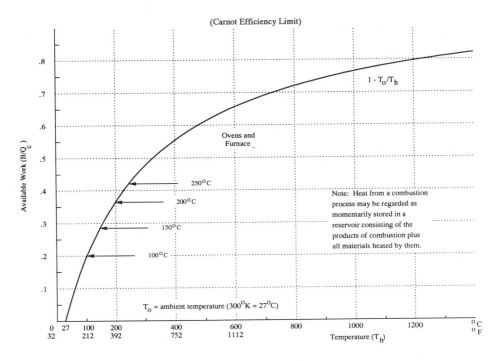

FIGURE 9.1. Available Work Obtainable from a Hot Reservoir vs. Temperature (Carnot Efficiency Limit).

by itself, the initially available useful work or essergy B is converted first (in part) to useful mechanical work, and subsequently back to heat and lost. Since thermodynamic information H is proportional to B, the original thermodynamic information content of the fuel is also lost.[3]

That is not the whole story, however. The energy and power used in manufacturing drives thermodynamic transformations of materials later used in manufacturing and construction. In effect, some of the available useful work (B) and/or thermodynamic information (H_{thermo}) content of fuels or other energy sources is ultimately embodied in these materials. In fact, it follows from equation (2.22) that the thermodynamic information embodied in metals and products, such as iron and steel, plastics and synthetic fibers, paper, aluminum, and cement, is also equal to B_{min}, the minimum amount of available work needed in theory to produce them. The most general definition of thermodynamic process efficiency η for an energy-conversion process is thus

$$\eta = \frac{B_{min}}{B}, \tag{9.1}$$

where B_{min} is the minimum quantity of available work theoretically required, and B is the quantity actually consumed. This definition applies just as well to

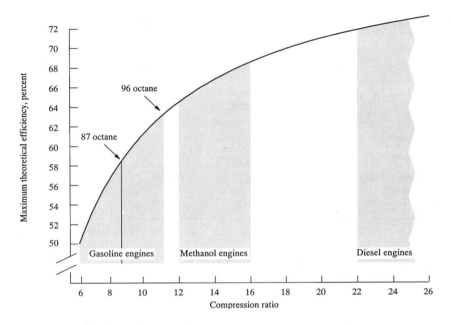

FIGURE 9.2. Theoretical Efficiency of Internal Combustion Engines.

metallurgical and chemical processes as to energy conversion *per se.*

For metals normally found in oxides, analysis of the theoretical requirements for the minimum amount of available work is fairly straightforward. In fact, as pointed out in Chapter 2, they can be determined from tabulated heats of combustion, with only a minor correction needed for the change in entropy due to dispersion of the combustion products and irreversible heating in the flame.

For metals found in nature as sulfides, the situation is only slightly complicated by the fact that the sulfide form is first converted to an oxide and the sulfur driven off as SO_2. In this case, one must also make use of tabulated heat of formation of the sulfide, then proceed as before. (The heat of formation of compound A is equal to the sum of heats of formation of the combustion products of A minus the heat of combustion of A.) In principle, the sulfur could be burned as fuel to recover available work, leaving the metals in pure form, but at present this is not done. The purified metals actually embody less available work than the sulfide minerals, but it is impractical to take advantage of this fact in an oxygen atmosphere.

Data on process efficiency for several industrial processes are shown in Table 9.2. In the case of paper, the figures given in the last column do not include the heat of combustion of the organic materials *per se,* which is also ultimately recoverable. However, the figures for steel and aluminum do roughly correspond to the heats of formation of the pure metallic ores from the metal and oxygen. For example, the heat of formation of ferric oxide Fe_2O_3 from pure iron and pure

TABLE 9.2. *Comparison of Specific Fuel Consumption (BTU/ton×10⁶) of Known Processes with Theoretical Minimum for Selected US Industries.*

Process	Actual 1986 Specific Fuel Consumption	Potential Fuel Consumption Using 1973 Technology	Theoretical Minimum Specific Fuel Consumpsion B_{min} Based on Thermodynamic Availability Analysis
Iron and Steel	26.5	17.2	6.0
Paper	39.0[a]	23.8[a] > −0.2[b], < +0.1	
Primary Aluminum production[c]	190	152	25.2
Cement	7.9	4.7	8.8

[a]Includes 14.5*10⁶ BTU/ton of paper produced from waste products consumed as fuel by paper industry.
[b]Negative values means that no fuel is required.
[c]Does not include effect of scrap recycling.
Source: [Gyftopoulos *et al.*, 1974: see also Hall *et al.*, 1975]

oxygen is − 198.5 kcal/gm-mole at 25°C. This translates to 4.46×10^6 BTU/ton of Fe_2O_3 or 6.4×10^6 BTU/ton of iron. It is the heat that would be released (as heat of combustion) if the pure iron were burned as a fuel in air.[4]

9.3. COST OF REFINING

There are two basic types of refining to consider, namely, physical separation/concentration (sometimes called "beneficiation") and chemical refining. Examples of the first type include screening, magnetic separation, gravity separation (e.g., by centrifuge), flotation, osmosis, reverse osmosis, solvent refining, crystallization, liquefaction (e.g., of oxygen and nitrogen), and fractional distillation. Examples of the second kind include smelting (with coke or natural gas), electrolysis, thermal or catalytic cracking (e.g., of petroleum), reforming, alkylation, hydrogenation, dehydrogenation, digestion, and other chemical processes.

As pointed out in Chapter 2, the information embodied per mole by physical concentration, or lost by diffusion, can generally be approximated by Boltzmann's ideal gas approximation (9.8):

$$H_c = R \ln \frac{X_c}{X_0} = R \ln C, \qquad (9.2)$$

where X_c is the mole fraction in the concentrated state, X_0 is the mole fraction in the diffused state, and R is the ideal gas constant (2 cal/mole). The ratio of mole fractions is equal to the concentration ratio C.

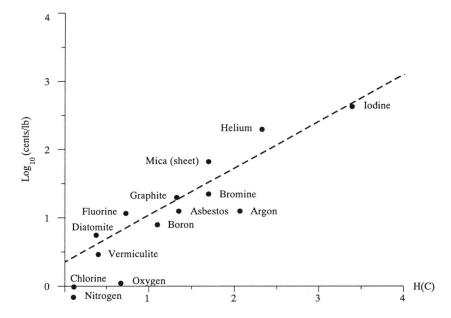

FIGURE 9.3. Cost of Physical Separation vs. Information Added.

It is important to note again that the incremental information added by concentration depends on the starting point. Not much is gained by starting from a highly concentrated source. If we are interested in comparing the information embodied in different materials in absolute terms, the best way is to calculate the information that would be lost if the material were completely dispersed (diffused) into the environment.[5] The difference between information lost by diffusion, and information added by concentration from high-quality natural sources, is, of course, a "gift from nature."

It has been observed, e.g., by Sherwood [1978], that costs (or prices) of many materials are inversely proportional to their concentrations in ore or other original form, and therefore proportional to the concentration factor needed to purify them. This relationship implies a linear relationship between the logarithms of price (cents/lb.) and concentration factors. Such a relationship is indeed observed for many materials, particularly those that have several different initial concentrations, such as vermiculite, diatomite, graphite, asbestos, sheet mica, and gold. It also applies to atmospheric gases (oxygen, nitrogen, argon) and to various chemicals found in brine. Data compiled by Sherwood is plotted in Figure 9.3. It compares a number of materials requiring only physical separation, which eliminates one of the complicating factors.

For materials requiring both physical and chemical processing, of course, the situation is much more complex. As an example of the latter, consider the multi-stage chemical process for refining platinum group metals from their ores.

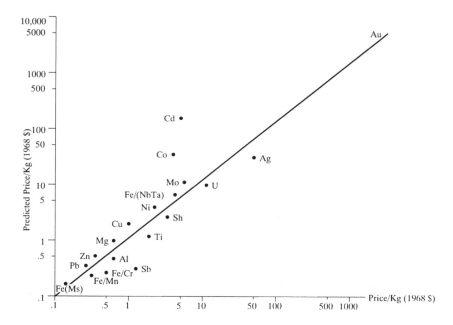

FIGURE 9.4. Cost of Chemical Separation. Source: Philips and Edwards, 1976. Reproduced with the Permission of the Controller of Her Britannic Majesty's Stationery Office.

Refining copper-lead-zinc ores and separating all the important by-products is comparable in its complexity. Consider also the different problems of recovering by-products, such as arsenic, antimony, cadmium, tellurium, thallium or silver vis–vis primary products in this regard. Contrast this with the extraordinarily simple process for separating and refining mercury from its sulfide ore (simple low-temperature retorting).

In view of this diversity and the enormous differences, for instance, in scale of production or use among different substances, it is more than a little surprising to observe a relationship between cost (price) and heat of formation (or heat of combustion). Figure 9.4 shows that metal prices (in relative terms) can be predicted quite well from heats of formation or heat of combustion alone, disregarding all other factors, including the relative concentration of the metal in its ore. Figure 9.4 neglects the benefits of economies of scale, which would presumably result in lower costs for metals used in large quantities *ceteris paribus* than for metals with only minor uses.[6]

9.4. THERMODYNAMIC ANALYSIS OF ALTERNATIVE PROCESSES

As pointed out in Chapter 2, the generalized thermodynamic potential B is a measure of the ability of a subsystem to perform useful work. Thus, when an "uses" energy, it is really thermodynamic potential being consumed. The science of thermodynamics enables us (in principle) to utilize the available useful work efficiently, or even to minimize the quantity of B used to achieve a given result.

In the 19th century, thermodynamic analysis of this sort was devoted largely to heat engines. In recent times, the techniques originally pioneered to analyze prime movers have been extended to the analysis of refrigeration cycles, heat pumps, air conditioning and heating systems, chemical processes, and metallurgical processes. By describing processes as sequences of steps carried out at constant temperature, constant pressure, constant volume, or other suitable simplifying assumptions, it is possible to construct idealized processes that can be compared with real ones. Such comparisons provide insight into the sources of losses and the potential for improvement. Quantitative examples of this sort of analysis have been presented, in particular, by Gyftopoulos et al. [1974], Ross et al. [1975] and Tribus et al. [1971]. It has been demonstrated that there are general design principles for efficient processes, and significant opportunities to conserve energy (that is, B) by redesigning all kinds of industrial processes.

Without underestimating the value of detailed analyses of the thermodynamic process at the micro-level, I think it safe to say that the larger implications are even more important. Consider that every product of the economy, whether a consumer good or a producer good (capital equipment or structure) has a *life cycle*. (Recall Figure 8.2). It begins "life" as raw materials extracted from the earth. These are concentrated, refined, and processed into finished materials. Thence the materials are shaped and formed into components, which are subsequently assembled into finished goods and sold to consumers. But, of course, *consumption* is only a word used by economists to contrast with *production*. When goods have exhausted their utility to users, they do not disappear. On the contrary, they become waste matter or "garbojunk" and must be disposed of in some fashion.

Figure 8.2 showed that the local entropy of the material good is actually decreased by the various processing steps, although this occurs at considerable cost as the various processes involved generate much larger amounts of entropy in the environment. When the good becomes waste matter or a piece of junk, this investment in local entropy reduction is lost too. This is the scientific basis of the argument for increasing the useful lives of material goods and enhanced reuse and recycling, especially of those materials (steel, aluminum, copper, etc.) that cost a lot of thermodynamic potential to extract, concentrate, and refine [e.g., Berry, 1972].

ENDNOTES

1. Material in this chapter has appeared previously in [Ayres, 1987].

2. Nevertheless, finished fuels, such as gasoline, are more information-intensive than crude fuels such as coal or petroleum. This is a consequence of the distillation or

other separation processes used in refining.

3. The constant of proportionality is the 'standard' temperature of the reference system, namely the average temperature of the earth's atmosphere. Recall equation (2.22).

4. In fact, most iron and carbon steel eventually oxidizes to rust, thus reversing the original reduction (smelting) process. Recycling is important because it avoids this unnecessary loss of energy. (The negative sign is a convention).

5. The appropriate definition of environment would be the earth's crust for most solids, the oceans for water-soluble salts or liquids, or the atmosphere for gases.

6. For this reason, a more accurate costing methodology would also have to take into account differences in inherent utility from one material to another. As an example, consider the great inherent utility of platinum as a catalyst compared to osmium, an equally rare metal of the platinum group with no known uses whatever.

REFERENCES

Ayres, Robert "Manufacturing and Human Labor as Information Processes" IIASA Research Report RR-87-19, Laxenburg, Austria, November 1987.

Berry, Stephen, "Recycling, Thermodynamics and Environmental Thrift," *Bulletin of Atomic Scientists* 27(5), May 1972 :8-15.

Gyftopoulos, E. P., L. J. Lazaridis and T. F. Widmer, *Potential Fuel Effectiveness in Industry* [Series: Ford Foundation Energy Policty Project], Ballinger Publishing Company, Cambridge, Mass., 1974.

Hall, E. H., W. H. Hanna *et al.*, *Evaluation of the Theoretical Potential for Energy Conservation in Seven Basic Industries*, Research Report (PB-244,772), Battelle Institute, Columbus, Ohio, July 1975.

Philips, W. and P. Edwards, *Metal Prices As Function of Ore Grade*, 1976. Resource Policy.

Ross, M. *et al.*, *Effective Use of Energy: A Physics Perspective*, Technical Report, American Physical Society, City, State, January 1975. [Summer study on Technical Aspects of Efficient Energy Utilization]

Sherwood, Martin, *New Worlds in Chemistry*, Basic Books, New York, 1978. [Revised edition.]

Tribus, Myron and Edward C. McIrvine, "Energy and Information," *Scientific American* 225(3), September 1971 :179-188.

Chapter 10

Morphological Information[1]

10.1. SURFACE INFORMATION: GENERAL CONSIDERATIONS

A discussion of morphological information would naturally start with a consideration of the thermodynamic information associated with a shape. To be as precise as possible, let us first assume the existence of a regular three-dimensional lattice of geometric points fixed in Cartesian space. If every one of the points were occupied by molecules, it would be a perfect crystal.

The next logical step is to define a plane surface through the lattice. Because all crystals have a number of natural planar surfaces, let us begin with a plane that exactly coincides with one layer of the lattice. That is to say, a subset of lattice points are in the plane. The plane defines a surface if all lattice sites on one side of the plane (e.g., above it) are empty and at least some of the sites in the plane itself are occupied. Because the surface is associated with a solid, it can also be assumed that most of the sites on the non-empty side of the plane are occupied.

A D-information value associated with the occupancy of this surface can now be determined. Assume the plane surface has a finite area enclosing M lattice points. There are just two possible states: *occupied* and *empty*. Assuming independence, the total number of possible occupancy-complexions W_{occ} for the surface is exactly 2^M. If all complexions are equally likely, the information gained by any observation that determines exactly which lattice sites in the plane are occupied is

$$H_{occ} = K \ln W = KM \ln 2 \qquad (10.1)$$

or $K\ln2$ (1 bit) per lattice site. This is inherently a small number compared to any entropic change of thermodynamic origin associated with the entire mass of the solid, because the number of particles (or sites) on any surface M is very small compared to the number N in the volume. In fact, since the surface area (in any unit) is roughly the 2/3 power of the volume, it follows that

$$M \approx N^{2/3}.$$

225

If the solid has a mass of the order of mole, then N is of the order of Avogadro's number $(A = 6 \times 10^{23})$, whence M is somewhat less than 10^{16}. In other words, for a 1-mole solid, the surface entropy appears to be smaller than the volume entropy by a factor of at least 10^8. This discrepancy in magnitudes makes it appear, at first glance, that the thermodynamic information associated with a surface can be neglected in comparison with the thermodynamic information associated with a volume, which is always proportional to the number of molecules in the volume (see Chapter 2, equation 2.8).

The foregoing argument is incomplete, however. On deeper reflection, it is clear that surface D-information is not completely defined by occupancy. It is true that all three-dimensional shapes must have surfaces. It is also true that a given (plane) surface with M possible lattice sites must have a D-information content of $MK\ln2$, as argued above. In fact, the assumption of a plane surface can be relaxed, since it was not required to derive the result. It is important to emphasize that the surface D-information derived above relates only to lattice site occupancy and not to geometry. From the lattice-occupancy perspective, all geometries (or shapes) are the same. It follows that the information gained by a choice of one particular surface geometry from among all possible *non-overlapping* surface geometries with a common perimeter has not yet been taken into account. (Surfaces that overlap, in the sense of differing only at a small subset of points, must be excluded, since they have different possible occupancy states *only* to the extent that they include different lattice sites.)

The next question is: How many different non-overlapping surfaces with the same perimeter must be considered? The number of distinguishable surfaces with a common perimeter is extremely large if the non-overlapping condition is relaxed. It can be shown that the number of common-perimeter surfaces containing M lattice sites from only two adjacent layers (with M sites each) is slightly less than 2^M due to edge effects. Most of these surfaces have many lattice sites in common with others, however, and the total number of possible different occupancy complexions encompassed by all the surfaces limited to two layers is roughly 2^{2M}. Notice that the total number of possible sites on the two layers appears in the exponent. By a logical extension of the reasoning above, virtually all lattice points in the volume can belong to at least one (distorted) geometric surface passing through the specified perimeter, and the sum total of different points in all such surfaces—subtracting overlaps—is just the number of lattice points in the volume itself. Following this argument to its limit, it is logical to conjecture that the number of different occupancy complexions encompassed by all possible surfaces passing through the volume with a common perimeter is 2^N, where N is the total number of lattice sites in the volume as a whole. Obviously, N corresponds essentially to the number of molecules, since most sites are occupied. Hence, when shape uncertainty is added to surface-occupancy uncertainty, the information gained by a particular choice is roughly

$$H_{\text{shape}} = KN \ln 2 \qquad (10.2)$$

or *N* bits, given the assumption of equal probability of all surfaces as well as all occupancy states. It follows that D-information associated with a particular choice of surface is roughly comparable in magnitude to the thermodynamic information associated with a particular composition and distribution of quantum states.

The assumption of equal probability clearly does not apply to shapes in the economic context. Some shapes are far more probable than others, resulting in a vast reduction in the D-information value of a particular choice in the real world, as compared with the theoretical possibilities. As a consequence, the amount of morphological information *embodied* in the shape of a real object (such as a washer, bolt, piston, or bearing) is comparatively small in magnitude, and of a different kind.

In fact, most surface D-information is not useful. The useful information about shapes relates forms with functions. As noted in passing in Chapter 2, S. Kullbach suggested many years ago that the appropriate reference state for shapes is the *ideal* shape that best facilitates some specified function. Kullbach entropy, so-called, is a measure of the deviation of the real from the ideal.

10.2. MORPHOLOGICAL INFORMATION EMBODIED IN MANUFACTURED SHAPES

To estimate the magnitude of *useful* shape information in contrast to the information associated with all possible shapes one must abandon the combinatorial methods of statistical mechanics and approach the question from another direction. To see the nature of the problem, David Wilson offers the following example: "Consider the complexity of a simple control valve. Its effective information content would differ in the following situations:

1. Selection from three stock models.
2. Assembly from standard parts.
3. Fabrication from scratch.

... the three situations represent increasing information content for the same function. The total information content of the valve is the sum of many components: those associated with fabrication, assembly, and selection, to name a few" [Wilson, 1980, p. 131]. The reference case, in each situation, is obviously different. In the first and second situations, external constraints (models in stock, configurations of standard components) limit the choices.

In the third case, which is the most general, the choices are also limited by the capabilities of standard machine tools and materials. This constraint can be generalized to a choice among certain relatively simple geometric surfaces (shapes) and dimensional precision. These two can be discussed together, since dimensional precision is a part of parametric specification.

Most part shapes are quite simple or are constructed from simple geometric elements, such as straight lines, circles, and angles. It is convenient to divide part shapes into three basic groups, namely, two-dimensional planar (flat), and three-dimensional (cylindrical or prismatic). Each group can be further subdivided, based on symmetries and whether or not the shapes can be obtained from simpler symmetrical shapes by analogs of such industrial operations as bending, winding, or by stretching/shrinking. Nine major shape categories relevant to manufacturing operations are shown in Table 10.1.[2]

It can be seen that simple shapes are generally constructed by sequences of geometric operations like rotations, translations, and intersections. (The theory of simplexes in mathematics deals with such sequences.) Many of these geometric operations have counterparts in physical manufacturing processes, though physical processes do not always correspond exactly to geometric ones.

The information embodied in a complex geometric shape defined by the intersection of surfaces consists of two components:

1. Dimensional specifications for each surface.
2. Construction instructions for combining several surfaces of specified types (planes, conic sections, etc.).

Each surface of the shape must be completely defined and oriented with respect to a single common Cartesian (or other) coordinate system. The orientation requirement applies to surfaces that are symmetric with respect to rotation around an axis (or three axes in the case of spheres). To orient such a surface, simply imagine that coordinates are fixed and imprinted on its surface (as a map of the world is imprinted on a globe). A plane or flat surface is completely defined by the four parameters of a first-order (linear) equation of the form:

$$ax + by + cz + d = 0. \tag{10.3}$$

Only three of these parameters are independent. A point in space or a vector from the origin is also defined by three parameters. A **conic section** (ellipsoid, paraboloid, or hyperboloid) is completely defined by the ten parameters (nine independent) of a second-order (quadratic) equation of the form:

$$ax^2 + by^2 + cz^2 + dxy + exz + fyz + gx + hy + kz + 1 = 0. \tag{10.4}$$

By extension, more-complex surfaces are defined by cubic, quartic, and higher-order equations with 20 (19), 35 (34), or more parameters, respectively. Note that many of the simpler and more familiar surfaces are defined by specialized quadratic forms in which many of the nine parameters of a quadratic can be collapsed into a smaller number. For instance, a sphere with a definite location in Cartesian space is fully defined by its radius and the distance and direction of its center from the origin (four parameters). A located and oriented cylinder of infinite length is defined by its axial vector (the intersection of planes) plus a radius, or seven parameters in all. Indeed, a plane is also a special (three-parameter) case of a generalized quadratic form in which most of the parameters are set equal to zero. Similarly, a quadratic is a special case of a cubic, and so on, to higher and higher orders. By this logic, an arbitrary shape can be regarded as

TABLE 10.1. *Parts-Shape Taxonomy, Based on Symmetry.*

	Category	Shape Specification	Examples
1.	Flat shapes (x-y plane); made by cutting, shearing, or punching	Thickness Edge profile is in x-y plane	Fabric parts, can tops, washers, generator core laminations
2.	Distorted x-y planar shapes (with extension in x direction), invariant under any rotation around x axis; made by drawing or stamping	Like category 1, plus Profile of intersection with x-y plane	Metal cups, cans, tire rims, spherical or parabolic reflectors, roller-bearing races, stamped wheel hubs
3.	Flat shapes made by folding and/or bending around x axis without surface distortion	Like category 2	Paper or metal boxes or origami shapes, tubes wound from strip, metal gutters
4.	Non-symmetric; made by stamping	Thickness Specification of non-flat surface Edge profiles on surface	Auto body shapes, etc.
5.	Cylindrical (3-D) shapes that are invariant under rotation around z axis; made by molding, rolling, grinding, turning, drilling, or boring	Profile of intersection with x-z plane Specification of end plane(s) or surfaces	Ball bearings, pins, nails, bushings, piston wheels, axles, tires
6.	Cylindrical (3-D) shapes that are invariant under a set of finite rotations around or translations along z axis; made by rolling, turning, or milling	Like category 5, plus Profile of intersection with x-y plane	Threaded connectors, worm gears, helical gears, bevel gears, spur gears, sprockets
7.	Prismatic (3-D) shapes that are invariant under pure translation along z axis	Like category 5	T-beams, H-beams, rails, piles, decorative moldings, window-frame extrusions, wheel spokes
8.	Prismatic (3-D) shapes made from translationally invariant shapes by winding or bending without distortion	Like category 5, plus Specification of curve in some plane or 3-D space	Electrical windings, wire springs, hangars, hooks, pipe systems, crank handles, horse-shoes, chain links
9.	Prismatic (3-D) shapes that are symmetrical under reflection or non-symmetrical; made by casting, molding, hammer forging and/or milling	Specification of intersecting planes or curved surfaces	Engine blocks, base parts for machines, turbine blades, crank-camshafts, connecting rods, cutlery, hand tools

a locus of intersecting surfaces of nth order.

It is worth mentioning, at this point, that geometric shapes can be characterized in terms of Fourier descriptors. This idea was first introduced by Cosgriff [1960] and has been used in pattern recognition [Zahn and Roskies, 1972] and parts classification in Group Technology [Persoon and Fu, 1974]. Wilson [1980] has applied it in the context of measuring complexity.

Without some *a priori* limitation, the information embodied in a material shape would be of the order of the number of molecules in the volume of material, or N bits, as pointed out above. This reflects the fact that any given shape is, in a sense, selected from among an enormous variety of possible shapes. In practice, however, any surface likely to be required could probably be adequately approximated piecewise by a finite number of plane or quadratic surfaces.

Recalling the importance of external constraints, one should note that the useful information content of a plane surface when it is regarded as a special case of a general nth-order equation differs greatly from when it is preselected from the category of plane surfaces. In the latter case, the surface is defined in terms of only three independent parameters, while for the nth order it is defined by

$$N = \sum_{j=1}^{n} \left[\sum_{i=0}^{j} (i+1) \right] - 1 \qquad (10.5)$$

independent parameters (of which all but four are numerically set equal to zero). Assuming, for purposes of argument, that the second point of view is the appropriate one, i.e., that many surfaces can be prespecified as planes and hence defined by only three independent parameters, a similar problem arises when a second plane surface exists parallel to the first. Should this be regarded as requiring three more independent parameters or only a single parameter corresponding to the distance between them? Again, the answer depends on whether the parallelism is a condition of belonging to a particular shape category or not. Note that strict surface parallelism is a condition of belonging to the category of flat parts (category 1 in Table 10.1). This, in turn, depends on the *function* of the shape, that is, its form/function relationship. An analogous situation arises for some curved surfaces. For cylinders and spheres, with their common centers or axes of rotation, it can be termed *concentrism,* but it is effectively equivalent to local parallelism. Note that local (as opposed to strict) parallelism is a defining characteristic of the major surfaces of all the shapes in categories 2, 3, and 4 of Table 10.1.

Similarly, since rotational symmetry is a condition of belonging to category 5, all the rotational surfaces are (by definition) concentric cylinders. Once the central axis is located and oriented, each specification requires only one additional parameter. Incidentally, the same argument holds for cylindrical holes, given that the shape category can be prespecified. The argument applies, with some modification, to rotational surfaces in category 6.

Finally, translational invariance is a condition of belonging to category 7. This implies that the entire external surface (except for the ends) is defined by a cross-section, i.e., a closed perimeter defined on a two-dimensional plane. The perimeter may be defined by intersections of straight lines, quadratics, or higher-order curves. (One can assume that virtually any higher-order curve can be adequate, approximated piecewise by several quadratics.) The information required to specify any shape in category 7 is, therefore, the information required to specify the perimeter of the translationally invariant (long) segment plus the two bounding surfaces at the ends. The simplest case is, of course, a cylinder also belonging to category 5 for which the cross-sectional perimeter is a circle.

If each shape category in Table 10.1 were equally probable, the information equivalence of a choice of category *per se* would be $\log_2 \simeq 3.2$ bits. In practice, simpler shapes predominate, so the information embodied in a shape choice such as category 1 (flat), 5 (rotationally invariant), or 7 (translationally invariant) is even less—probably of the order of 2 bits or so. On the other hand, the truly non-symmetrical shapes belonging to categories 4 or 9 are much less common, i.e., much less probable. For instance, the information equivalence of such an *a priori* choice is greater than 10 bits if the *a priori* probability is less than 10^{-3}.

Of course, the classification in Table 10.1 is not the only possible one. (In fact, to a mathematician it will appear arbitrary, inconsistent, and quite unsatisfactory). Nor is it sufficiently detailed for practical application. As it happens, a number of parts-classification coding schemes have been implemented over the past 20 years under the rubric Group Technology (or GT) to facilitate the design of manufacturing systems. Herman Opitz's scheme, for example, is a five-digit (decimal) code [Opitz, 1970]. The Opitz code for a hexagonal nut is 30500. In this case the first digit (3) implies that the part is roughly rotational, with deviation from perfect symmetry, with a length/diameter ratio <2. The second digit (0) implies hexagonal shape. The third digit (5) implies a rotational internal shape with screw threads. The fourth digit (0) implies flat, unstructured, plane external surfaces; and the fifth digit (0) implies an absence of auxiliary holes in gear teeth (see Figure 10.1). Again, it is evident that the 10^5 different Opitz code possibilities are not equally likely to occur. The notion of a *random* part is itself not well-defined. As a *Gedanken* experiment, one can suppose that all manufactured products produced on a typical day next year are collected in a large warehouse. Suppose that each product is then dismantled and reduced to its individual component parts. The pile might contain billions of individual parts. Finally, suppose that all the parts are coded and sorted into 10^5 file drawers by means of the Opitz code. Some code numbers (such as 30500) would be used a great many times, whereas some others would be used very seldom.

The frequency distribution of parts across all possible code numbers should be proportional to the number of parts in each drawer. This frequency distribution describes the current relative probability of each code number. The information

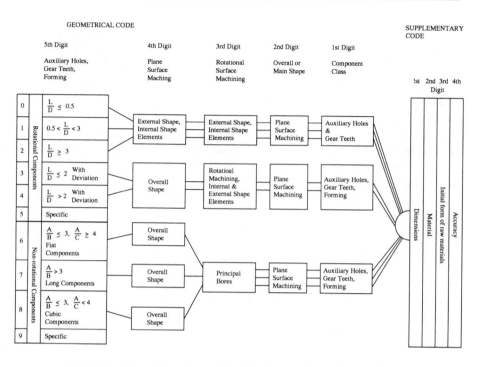

FIGURE 10.1. Workpiece Classification. Source: Optiz, 1970.

contained in the ith code number is the negative logarithm (base 2) of its probability of occurrence:

$$H_{\text{code}(i)} = -\log_2 P_i. \qquad (10.6)$$

Frequently encountered shapes have a relatively high probability and relatively low (3-8 bits) information value. Conversely, for very infrequently encountered or unique shapes, the information value can be arbitrarily high.

The problem of determining the information content associated with a unique or unusual shape can probably best be approached in another way. One suggestion is to count the number of instructions required to program a numerical control (NC)-machine tool to cut the shape. For a turbine blade, for instance, the program might require several thousand instructions (bits).

Once the shape category is fixed, the range of possibilities for parameter specification is much reduced. One can suppose, for convenience, that each of these parameters is a number (in some system of measuring units) specified to a standard accuracy of 1 part in 1,000. This level of accuracy, or tolerance, is essentially equivalent to 10 bits of information per parameter, because $2^{10} = 1024 \approx 1000$. Hence $\log_2 1000 = 10$ bits. (In cases where lesser or greater accuracy is needed, a suitable correction term is subtracted or added.) It follows that a localized plane surface defined to standard accuracy "embodies" about 30 bits, while two parallel planes separated by a fixed distance embody 40 bits. An intersection of N localized planes ($N \geqslant 4$) therefore embodies $30N$ bits. An intersection of M pairs of parallel planes ($M \geqslant 2$) embodies $40M$ bits. The intersection of three localized pairs of parallel planes embodies $3 \times 40 = 120$ bits of information (assuming standard accuracy).

The foregoing is actually an overspecification, since it also includes both location and orientation information that is not needed to define the shape itself. A pure shape is (by definition) invariant under all translations of the center of mass and all rotations around its center. Location and orientation in Cartesian space require six parameters to be specified, or 60 bits of information. This must be subtracted. The total amount of information embodied in simple shapes is therefore determined by (1) the shape category, (2) the number of surfaces, and (3) the form or order of the surface-defining equation (e.g., quadratic, cubic, etc.).

A tetrahedron defined by the common intersection of four nonparallel planes, independent of position or orientation, therefore embodies (4×30) 60 = 60 bits, or 10 bits for each of six edges that can be independently fixed. The parallelopiped above also embodied 60 bits of information, corresponding to 10 bits for each of three independent edges and three independent angles. For the set of shapes defined by the intersection of a plane surface and a conic section, the embodied information is, again, $30 + 90 - 60 = 60$ bits. (For an ellipsoid, such an intersection defines two or more shapes, like those of an orange cut into two parts. To specify which of the two parts is intended requires one additional bit of information.) By similar logic, a shape defined by a localized conic section (90 bits) intersected by two non-parallel planes (60 bits) embodies $60 + 90 - 60 = 90$ bits. A generalized conic section intersected by two parallel planes (40 bits) embodies $40 + 90 - 60 = 70$ bits. Both of these last two cases may have up to four distinct regions of intersection. To select one of the four possibilities requires an additional two bits ($\log_2 4$).

In practice, it would normally be most convenient to use a GT code, such as the Opitz code. Once a part is classified, the number of independent dimensional parameters is easily determined from the code. For example, the "hex nut" has at least six dimensional specifications altogether: external diameter, thickness, internal radius, and depth, width, and pitch of the thread groove. (In practice, others—e.g., bevels—may also be needed.) Each of these six parameters corresponds to 10 bits of information in the case of "standard" precision. In addition, the screw thread may be right-handed or left-handed. Thus, in addition to the code itself, the complete specification involves exactly 61 bits of information. To recapitulate: These are the minimum numbers of bits of information needed to specify the parameters of a given shape to a given level of precision.

One final clarification: The information required to specify a parameter depends on the specified precision. The most general expression for the information content of a single parametric specification is

$$H_{\text{par}} = \ln_2 \frac{m}{\Delta m},$$

where m is the dimensional measure and Δm is the allowable tolerance or uncertainty. Thus, when a tolerance is expressed as one part per thousand, one has immediately

$$H_{\text{par}} = \ln_2 10^3 \approx \ln_2(2^{10}) = 10 \text{ bits.}$$

While if the tolerance is one part per million, it immediately follows that

$$H_{\text{par}} = \ln_2 10^6 \approx \ln_2(2^{20}) = 20 \text{ bits.}$$

To summarize: the amount of useful morphological information embodied in a simple manufactured shape can be estimated by the following procedure:

1. Determine the applicable external constraints and the category specification by using, for example, any general-purpose GT code.
2. Determine the required precision of parameter specification. In general, it is reasonable to assume 10 bits per parameter, but this depends on the intended function of the item.
3. Determine the frequency distribution of part shapes among possible categories defined by the code. The code information is the logarithm of the (inverse) frequency of the code for the shape. (This could be a major research project, of course.)
4. Determine the number of parameters needed to select a specific shape *within* the designated code category, taking into account parallelism and concentrism, as appropriate, and multiply by the appropriate number of bits per parameter (based on required precision; see number 2 above). Add this number to the code information.

More-complex shapes can be constructed geometrically by combining or superimposing simpler shapes, either positive or negative (holes). The combination process is closely analogous to assembly, as will be seen. A rivet, for instance, is a simple cylindrical shape with a hemisphere at one end. A cylindrical hole is a negative cylindrical shape superimposed on a positive shape. Complex shapes in the real world may begin as simple shapes that undergo a sequence of forming operations. Alternatively, many complex shapes can be assembled or constructed from simpler shapes. This is discussed further in the next section.

As the acute reader will have realized, the main approach that has been discussed up to now, which can be characterized as the *decomposition* of complex shapes into simpler ones, is still comparatively limited in its practical utility.[4] In the first place, decomposability implies a pre-existing underlying classification of recognizable parts. Faces can be decomposed into *features*, such as eyes, nose, mouth, hairline, etc. But how do you decompose a lump of clay? In the second place, even within a classification system, the task of recognizing and classifying components is clearly limited by the precision of the representations and the idealized shapes.

The foregoing problems have been avoided, so far, by considering only externally constrained cases, e.g., where *ideal* surfaces are defined by equations of some order. One might postulate that all surfaces that would ever be of interest for manufacturing (if not sculpture) should, in principle, be definable by such an equation of some sort. But this is conjecture.

10.3. INFORMATION OF ORIENTATION AND ASSEMBLY

The process of creating complex shapes from simpler components (including holes) requires further elucidation. In such an assembly (or composition), each component must be correctly positioned (that is, oriented *with respect to all others*). For convenience, one may assume that the core shape is fixed in space, and another shape is brought to it, just as a component is added to an assembly. The information embodied in a specified combination of the two shapes is the sum of the information embodied in each shape separately plus a relative location and orientation specification that depends on the symmetry of each component. Information may also be lost if an interior surface is eliminated by the joining.

Assuming that complex geometric shapes can always be constructed by superposing simpler ones (both positive and negative) on one another, a straightforward decomposition strategy can be adopted:

1. Identify and classify the core shape, as distinct from any minor exterior projections or interior spaces, holes, slots, or grooves. There can be ambiguity in a few cases. For instance, is a gear tooth an exterior projection, or is the space between two gear teeth a slot? Most of these special cases are included under category 6 in Table 10.1 and can be treated separately.
2. Identify the core shapes of exterior projections, such as knobs, heads, etc.
3. Identify the core shapes of interior spaces within the main core shape.
4. Identify the core shapes of exterior projections within interior spaces, if any.
5. Identify the core shapes of interior spaces within external projections, if any.

This decomposition scheme is applicable to either 2-D or 3-D shapes. It can be continued indefinitely, although three or four steps will suffice for almost any part used industrially. (If more than five steps are required, the design is probably unnecessarily complex.) The core shapes, in turn, can be decomposed into surfaces. When the decomposition is complete, the vast majority of the surfaces will be either flat, cylindrical, or derivable from one or the other. A very few (but important) surfaces will have more-complex profiles, such as involute curves, parabolas, ellipses, etc. In the most general case, one may assume the major

TABLE 10.2. *Symmetry Information Embodied in Simple Shapes.*

Dimension Shape		Number of Equivalent Directional Orientations	H_{sym} (Bits)
2	Rectangle	2	$\log_2 2 = 1$
2	Equilateral triangle	3	$\log_2 3 = 1.585$
2	Square	4	$\log_2 4 = 2$
2	Equilateral pentagon	5	$\log_2 5 = 2.322$
3	Equilateral tetrahedron	12	$\log_2 12 = 3.585$
3	Rectangular parallelopiped	12	$\log_2 12 = 3.585$
3	Cube	24	$\log_2 24 = 4.585$
3	Right cone*	1,000	$\log_2 1,000 = 10$
3	Ellipsoid	1,000	$\log_2 1,000 = 10$
3	Right cylinder	2*1,000	$10 + \log_2 = 11$
3	Sphere	$(1,000)^3$	$3\log_2 1,000 \approx 30$

*By the convention on standard precision (see text), we assume 1,000 distinguishable angular orientations around one axis. If a greater precision is assumed, the number of bits associated with orientation of a cone, ellipsoid, cylinder, or sphere is correspondingly greater.
Adapted from Sanderson, 1984.

(core) piece to be fixed in space and an asymmetric projection or hole to be located and oriented exactly with respect to a coordinate system embodied in it. This requires 6 degrees of freedom and corresponds roughly to 60 bits of information, as argued above. Symmetries can reduce the orientation-information requirement, however. This is clear from the observation that a non-symmetric shape looks different from every orientation, whereas a symmetrical shape looks exactly the same in two or more different orientations. This is important when a symmetrical shape is added to (or subtracted from) an oriented core shape: *The more symmetrical the second shape is, the more equivalent ways there are to bring them together and the less information is needed to specify the superposition.* It is convenient to define a symmetry correction term H_{sym}, which has to be subtracted from the total information embodied in the combined shape. A few illustrative values of H_{sym} are listed in Table 10.2. If the piece to be added or subtracted is rotationally symmetric about one axis, for instance, there is a reduction of 10 bits in the amount of additional orientation information required. In practice, most holes are round.

It is also important to think about the case where the core shape also has symmetry. It is clear that the two situations are conceptually equivalent: Either piece could be designated as the core. The symmetry information must again be subtracted. If both pieces are symmetric, however, it does not follow that there must be a subtraction for each. On the contrary, whichever piece is selected as the core can be oriented arbitrarily, by assumption. Its symmetry does not reduce the information required for superposition.

When physical shapes are permanently joined, the attached surfaces disappear and become indistinguishable. This results in lost information. To take a simple example, suppose we join two right cylinders of the same diameter together, end to end, to form a longer cylinder. The two joined plane surfaces simply disappear as far as the final shape is concerned. The information embodied in the two adjoining surfaces is lost.[5]

Moreover, the two cylindrical surfaces also merge and become one. Suppose the two original right cylinders each embodied $40 + 30 - 50 = 20$ bits of information in their shapes, the final merged cylinder also embodies 20 bits, even though an additional $50 - 2 = 48$ bits of information is subtracted from the orientation information. (Because each cylinder is symmetric end to end, there are actually four possible ways of joining them. Selection of any one of these four possibilities creates $\log_2 4 = 2$ bits of information, which, in principle, are required to position and orient the two original cylinders exactly end to end.) Thus, $20 + 50 - 2 = 68$ bits of information are actually lost in this particular assembly process. (Notice that the more symmetric the parts being joined, the less information lost in superposition or assembly.)

A different situation arises when one disk has greater diameter than the other. In this case, only one of the plane surfaces the side of the smaller disk completely disappears. The surface with the larger area does not disappear, although it is partly covered up, and only the information associated with flat end of the smaller disk is actually lost.

The simple shape in Figure 10.2 yields several observations. Given that it belongs to category 5 in Table 10.1 it can be defined as the intersection of four parallel planes at right angles with two concentric infinite cylinders. The first plane is specified by three parameters (30 bits), and the other three planes by one parameter each, or 30 bits in all, making a total of 60 bits. The first infinite cylinder orthogonal to the plane embodies 40 bits, and the second (concentric) cylinder adds 10 bits of information. Thus, the complete shape embodies $60 + 50 - 60 = 50$ bits of information (there are two radii and three lengths).

The shape can be decomposed in two ways: (1) as a superposition of three cylinders, merged end to end, or (2) as a superposition of two cylinders, one of which fits inside an axial hole in the other. In the first case, the three simple right cylinders each embody $40 + 40 - 60 = 20$ bits of information, or 60 in all. After the end-to-end assembly [requiring $(2 \times 50) - 3$ bits of positional and orientational information], one is left with a final shape embodying only 50 bits, as noted above, which means $60 + 97 - 50 = 107$ bits of information are lost in the assembly process. Now consider the other starting point: two cylinders, one hollow. The hollow cylinder embodies $40 + 50 - 60 = 30$ bits, while the solid cylinder embodies 20 bits, or 50 bits altogether. After combination, the resulting shape still embodies 50 bits, as noted, and $50 - 1 = 49$ bits of positional and orientational information are required to bring the two cylinders together. In this case, the information loss is 49 bits, less than half of loss in the first case.

The above example illustrates two key points:

1. There may be several ways to decompose a complex shape into simpler

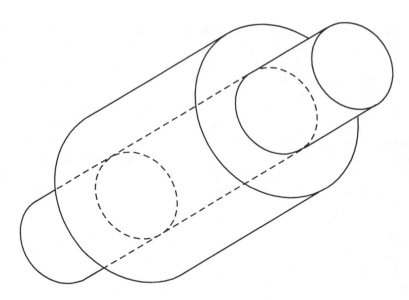

FIGURE 10.2. Superposition of Cylinders.

ones and, conversely, several ways of composing (i.e., assembling) a complex shape from simpler ones. In any decomposition into independent subsystems, the information embodied in the components must equal the information embodied in the final shape.

2. The amount of information lost in the composition (i.e., assembly) process depends strongly on the number of components. The fewer distinct components required, the less information lost.

The preceding discussion suggests a fundamental measure of efficiency: the ratio of information embodied in the final shape per se to information embodied in the component shapes plus information required for positioning and orientation. The information efficiency of constructing the shape in Figure 10.2 from three cylinders is evidently $50/157 \approx 0.32$, while the information efficiency of constructing it from two cylinders (one hollow) is $50/99 \approx 0.50$.

10.4. THE RELATIVE PRICES OF METABOLIC AND MORPHOLOGICAL INFORMATION

As previously suggested, manufacturing can be considered as consisting of two distinct information-conversion processes:

1. metabolic (or materials-processing) activities.
2. morphological (shaping or forming) activities.

It follows that manufacturing value added can be subdivided into two components, viz.,

$$V_{mfg} = V_{metab} + V_{morph} = P_{metab}H_{metab} + P_{morph}V_{morph}. \qquad (10.7)$$

The first (metabolic) component would seem, at first glance, to include most or all of the following processes:

- extraction or winning
- beneficiation (physical separation)
- digestion or leaching
- carbothermic or electrolytic reduction
- refining (including petroleum)
- alloying
- chemical synthesis
- food processing
- dehydration, calcination
- distillation and related separation processes

Industries using these processes are also the greatest users of energy, in relation to value added. The five most energy-intensive sectors in the United States are as shown in Table 10.3.

These five sectors together accounted for about \$152 billion in value added, and 8.55×10^{15} BTU (quads) in *purchased energy* consumption and about 13.6 quads in *total energy* consumption. The difference is accounted for by energy derived from waste materials, such as wood. Part of the energy is used to transform crude, petroleum, and chemical feedstocks into more useful forms, and part of it is used to transform fossil fuels into electricity. (Electricity is counted at its thermal-equivalent value: 3,412 BTU per kWh.) Detailed analysis of processes [e.g., Battelle, 1975] shows that most of the remainder is used to separate metals from ores and to increase the free energy or available useful work in structural materials prior to subsequent shaping, forming, and assembly. Of course, the free energy in the fuels and electricity used in the manufacturing processes is simultaneously dissipated and lost.

In the case of the chemicals industry, the biggest user of processing energy are simple minerals and hydrocarbon feedstocks, mainly sulfur, sodium chloride, nitrogen, oxygen, methane, propane, butane, and benzene, toluene, and xylene. They are first converted to more reactive chemicals, such as sulfuric acid, chlorine, caustic soda, hydrochloric acid, ammonia, acetylene, ethylene, propylene, methanol, ethanol, and so on. Except for the production of sulfur dioxide from sulfur, most of these first-stage reactions are endothermic, which means they require substantial amounts of processing energy from an external source. This is where the *purchased energy* in the chemical industry is largely used. In most cases, the chains of subsequent reactions to produce more-complex chemicals are actually exothermic, or self-energized. (There are, obviously, many exceptions. For instance, several important polymerization reactions are endothermic and

TABLE 10.3. *Metabolic Processing Activities.*

Sector	Purchased Energy 1980 (10^{12} BTU)	Total Energy* 1980 (10^{12} BTU)	Value Added 1977 ($ Billions)
Chemicals	2,717	3,354	56,721
Primary metals	2,277	3,712	37,568
Petroleum refining	1,178	3,061	16,378
Pulp and paper	1,278	2,328	22,171
Stone, clay and glass	1,132	1,132	19,130

*Excluding feedstock energy ultimately embodied in product, but including "waste fuels" derived from feedstock. Energy data for 1980 were complied by Doblin [1987] from the Census of Manufacturers and various special surveys. Unfortunately, comparable data for more recent years have not been published. Value added figures are also difficult to obtain, but are being published on a regular basis; 1977 was the nearest year I could find. Using figures for exactly corresponding years would not affect the conclusions significantly.

cannot proceed spontaneously.) Here the energy is extracted from the intermediates, whence the free energy of the products is less than the free energy of the intermediate inputs. One estimate [Burwell, 1983] puts the non-purchased fraction of total energy at 15 percent, implying that about 25 to 30 percent of feedstock energy is lost in conversion.

In the case of the petroleum industry, processing energy is used both for separation (distillation) and for cracking, reforming, alkylation, and other processes to increase the gasoline yield per barrel of crude oil and to purify the products, especially by removing sulfur. The industry both consumes feedstocks—mainly liquid propane gas from natural gas liquefaction plants—and produces them mainly ethylene, propylene, butylene, and B-T-X. Roughly 10 percent of the free energy originally in the crude oil is lost in these various conversion processes.

In the case of pulp and paper, processing energy is needed mainly to get rid of excess water and recycle the various leaching chemicals. About 40 percent of the total processing energy used in the industry is now derived from the burning of waste lignin and cellulose. In principle, this figure could be much larger, but more-efficient use of the biomass energy is inhibited by the large amounts of water used in all the pulping and digestion processes. An idealized papermaking process would produce net free energy, not consume it.

The primary-metals industry has four distinct branches: ferrous and non-ferrous, and primary and secondary for each. The primary ferrous branch extracts iron ore (Fe_2O_3, Fe_3O_4), smelts the ore with coke in a blast furnace, and then refines the impure pig iron by reacting it with oxygen in a basic oxygen furnace. The final product, pure iron or carbon steel, has a much higher free energy than the ore from which it was extracted. (It could actually be "burned" again as fuel.) The overall process, however, involves a loss of all the free energy in the various fuels, especially coking coal. Overall efficiency in these terms is currently around 30 percent of ideal efficiency [Gyftopoulos *et al.,* 1974]. So-called electric steel and ferro-alloys are the secondary branch of the ferrous-metals sector. It is based on remelting and repurifying scrap. In this case, there is essentially no gain or loss in free energy of the steel, although the electric energy (for melting) is lost.

Primary non-ferrous metals can be subdivided into those with oxide ores (Al) and sulfide ores (Cu, Pb, Zn). In both cases the ore beneficiation process is very energy-intensive. Aluminum ore (bauxite) is first converted to nearly pure alumina (Al_2O_3) by a chemical leaching/dehydration process. The dry alumina is then reduced to metal in an electrolytic cell. The process is highly endothermic. The free energy of the product aluminum is, of course, much greater than that of the ore. (In principle, it, too, could be burned as a fuel.) However, much more processing energy is lost.

Copper lead and zinc are *chalcophile* (sulfur-loving) metals. Their ores tend to be rather low in grade and must usually be finely ground and beneficiated by a physical process, such as flotation/filtration. This, again, is very energy-intensive. Subsequently, the concentrate is "roasted" to drive off sulfur and arsenic subsequently recovered in modern plants and the concentrate is then smelted in a furnace. A final electrolytic purification stage is needed for copper. The first (roasting) stage is theoretically exothermic, although fuel is used to speed it up, but the second stage is endothermic. In principle, the combined roasting–smelting process with sulfur recovery ought to produce net free energy; in practice it never will. The final purification steps to eliminate or recover minor impurities, such as gold and silver, cadmium, tellurium, and selenium, are quite energy-intensive in the aggregate. In fact, pure copper requires nearly as much energy to produce, in practice, as aluminum.

The stone, clay, and glass sector consumes energy mainly in the manufacture of quicklime (CaO), hydraulic cement, and plaster of paris, and in the melting of glass. The first and second of these involve calcining—the use of heat from fuel to drive CO_2 and H_2O away from hydrated calcium carbonate (limestone). The third process (to make plaster) is a simple dehydration—the use of heat to drive H_2O away from a mineral calcium sulfate (gypsum). The resulting materials are eager to recombine with water, yielding an inert mineral solid and releasing heat in the process. This actually occurs when these building materials are used by the construction industry. Thus, the free energy in both the initial and final materials is equally zero. The processing energy used in this materials sector has only one practical function, namely, to enable the materials to be "fluidized" for purposes of forming and shaping. The same thing is also true for glass. Brickmaking, also in this sector, is essentially a forming/shaping activity.

Evidently, part (perhaps 20 percent) of the energy used in the primary metals sector for melting and casting and all the energy used in stone, clay, and glass sector are really attributable to forming and shaping, not extraction or refining.

By comparison, the five sectors covered in Table 10.3 accounted for $246.2 billion in value added and 1.4 quads of energy consumed, mostly as electricity

TABLE 10.4. *Morphological (Forming Shaping) Activities.*

Sector	Purchased Energy 1980 (10^{12} BTU)	Value Added 1977 ($ Billions)
Fabricated metal products	362	45,512
Non-electrical machinery	337	67,223
Electrical machinery and electronics	240	50,366
Transportation equipment	344	64,291
Instruments and related products	80	18,762
Total	1,363	246,200

(see Table 10.4). The products of these sectors are metal components or machines and instruments of varying degrees of sophistication. Many processes are used in these sectors, but most of the energy is used for the following:

- casting (foundry)
- stamping, bending
- cutting (drilling, boring, machining)
- grinding
- welding and soldering
- assembly

Much less energy is consumed per dollar of valued added, and the free-energy content of the final products is invariably *less* than the free-energy content of the purchased materials from which they are made. If the energy used in the stone, clay, and glass sector and 20 percent of the energy used in the primary metals sector (and 30 percent of the valued added) are attributed to forming and shaping, then we have roughly the following comparisons (considering only ten sectors):

1. Metabolic activities: separation, reduction, refining, purification, and synthesis of materials:

   a. Value added (1977): $141 billion
   b. Energy (1980): $12.1*10^{15}$ BTU
2. Morphological activities: melting or liquefaction of materials for purposes of forming/shaping, forging, bending, pressing, cutting, grinding, joining, and assembly:

   a. Value added (1977): $257.5 billion
   b. Energy (1980): $2.87*10^{15}$ BTU

For completeness, it may be noted that the remaining ten manufacturing sectors normally included in manufacturing had a total value added of $186.5 billion and a total energy consumption of less than 2×10^{15} BTU. Nearly half of this energy was used in the food-processing sector, which is more nearly metabolic than morphological.

Summarizing, it is clear that in the manufacturing sectors

$$V_{\text{metab}} < V_{\text{morph}}, \tag{10.8}$$

whence

$$P_{\text{metab}}H_{\text{metab}} < P_{\text{morph}}H_{\text{morph}}. \tag{10.9}$$

It was established in the preceding sections that, if H is always measured in bits, the useful shape information for a standard machine part is of the order of magnitude of 10 bits per parameter (for precision of 10^{-3}) plus a few more bits for the code specification. The total would usually be less than 100 bits and seldom more than 1,000 bits. In the assembly process, information is lost, not gained, so the morphological information embodied in a machine with 1,000 parts would be of the order of 10^5 to 10^6 bits, even allowing for a few specialized parts with moderately complex shapes.[6] A highly sophisticated machine, such as a helicopter, might have 10^5 or possibly 10^6 parts, many of them individually complex and requiring high precision. Yet the total morphological information content could scarcely exceed 10^8 or 10^9 bits. By comparison, the thermodynamic information embodied in any metal alloy or synthetic chemical is likely to be of the order of 10 to 100 kT per mole or 10^{24} to 10^{25} bits/mole. (A helicopter or jet engine would require 10^3 or 10^4 moles of mass.)

An obvious implication of these facts is that the ratio of metabolic to morphological information in the economic system is currently in the neighborhood of 10^{20}. Hence, for the manufacturing sectors of the economy, it is certainly accurate to say that

$$H_{\text{morph}} \ll H_{\text{metab}}. \tag{10.10}$$

It follows from (10.9) and (10.10) that

$$P_{\text{morph}} \gg P_{\text{metab}}. \tag{10.11}$$

The foregoing analysis can also be used to estimate, at least roughly, the actual value of P_{metab}. Here I make use of the assumption that H_{metab} is proportional to the free energy or available work B consumed or dissipated (Chapter 2, equation 2.35). This, in turn, is essentially equal to the free energy stored in the fuels used up or, using equation (9.4) in Chapter 9.

$$\Delta H_{\text{metab}}(\text{bits}) = \frac{\Delta B}{T_0}. \tag{10.12}$$

Here T_0 is the temperature of the ambient environment (i.e., the earth's surface), and ΔB is the change in available free energy. Thus,

$$P_{\text{metab}}(\$) = \frac{V_{\text{metab}}(\$)}{H_{\text{metab}}(\text{bits})} = \frac{V_{\text{metab}} * T_0}{B}, \qquad (10.13)$$

where P_{metab} is given in \$/bit.

Substituting $V_{\text{metab}} \simeq 140 * 10^9$, $T_0 = 300$ K, and $B \simeq 12 * 10^{15}$ BTU $= 1.266 * 10^{19}$ joules, one finds for metabolic inputs

$$P_{\text{metab}} \simeq 3.3 * 10^{-6} \quad (\%/\text{joule K}).$$

Each joule/K is equivalent to 10^{23} bits, so the price per bit of metabolic inputs is, very roughly

$$P_{\text{metab}} \simeq 3.3 * 10^{-29} \quad (\$/\text{bit}).$$

Assuming processing industries are operating at 5 percent of theoretical (second law) efficiency, this translates to 6.6×10^{-28} \$/bit for information embodied in finished materials.

By the arguments above, we see that P_{morph} is of the order of 10^{19} to 10^{20} larger than P_{metab} or, roughly,

$$P_{\text{morph}} = 3 * 10^{-9}. \quad (\$/\text{bit})$$

In words, it costs 10^{19} to 10^{20} times as much to embody a bit of *useful morphological information* in a manufactured product as it does to embody a bit of *useful thermodynamic information* (as free energy) in finished materials. These results are highly insensitive to the exact years for which the data were taken.

10.5. THE PRINCIPLE OF MINIMUM MORPHOLOGICAL INFORMATION

Four conclusions follow immediately from the fact that morphological information is extremely expensive compared to thermodynamic or metabolic information.

First, whenever there is a clear tradeoff between the composition of a product and its design (form and shape), the former should always be preferred. In practice, this means materials should be selected, wherever possible, to simplify design and, in particular, to minimize forming/shaping and assembly operations.

Second, it appears to follow that materials research and development should focus on two objectives:

1. Making materials that are easier to form and shape, consistent with the desired properties of the object and reflecting all stages of the product life cycle, from assembly to disposal and recycling. (There are tradeoffs among these requirements.)

2. Giving materials new physical properties, especially by deliberately introducing inhomogeneities (such as alternate layers of conductors and insulators) or properties that can be altered reversibly by external influences, thus permitting *monolithic* materials to substitute for assemblies of many parts.

Third, manufacturing research should focus on ways to increase the overall efficiency of forming, shaping, and assembly operations. Given that the design of a product specifies its shape, dimensions, tolerances, and the choice of material, an optimal method must still be determined for shaping and dimensioning to the required tolerances. This involves the selection of a sequence of unit operations that will convert a "formless" piece of the desired material into its final shape. To achieve this conversion, there are three generalized strategies, which may be termed (1) material addition or build-up, (2) material subtraction, and (3) "pure" deformation.

In practice, all three strategies and combinations of them are used in different situations. One cannot eliminate any of them from consideration *a priori*, because the functional requirements of the product may dictate constraints that make it impossible to select one strategy for all purposes. For example, a build-up or assembly approach is, to some extent, unavoidable if the product necessarily involves combinations of different materials, such as conductors and insulators. It is also unavoidable if there are moving parts in juxtaposition with fixed ones, especially if as is often the case the moving parts are largely contained within a fixed shell, as they are in a motor, compressor, pump, engine, or transmission.

Fourth, design research should seek to identify designs with *minimum morphological information*. This is consistent with some of the familiar rules of thumb, such as minimizing the number of distinct parts, especially connectors, and the number of high-precision interfaces. It is also essentially identical to the design axioms formulated by Suh *et al.* [1978]; and Nakazawa and Suh [1984].

Nevertheless, until recently, designers have sought relatively few opportunities to minimize information lost in manufacturing. Boothroyd and Dewhurst [1983] may have been the first to see the potential. They have been able to demonstrate impressive savings at the design stage by reducing the number of distinct parts, such as connectors. The principle of parts minimization is clearly implied by the larger principle of minimum information in design. At the same time, it is also compatible with the principle of minimum information loss in manufacturing. Fewer parts mean fewer manufacturing operations, fewer assembly operations, and less cost.

To determine the minimum (theoretical) number of parts in a product, Boothroyd and Dewhurst suggest the following criteria:

1. Does the part move relative to all other parts already assembled?
2. Must the part be of a different material than, or be isolated from, all other

parts already assembled? (Only fundamental reasons concerned with material properties are acceptable.)

3. Must the part be separate from all other parts already assembled because, otherwise, necessary assembly or disassembly of other parts would be impossible?

The number of distinct parts can be reduced, in some cases, only by making some individual parts more complex and costly. This may require more-sophisticated methods in another manufacturing domain. Each case certainly requires detailed analysis, but the foregoing analysis strongly suggests that, on balance, the extra cost of manufacturing some components will be more than justified by savings at the assembly stage.

The principle of minimizing embodied design information goes beyond the formulation of Boothroyd and Dewhurst. It also implies that the designer should seek to minimize the number of dimensional parameters needed to describe the product, as well as the precision with which those parameters need to be specified. Again, the designer must take into account tradeoffs between requirements for the different stages of the product life cycle (from assembly to disposal).

Similarly, it can be argued that any process involving the creation of a complex shape by subtraction (as a sculptor creates a shape from a block of stone) is likely to be inherently inefficient. This is so particularly where the starting point is a precisely dimensioned metal block or cylindrical rod, much of which is subsequently cut away. Again, morphological information is lost.

The ideal forming process, by the foregoing criteria, would require neither the addition nor the subtraction of materials from the net final shape. (This is called *near net shape forming*). Either addition or subtraction involves waste. Examples of pure deformation processes include injection molding (for plastics), die casting, and investment casting; a variant substitutes metal or ceramic powders for molten materials, followed by sintering. Solid deformation processes, such as rolling, extrusion, bending, and forging, are also inherently efficient to the extent that they can be precisely controlled.

The loss of position/orientation information in assembly is a major cause of inefficiency. The problem is most acute in a traditional multi-purpose machine shop, where individual machines are used successively to process batches of components, and each batch of parts is piled in a bin prior to the next operation. As the part comes off the machine, its position and orientation are precisely determined. Yet the information is lost when the parts are dropped into the bin and must be reestablished later when a worker picks up the part, orients it, and positions it for the next operation. The benefits of mechanical transfer systems are easily explained in terms of *not losing* this information. An automatic transfer system, such as a conveyor belt, delivers the parts to the next station with reduced positional and orientational indeterminacy and, hence, less information is needed to reorient the part. A synchronous transfer machine can deliver parts in *exactly* the required position and orientation, with no loss of information.

Of course, such machines (like all forms of physical capital) embody a great deal of morphological information in their design and construction, which is gradually lost as the machine wears out. (In fact, a very small loss of morphological information, in a critical place, can render a machine non-functional.) Unlike materials *per se,* morphological information is not conserved. The major drawback to a more-widespread use of such machines, however, is their inherent inflexibility. Technological advances may ameliorate this drawback in the near future.

Thus, the trend toward hard automation in materials-handling can be seen as a response to the imperative of reducing morphological information losses. Much the same can be said of recent trends to improve product quality by substituting computers for human workers in certain operations. Clearly, defects are a costly form of lost information, and it is increasingly easier to introduce machines than to detect and correct errors once they are embodied in a faulty component or assembly. The principle of minimum morphological information loss seems to explain much of what is going on in industry today.

ENDNOTES

1. Material in this chapter has previously appeared in [Ayres, 1987].

2. To a mathematician, this classification will inevitably be unsatisfying if not worse. It is not a classification based on group-theoretic concepts. It is based, rather, on industrial experience and practice. This underlines the point made above, that useful information about shapes must include (or reflect) the underlying relationships between form and function.

3. See, for example, Burbridge [1975]; Devries *et al.* [1976]; Edwards [1971]; Gallagher and Knight [1973]; Ham and Ross [1977]; Mitrafanov [1966]; and Opitz [1970].

4. These problems are currently at the research frontier for both computer-aided design (CAD) and machine vision (visual pattern recognition).

5. In this context, note that D-information (or entropies) for subsystems are additive if and only if the subsystems are independent. Joining geometric shapes violates the independence condition.

6. Thoma estimated the morphological information in a steam locomotive to be $8*10^4$ bits (assuming 8,000 different specifications); for a diesel locomotive his estimate was $3.3*10^5$ bits.

7. It hardly needs to be pointed out that integrated circuitry and silicon chips exemplify the monolithic approach. There are, however, a number of other, less-familiar examples.

REFERENCES

Ayres, Robert U., "The Entropy Trap: Is It Avoidable?" in: Kaplan (ed.), *East Asian Culture and Development*, Paragon House, New York, 1986. [An ICUS book]

Ayres, Robert U. "Manufacturing and Human Labor as Information Processes" IIASA Research Report RR-87-19, International Institute for Applied Systems Analysis, Laxenburg, Austria July 1987.

Boothroyd, Geoffrey and Peter Dewhurst, *Design for Assembly*, University of Massachusetts Department of Mechanical Engineering, Amherst, Mass., 1983. 2nd edition.

Burbridge, J. L., *The Introduction of Group Technology*, John Wiley and Sons, New York, 1975.

Burwell, C. C., *Industrial Electrification: Current Trends*, Technical Report (83-4), Oak Ridge Associated Universities Institute for Energy Analysis, City, State, February 1983. [ORAU/IEA]

Cosgriff, R. L., *Identification of Shape*, Report (820-11), Ohio State University Research Foundation, Columbus, Ohio, December 1960. [ASTIA AD254-792]

Devries, M. F., S. M. Harvey and V. A. Tipnis, *Group Technology: An Overview and Bibliography*, Technical Report (MDC 76-601), Machinability Data Center, Cincinnati, Ohio, 1976.

Doblin, Claire P., *The Impact on Energy Consumption of Changes in the Structure of US Manufacturing: Part I: Overall Survey*, Working Paper (WP-87-04), International Institute for Applied Systems Analysis, Laxenburg, Austria, February 1987.

Edwards, G. A. B., *Readings in Group Technology*, Technical Report, Machinery Publishing Company, London, 1971.

Gallagher, C. C. and Knight, W. A., *Group Technology*, Butterworths, London, 1973.

Gyftopoulos, E. P., L. J. Lazaridis and T. F. Widmer, *Potential Fuel Effectiveness in Industry* [Series: Ford Foundation Energy Policy Project], Ballinger Publishing Company, Cambridge, Mass., 1974.

Ham, I. and D. T. Ross, *Integrated Computer-Aided Technology (ICAM) Task-II Final Report*, U.S. Air Force Technical Report (AFML-TR-77-218), Wright-Patterson Air Force Base, Dayton, Ohio, December 1977.

Mitrafanov, S. P., *Scientific Principles of Group Technology*, (Russian), Mashinestroyesle, Moscow, 1970.

Nakazawa, Hiromu and Nam P. Suh, "Process Planning Based on Information Concept," *Robotics and Computer-Integrated Manufacturing* 1(1), 1984 :115-123.

Optiz, Herman, *A Classification System to Describe Workpieces* 1and2, Pergamon Press, London, 1970.

Persoon, E. and K. S. Fu, *Shape Discrimination Using Fourier Descriptors*, Second International Joint Conference on Pattern Recognition, Copenhagen, Denmark, August 1974.

Sanderson, Arthur C., "Parts Entropy Method for Robotic Assembly System Design," in: *IEEE International Conference on Robotics*, IEEE, City, State, 1984.

Suh, Nam P. *et al.*, "On an Axiomatic Approach to Manufacturing Systems," *Journal of the Engineering Industry (Transactions ASME)* 100, 1978 :127-130.

Thoma, Jean, *Energy, Entropy and Information*, Research Memoranda, International Institute for Applied Systems Analysis, Laxenburg, Austria, June 1977.

Wilson, David R., *An Exploratory Study of Complexity in Axiomatic Design*, Ph.D. Thesis [Massachusetts Institute of Technology, Cambridge, Mass.], August 1980.

Zahn, C. T. and R. Z. Roskies, "Fourier Descriptors for Plane Closed Curves," *IEEE Transactions on Computers* C21(3), March 1972.

Battelle-Columbus Laboratories, *Energy Use Patterns in Metallurgical and Nonmetallic Mineral Processing (Phase 4: Energy Data and Flowsheets, High Priority Commodities)*, Interim Report (S0144093-4), Battelle-Columbus Laboratories, Columbis, Ohio, June 17, 1975. [Prepared for U.S. Bureau of Mines].

Chapter 11

Labor as an Information Process[1]

11.1. ERGONOMIC BACKGROUND: THE WORKER AS INFORMATION PROCESSOR

With certain obvious exceptions, human-labor outputs are, in the first instance, *motions*. This chapter is concerned specifically with motions involved in manufacturing processes. For my purpose here, it is appropriate to consider workers as information processors who convert information inputs (mainly sensory data) to some kind of output (generally, decisions). Output rates of human workers differ widely in various situations. The information *output* of a worker in a manufacturing environment is, by definition, his or her information *input* to the process itself, but it is not necessarily the amount of morphological information embodied in the workpiece being processed. The latter would have to be considered independently, as discussed in Chapter 10. In general, the control information output of a worker is much smaller than the information content of sensory input. In effect, a great deal of information is wasted in every process, as schematically indicated in Figure 11.1.

Information inputs to humans in all situations—not only the workplace—are initially conveyed by the five senses: vision, hearing, touch, taste, and smell. It is known from a variety of evidence, including the amount of brain tissue devoted to each type of sensory-signal processing, that, for humans, vision accounts for as much as 90 percent of total information inputs, while touch and hearing account for most of the remainder (see Table 11.1). This does not imply that vision is the only sense that must be considered. In a number of workplace situations, audio or tactile information is at least as important as vision. One example is inserting a nut on a bolt, or tightening it. Indeed, in some industrial tasks other senses (hearing, smell, even taste) are critical. Many assembly operations are difficult or impossible without tactile feedback. However, the inability of humans to turn off their senses (except vision) means that significant amounts of input information are wasted in most cases.

Estimates of the (input) transmission capacity of the human eye vary consid-

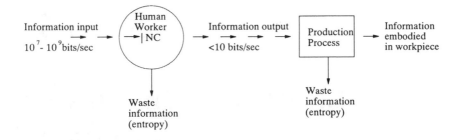

Inputs are variously estimated. The information INPUT rate
to human eyes has been estimated as 4 x 10^{6}bits/second [Flohrer 72],
10^{7}bits/second [Kalff 74], and 10^{9}bits/second [McCormick 70]

FIGURE 11.1. Sensory Information.

erably, depending on the measurement technique used. For instance, combining
data on monocular visual acuity and so-called flicker-fusion frequency leads to
an estimate of 4.3×10^6 bits/sec [Jacobson, 1951]. This implies an estimate of
4.3×10^6 bits/sec for each optic nerve fiber. Other data have suggested greater
input capacities, e.g., 5×10^7 bits/sec [Marko, 1967]. Visual-input capacity is
evidently a function of luminance (light intensity) and may be as high as 109
bits/sec at high luminance levels [Kelley, 1968]. The quantity of information
that can be extracted from visual or audio inputs (i.e., **channel capacity**) can be
estimated in terms of human ability to discriminate among stimuli and respond
suitably. Response normally involves a physical action. The ability to discriminate varies according to the nature of the sensory input, as shown in Table 11.2,
which assumes no time limitations.

The speed with which a single cell can detect and respond to a stimulus is

TABLE 11.1. *Sensory Inputs (Bits/Second).*

Sense	Number of Receptors	Number of Nerve Fibers
Vision	2*10^8 (retina)	2*10^6 (optic nerve)
Hearing	3*10^4	2*10^4
Touch	5*10^5	1*10^4
Smell	1*10^7	2*10^4
Taste	1*10^7	2*10^3
Pain	3*10^6	
Heat	1*10^4	1*10^6
Cold	1*10^5	

Source: Marko, 1967.

TABLE 11.2. *Information in Sensory Stimuli.*

Sensory modality and stimulus	$W=$ # of Levels that Can Be Discriminated on Absolute Basis	$H=\ln W=$ # of Information Bits Transmitted*
Vision, single dimensions:		
Pointer position on linear scale	9	3.1
Short exposure	10	3.2
Long exposure	15	3.9
Visual size	5-7	2.3-2.8
Hue	9	3.1
Vision, combined dimensions		
Size, brightness and hue	17	4.1
Hue and saturation	11-15	3.5-3.9
Audition, single dimensions		
Pure tones	5	2.3
Loudness	4-5	1.7-2.3
Audition, combined dimensions		
Six variables	150	7.2
Odor, single dimension	4	2.0
Odor, combined dimensions		
Kind, intensity and number	16	4.0
Taste		
Saltiness	4	1.9
Sweetness	3	1.7

*Amount of information in absolute evaluations on different stimulus dimensions.

limited by the rate at which certain chemical reactions take place and chemical signal carriers diffuse physically through the cell. This translates into a maximum rate at which a cell can process information. For most single cells, the maximum output rate is about 4,000 bits/sec.

In the case of an organ consisting of a functional structure with several kinds of different cells, stimuli are initially received by one type of cell (and not, in general, simultaneously). As the cells individually respond by producing a chemical signal carrier, the concentration of that chemical gradually builds until some threshold is reached, at which point a response can be defined for the organ as a whole. Animal organs differ in sensitivity and response rate. Eyes and ears are obviously specialized for large capacity and rapid response; but the effective range seems to be 340 to 460 bits/sec, depending on the nature of the input stimulus and the output response.

Within a single cell, the processes of detection and response are essentially indistinguishable, because the internal processes are all chemical reactions. An organism, however, undergoes a sequence of internal stimulus-response stages culminating in a physical motion. Again, an inherent latency period characterizes each step in the sequence. The longest latency period is probably associated with initiating a muscular contraction. Each muscle involved in a motion must relax and return to its original condition and position before it can repeat that motion. The latency period for hand motions is typically about 0.3 seconds. Taking into account relaxation or latency times, information outputs (responses) characteristically rise linearly for low input rates, reach a maximum when the channel capacity is saturated, and decline as problems of interference and overloading increase.

11.2. EXAMPLES: THEORY AND EXPERIMENT

Typing exemplifies a task in which information-output computations are comparatively straightforward. A very good typist can sustain a rate of more than 100 words per minute, or around 8 characters per second, for short periods, or 5 to 6 characters per second for extended periods. If the typist were selecting each character from equally probable possibilities, the information content of each character in a 26-letter alphabet (plus a space) would be $\ln_2 27$, or 4.75 bits. However, taking into account unequal probabilities of occurrence of various letters alone, the information content drops to 4.03 bits.

Taking into account preferential combinations of 2 letters, 3 letters, and N letters brings the figure down to less than 3.3 bits. Further taking into account the fact that language consists of words, sentences with unequal word frequencies, and strings of words subject to grammatical rules adds significantly to the redundancy. Shannon [1951] calculated some of these factors from published frequency tables and estimated others (allowing, also, for the information content of punctuation, capitalization, etc.). Shannon concluded that the information content of English is about 1.5 bits/character [see also Cherry, 1978]. This implies that an expert typist (100 wpm) can achieve a maximum short-period output rate of 12 bits/sec and a sustained output of 7.5-9 bits/sec.

Direct measurements, based on various experimental arrangements, have been made to determine maximum channel output capacities, or output rates, of subjects doing various activities for short periods. Data from a number of experiments involving human response to stimuli are summarized in Figure 11.2. Reaction times per bit of information processed range from 0.2 to 1.2 seconds for 1 bit, and from 0.5 to 2.8 seconds for 4 bits. These data imply human processing rates of 0.8 to 5 bits/sec for 1 bit and from 1.4 to 8 bits/sec for 4 bits, depending on the experimental situation. When specialized learned skills, such as typing, reading or piano playing, are called upon humans can respond at rates up to 24 bits/sec (see Table 11.3). The maximum response rate for human groups or organizations is lower (3-4.5 bits/sec). Whether the information processor is a cell, an organ, or an organism, there is a tendency toward *overload* if the input rate is too high. Successive stimuli begin to interfere with one another and cause a saturation effect. One cause of this difficulty is that the detector, whether it is a single molecule or an organ, necessarily changes its state in the process of

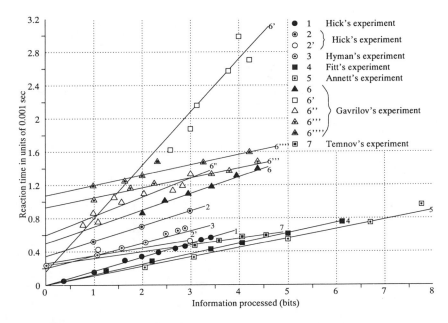

FIGURE 11.2. Experimental Data on Human Response VS. Information Input Rate. Source: Yaglom and Yaglom, 1983. Reprinted by Permission of Kluwer Academic Publishers.

detection. After responding to one impulse, it is not immediately ready to receive another until returning to the original "ready" state. In the case of humans, the latency period ranges from 0.1 to 0.2 seconds [Miller, 1978]. Results are shown in Table 11.3.

The numbers in Table 11.3 are in reasonable agreement with the previous calculation. An important point emerging from the experimental research on keyboard operation is that error rates tend to be constant but not zero at speeds up to about 3.2 keys/sec, at which point channel capacity reaches 9.6 bits/sec for 8 keys, 12.8 bits/sec for 16 keys, and 14.5 bits/sec for 32 keys. At slightly faster speeds (up to 4 keys/sec or so), errors increase in proportion to speed, keeping output constant. Beyond this point, errors increase non-linearly and effective

TABLE 11.3. *Output-Information Capacities.*

Activity	Maximum Rate (Bits/Sec)
Typing random sequences of letters	9.6-14.5
Piano playing	22
Reading	24
Mental arithmetic	24

Source: Quastler and Wulff, 1955.

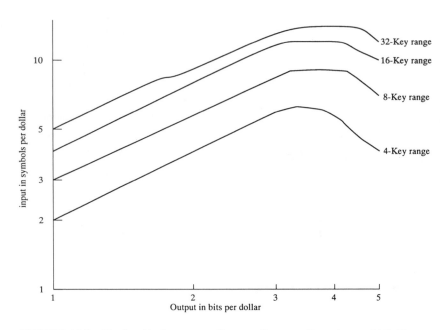

FIGURE 11.3. Typing Performance Curves. Source: Quastler and Wulff, 1955.

output drops quite sharply, as shown in Figure 11.3. Some implications of these data are discussed later in connection with the error-defect problem.

Other experiments have led to comparable results. For instance, experiments requiring subjects to respond verbally to visual stimuli gave output rates of up to 30 bits/sec [Licklider *et al.*, 1954]. When the required response was to point a finger at a target, the maximum rate dropped to 15 bits/sec. But when verbal and manual responses were permitted simultaneously, the maximum output rate actually increased to 45 bits/sec the sum of the individual rates. This (and other evidence) suggests that several parallel independent processing channels exist in the human brain.

11.3. OUTPUT OF A WORKER: TIME AND MOTION

The question now arises: How can the information output of a worker be estimated for typical tasks involving arm (or other body) movements? The pioneer of scientific management, Frederick W. Taylor [1911] argued that each particular industrial task can be optimized from a time-motion point of view, so that it is accomplished with minimum effort. In Taylor's words:

> "For each job there is the quickest time in which it can be done by a first-class man. This time may be called the quickest time or the **standard** time for the job. The best

way ... to determine how much work a first-class man can do in a day ... is to divide the man's work into its elements and time each element separately."

Harold and Lillian Gilbreth subsequently pioneered the development of micro-motion study and postulated that elementary motions could be timed as multiples of a fundamental time unit the *therblig*. The basic empirical "law of fundamental times" states that, "within reasonable limits, the time required by experts to perform a fundamental motion is a constant" [Steffy, 1971]. If each task can be broken down into fundamental motions, and each motion requires a certain minimum amount of time, then it follows that each well-defined task also requires a minimum amount of time that can be determined from the motions themselves.

A number of practical schemes for subdividing tasks into work elements have been proposed [see Steffy, 1971]:

MTA:	(Methods-Time-Analysis)	Segur, *c.* 1925 [Segur, 1973]
WF:	(Work Factor)	Quick, Duncan, and Malcolm, *c.* 1938 [Quick *et al.*, 1962]
ES:	(Engstrom System)	Engstrom and General Electric, *c.* 1940
400:	(400 System)	Western Electric Co., *c.* 1944
MTM:	(Methods-Time-Measurement)	[Maynard *et al.*, 1948]
MTS:	(Methods-Time-Standards)	General Electric Co., *c.* 1950
BMT:	(Basic Motion-Time Study)	J.D. Woods and Gordon Ltd., *c.* 1951 [Presgrave and Bailey, 1963]
DMT:	(Dimension-Motion-Time)	General Electric Co., *c.* 1954

Industrial ergonomists have classified at least ten elementary arm, hand, and eye motions, each with several subcases depending on situation variables. One well-known list, known as the Methods-Time-Measurement, or MTM system (the standard references for which are Antis *et al.* [1979] and Maynard *et al.* [1948]), is shown in Table 11.4.

Any standard manufacturing task can be decomposed into a series of such elementary motions, given information on the location and orientation of work-pieces, tools, parts, etc. An example of such a decomposition is shown in Figure 11.4. Each elementary motion requires a characteristic length of time for humans, tabulated in standard manuals [Antis *et al.*, 1979]. For example, Table 11.5 displays the average time required for various cases of "reach" as a function of distance moved. (All MTM measurements use a standard time unit or TMU corresponding to 10^{-5} hours or 0.036 seconds). Given a task decomposition, such as that shown in Figure 11.4 (assumed to be optimal), and a set of elementary motion timetables, such as Table 11.5, the ideal time (in Taylor's sense) for the task can be easily computed.

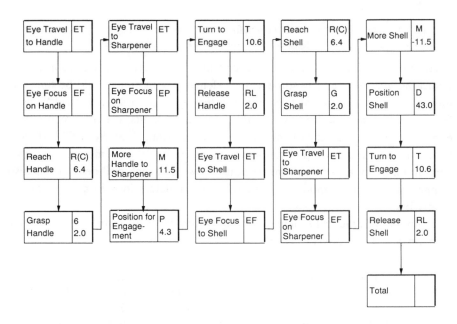

FIGURE 11.4. Pencil Sharpener Assembly (1 Hand, Full Sight) MTM.
Source: Ayres *et al.*, 1984.

11.4. OUTPUT OF A WORKER: TIME AND INFORMATION

One can subdivide each physical (i.e., manipulative) task into two kinds of mental activity. Each requires a characteristic increment of time to:

TABLE 11.4. *Method-Time-Measurement (MTM) system.*

Motion	Abbrev.	Cases
Reach (empty hand), gross arm	R	A through E, depending on exactness of location and size of object. Function of distance.
Move (loaded hand), gross arm	M	A,B,C, depending on exactness of location. Function of distance.
Turn (wrist)	T	Function of angle.
Apply pressure (thumb or wrist)	AP	A,B, depending on whether pressure is released.
Grasp	G	A,B,C, depending on size of object.
Position (orient)	P	Function of symmetry, fit.
Release	RL	
Disengage	D	Function of fit, recoil, etc.
Eye travel	ET	
Eye focus	EF	

TABLE 11.5. *MTM Values for Various Types of "Reach."*

Distance Moved (inches)	RA[a]	RB[b] or RC[c]	RD[d]	RE[e]	RA	RB
		Time in TMU (1 TMU=.00001 hr=.036 sec)				
1	2.5	2.5	3.6	2.4	2.3	2.3
4	6.1	6.4	8.4	6.8	4.9	4.3
12	9.6	12.9	14.2	11.8	8.1	10.1
30	17.5	25.8	26.7	22.9	15.3	23.2

[a]Reach to object in fixed location, or to object in other hand or on which other hand rests.
[b]Reach to single object in location that may vary slightly from cycle to cycle.
[c]Reach to object jumbled with other objects in a grup so that search and select occur.
[d]Reach to a very small object, or where accurate grasp is required.
[e]Reach to indefinite location, to get hand in position for body balance, or for next motion, or out of the way.

1. decide on a course of action by evaluating information inputs (e.g., scanning and interpreting instrument dials), and
2. control arm and hand motions.

For the first type of mental activity (decision), the time required is given by Hick's law (see Figure 11.2):

$$T_d = K_p + \frac{H_d}{C_d}, \tag{11.1}$$

where K_p is the minimum delay time associated with sensory perception, C_d is the effective perceptual response channel capacity, and H_d is the amount of information output (decisions) in bits [Hick, 1952]. Depending on the sensory mode, $K_p = 0.15$–.225 seconds for visually presented information, 0.12–0.18 seconds for auditory information, and 0.115–0.19 seconds for tactile information [Salvendy and Knight, 1982]. Hick estimated C_d at 4.5 bits/sec, which is well below the maximum channel capacity figure suggested by others.

The time required for physical motion per se is analogously given by

$$T_m = K_m + \frac{H_m}{C_m}, \tag{11.2}$$

where K_m is the minimum delay time associated with initiating a motion (0.177 seconds for hand motion), C_m is the channel capacity for motion control, and the amount of control information required is

$$H_m = \log_2\left(\frac{2A}{W}\right). \tag{11.3}$$

Here A is the amplitude of motion, and W is the target width [Fitts, 1954].

The motion-control channel capacity C_m is assumed by Salvendy and Knight [Knight, 1982] to be 5 bits/sec. This is probably reasonable for unpracticed motions, although, as noted previously, expert typists can achieve twice this rate. However, an alternative estimate is derived as follows. It is experimentally observed that most arm movements are accomplished in two stages:

1. An open-loop gross ballistic motion, which moves the arm into the vicinity of the target, with about 93 percent accuracy (7 percent error in amplitude), based on information stored in memory.

2. A series of closed-loop, visually controlled correction movements if higher precision is required. Each such correction requires about 0.3 seconds and reduces location error by 93 percent.

Thus, for a single-stage (open-loop) gross ballistic motion, setting $2A/W = (.07)^{-1}$ whence

$$H_m = \log_2 14.3 = 3.84 \text{ bits.} \tag{11.4}$$

Assuming the parametric values given above ($C_m = 5$ bits/sec)

$$T_m = 0.177 + 0.768 = 0.945 \text{ sec.} \tag{11.5}$$

However, for each successive (closed-loop) correction,

$$H_i = 3.84 \text{ bits;} \quad T_i = 0.3 \text{ sec.} \tag{11.6}$$

Plotting this relationship for two different values of C_m yields the results shown in Figure 11.5. It can be seen that the control-information channel capacity required for motion—hence, the mental load—increases sharply for the closed-loop search-correction phase:

$$\frac{1.84}{0.3} = 12.4 \frac{\text{bits}}{\text{sec}}. \tag{11.7}$$

This requires intense concentration at a level that is almost certainly not sustainable over long periods. It is unclear whether open-loop (gross) motion-control channel capacity is really less than closed-loop channel capacity or, if it is, why. The reverse would seem more plausible. The experimental data are ambiguous, and many unresolved puzzles remain.

There is evidence from other experiments, too, that the time required for certain tasks is a function of the information input rate. Such tasks are inherently sensory-intensive, if not sense-limited. A particularly direct example comes from Murrel [1965], who shows a relationship between illumination, accuracy, and time (Figure 11.6). The figure shows that for fixed levels of illumination, accuracy increases with time. If the available time is fixed, on the other hand, accuracy is an increasing function of illumination. Illumination level can be regarded as a rough measure of visual information received by the worker. Interesting

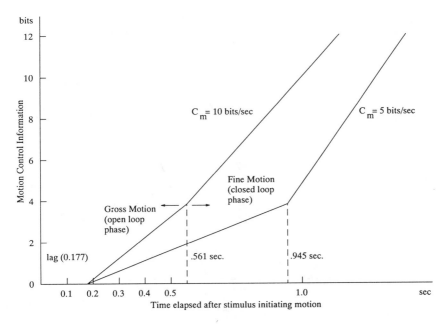

FIGURE 11.5. Information for Motion Control VS. Time.

results of a similar nature were obtained by McCormick and Niven [1952] in experiments with students. They found that examination scores (accuracy) were significantly improved by increased illumination, although this effect leveled off beyond a certain point. Kaswan and Young [Young, 1965] have obtained comparable results.

In recent experiments [see Ayres *et al.,* 1984; also reported in Ayres, 1986] We obtained similar results for a series of manual tasks requiring both visual and tactile sensing. Experiments such as these imply that various tasks can be characterized and ranked in terms of their relative sensory dependence. It is quite plausible that the feasibility of substituting a "smart sensor" for a human worker is essentially inversely proportional to the sensory dependence of the task, i.e., the degree to which sensory deprivation degrades performance.

11.5. OUTPUT OF A WORKER: MOTION AND INFORMATION

Two things related to the same thing must be related to each other. In the preceding two sections it was argued, first, that any set of human motions requires a minimum amount of time and, second, that the amount of information required for motion-control purposes can be estimated on the basis of the time required. From this it follows that a given sequence of motions requires (at

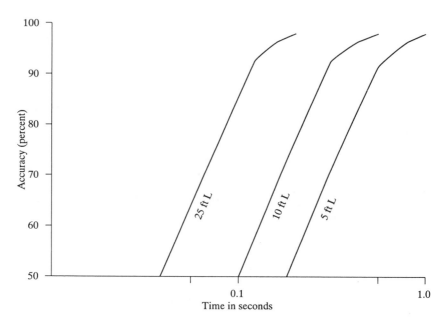

FIGURE 11.6. A Relationship Between Illumination, Accuracy and Time of Viewing. Source: Murrell, 1965.

least) a given amount of information output in the form of control signals. On the average, the information output of a manual worker performing a manipulative task is essentially proportional to the time required for the motions.

Not all motions are productive, of course, just as not all information is useful. The original purpose of Taylor's time-motion analysis (1911) was to compare alternative *methods* (i.e., sequences of elementary motions) of performing a given task, so as to select the best one. When the optimal method has been found, workplace layout can be optimized in its turn. Finally, operators can then be trained systematically. It is estimated that 80 percent of the benefits of time-motion study are attributable to the optimization of task sequences and to worker training, while 20 percent are attributable to better workplace layout [Steffy, 1971].

The cumulative effect of training can be significant. This phenomenon is generally known as the learning or "experience" curve. It is often used to predict unit-cost reductions as a function of cumulative production [see Cunningham, 1980]. Experience in operating cigar-making machines, for instance, showed that a speed-skill improvement factor of 2.5 was possible, but only after 2 years and 3 million machine cycles of experience [Crossman, 1956, 1961]. The speed-skill improvement presumably is due to an increase in the output of *useful* information, rather than in the *total* amount of information output. The gain in useful information can be interpreted as a decrease in the amount of waste motion, including errors.

It was pointed out in Chapter 10 that the amount of morphological information actually embodied in a product, such as an assembly, can be computed directly, at least in principle. It can be assumed that if such a task is carefully optimized from both time-motion and workplace layout perspectives, and if the worker is adequately trained, the motion-information output of the worker will be roughly proportional to the morphological information ultimately embodied in the product (assembly).

The ergonomic evidence presented up to this point clearly suggests that Taylor's "quickest time" principle of task optimization can be re-expressed in different language. In effect, Taylor implies that the optimal work rate is that which maximizes the output of (specified) motions, where each job is described as a sequence of motions. Relating motion to information processed (as above) further implies that information output is thereby maximized by Taylor's principle.

This interpretation is misleading, however. It fails to distinguish clearly between *useful* information outputs and *garbage* (or error) outputs. The experiments cited earlier (e.g., Figure 11.2 and many similar experiments) show maximum useful information output as a function of input rate, not maximum total information output (including wrong letters, wrong notes, etc.). Motion *per se* cannot be equated with useful information output, only with total information output. As pointed out earlier, the ergonomic evidence clearly implies that there exists a maximum rate of useful information output for human workers. This, by itself, is not inconsistent with Taylor's point of departure. However, Taylorism neglects the fact that useful information is often combined with non-useful information (errors or "garbage"). Seeking to maximize the former makes sense if, and only if, the associated errors, or garbage output, have no economic importance and can be neglected.

11.6. THE ERROR/DEFECT PROBLEM

Defects in products arise from one of two sources: design flaws, or operational errors by workers. The latter predominate, but both are ultimately attributable to human error. Industrial engineers and ergonomists attack the problem by seeking to understand and eliminate the environmental factors that increase *human error propensity*, or HEP.

Factors that tend to increase the error rate include:

- Emotional stress.
- Physical strain and discomfort.
- Poor illumination.
- Information load (overload).

The influence of these factors on human performance is discussed in a number of ergonomics and human-factors studies and textbooks [see, for example, Meister 1971, 1976; and Grandjean, 1980]. Most of today's "quality control" methods are predicated on the reduction or elimination of adverse factors.

Statistical quality control, now practiced with outstanding success by the Japanese, was based in large measure on the pioneering work of W. A. Shewhart of Bell Laboratories and systematized by the Western Electric Company [Shewhart, 1980, originally published in 1931]. Statistical methods were transferred to the Japanese after World War II by Bell System engineers at the request of U.S. Occupation authorities, who were anxious to get the Japanese telephone system functioning again [Trevor, 1986]. The concept of total quality control was first introduced in 1967 by A.V. Feigenbaum [1983], of the General Electric Company. But in recent years these methods have been implemented most successfully by Japanese manufacturers and have become a major competitive weapon. Japanese methods have been able to reduce defect rates and reject rates by factors of 10 to 100 from levels considered normal and acceptable in U.S. plants only a decade or so ago [see, for example, Garvin, 1983].

Nevertheless, organizational and "human factors" approaches to quality must ultimately approach natural limits of a fundamental sort. Defect rates cannot be reduced to absolute zero in any production system operated by human workers. Not all mistakes can be detected and, of those discovered, not all can be corrected or adjusted for. Beyond some point, the cost of additional inspection, testing, and rework must exceed the benefits.

When organizational and ergonomic strategies for error/defect control are exhausted, the only option left is to eliminate the remaining source of error from the production system. This source is the human workers themselves, especially those having direct contact with workpieces and production machines. Recognition of this dilemma provides the motivation for introducing computers in place of humans as machine controllers and operatives.

A 1961 study of the sources of 23,000 product defects in electrical assembly plants revealed that 82 percent were caused directly by worker errors while the remainder could be attributed to design flaws (see Table 11.6). It must be emphasized that these are *net* error rates computed by examining finished products, and therefore after ordinary inspection, error detection, and rework. The intrinsic human error probability is clearly higher than the net figures shown. More recent work by Swain [1977] and Swain and Guttmann [1983] suggests that intrinsic human error probabilities for various repetitive operations may be around an order of magnitude higher than Rook's number, or around 10^{-3} per opportunity. Error rates in some activities, such as computation and number transcription, approach one error in 30 opportunities [McKenney and McFarlan, 1982]. In this particular case, at least, electronic computers are now at least 100,000 times more reliable than humans.

It is true that these error rates (HEPs) are not absolute. They are subject to environmental conditions, for instance. But stress, discomfort, and distraction tend to increase the HEP above its normal level. In fact, under life-threatening

TABLE 11.6. *Imputed Error Rates for Certain Manufacturing Operations.*

Error	No. of Observations	No. of Observed Defects	Calculated Human Error Probability HEP
Two wires transposed	13,083	22	0.0006
Component omitted	103,880	213	0.00003
Solder joint omitted	47,075	596	0.00005
Operation (e.g., applying, staking) omitted	59,435	11	0.00003
Component wired backward (diodes, capacitors, etc)	2,610	27	0.001
Wrong value component used	103,880	213	0.0002
Lead left unclipped	33,000	551	0.00003
Component damaged by burn from soldering iron	103,880	213	0.001
Solder splash	47,075	596	0.001
Excess solder	47,075	596	0.0005
Insufficient solder	47,075	596	0.002
Hole in solder	47,075	596	0.07

Source: Rook, 1962.

stress conditions, the HEP can approach 25 percent [Swain and Guttmann, 1983].

There is also a learning curve, as mentioned previously. Practice does tend to make perfect, although even with practice the error rate does not approach zero for any production process. Training can increase accuracy as well as efficiency; but, as noted previously, the number of repetitions (cycles) needed to reach maximum accuracy tends to be rather large.

11.7. ERRORS AND INFORMATION OVERLOAD

There is compelling empirical evidence that human errors beyond the minimum level are due in large part to *information overload* inability of the organism to respond to input signals as fast as they are received. Many studies have found the following pattern: When information output (i.e., human responses or actions) is plotted as a function of the rate of information input, the curve rises linearly at low input rates, reaches a maximum at some input rate where the channel is said to be *saturated* (depending on the type of input and the type of output), and then drops off as input rates rise further (Figure 11.7). For the sake of clarity, the input information may be of any kind, including noise, whereas the output (in the experiments) is assumed to be useful. The literature has been summarized by Miller [1978], who notes that the same pattern applies

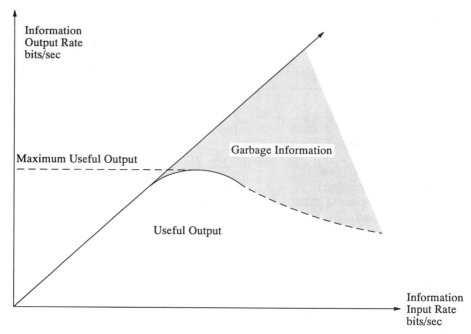

FIGURE 11.7. Information Overload Phenomenon. Source: Author [Adapted from Miller, 1978].

to cells (e.g., neurons), sensory organs, organisms, groups, and organizations. In general, the maximum information-output level occurs at lower and lower input levels as one proceeds along this sequence. For human organisms, the maximum output level is of the order of 10–15 bits/sec. As information output rate falls below a simple proportion of input, the information loss (i.e., error rate) rises sharply. This implies a strongly nonlinear relationship between output rate and error rate (Figure 11.8). It also implies that, to maximize the output rate of any information-processing activity, one must accept a significant non-zero error rate. To reduce the error rate to near zero also necessitates sharply reducing the (useful) output rate.

The economic optimum obviously depends on the cost of discovering and correcting errors and defects vis—vis the implicit cost of reducing errors by cutting output rates or cycle times. There is good reason to suppose, however, that work rates adopted in most modern factories are based largely on job descriptions and standards set in the early years of method-time-motion studies during the heyday of Taylor's scientific management (1900–1925). At that time there was very little understanding of fundamental ergonomic relationships. It is likely that, in most cases, rates were adjusted to *maximum* output levels rather than to levels that would maximize the net useful output of the entire production process (i.e., after error detection and correction). Given the increasing mechan-

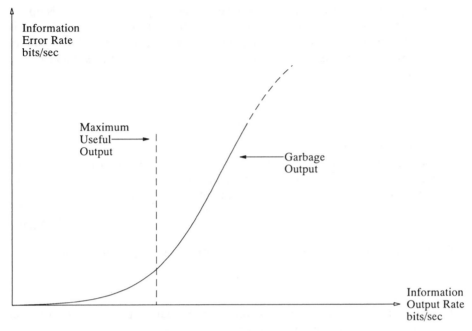

FIGURE 11.8. Error Rate VS. Output Rate.

ical complexity of most products and the increasing cost to manufacturers of uncorrected defects, it would appear that the optimum work rate (if human labor is used) should actually be adjusted downward from levels set early in the 20th century. The amount of the downward adjustment depends on the complexity of the product, because the cost of eliminating defects (or tolerating them) increases sharply with product complexity. A simple economic model of the optimum output rate is given in the next section.

11.8. OPTIMUM WORK PACE

We can assume, for purposes of analysis, that the worker's output response curve for a set of given conditions is a known function of input rate, all other factors remaining constant (see Figure 11.7). Define y as useful output rate (in bits/sec) and x as the input rate, also in bits/sec.

For very low rates of input (small x) we can also safely assume that

$$y = kx, \quad 0 \leqslant k \leqslant 1, \tag{11.8}$$

where the constant k reflects only the technical nature of the task, i.e., the degree of information reduction involved in converting sensory stimuli into useful output (motions).

For higher rates of input, $y(x)$ falls below the simple linear relationship. In fact, the conditions are $dy/dx > 0$ and $d^2y/dx^2 < 0$. This characteristic behavior is usually interpreted as an overload phenomenon of some sort. The point of maximum output corresponds to an input rate x_m, which can be found by the condition

$$\frac{dy}{dx} = 0 \tag{11.9}$$

assuming the function $y(x)$ is known.

The difference between potential useful output (if there is no overload or fatigue factor) and realizable output is, in effect, lost information or garbage output. For convenience, define "garbage output" $g(x)$ as follows

$$g(x) = kx - y(x). \tag{11.10}$$

By previous assumption, at very low input rates, $g(x)$ approaches zero:

$$\lim_{x \to 0} g(x) = 0. \tag{11.11}$$

Rewriting (11.10), we can reinterpret the potential output as *total* output of useful information plus garbage, namely,

$$kx = y(x) + g(x). \tag{11.12}$$

$$\begin{array}{ccc} \text{total} & \text{useful} & \text{garbage} \\ \text{output} & \text{output} & \text{output} \end{array}$$

Now suppose the worker is employed by a firm and his or her useful output has an economic value, or shadow price P_y per unit. For generality, we can also assign a price P_g per unit to the garbage. Of course P_g is non-positive, but it may or may not be zero. It can be interpreted as the (shadow) penalty cost of garbage to the firm employing the worker. Garbage, in this context, can be thought of as defective or spoiled output that must (at the very least) be separated and disposed of or (more likely) repaired or reworked.

If the garbage output is simply waste motion of some sort, then P_g can be neglected and simply set equal to zero; but this is not likely or important in a realistic situation. If the object of the work is to add value to a material work-piece, an error on the part of the worker has a finite probability, depending on the case, of introducing a defect to the workpiece. If the defective piece is immediately detected and discarded, the minimum loss is either (1) the value of all work done on the material in earlier stages of production, plus the value of the purchased material, less any salvage value; or (2) the cost of repairing the defect whichever is less. On top of this must be added the *pro rata* cost of the inspection (defect detection), since inspection would not be required if there were no worker errors and/or resulting defects.

It follows immediately from the analysis above that, in general, P_g is negative, and non-zero:

$$P_g < 0. \tag{11.13}$$

Moreover, the absolute value of P_g is clearly an increasing function of the complexity of the entire production process and of the stage of the process where the defect occurs. This reflects the well-known point that repair and rework are far more expensive per unit output than original production because of the high labor intensity involved.[3]

The value added per unit time by the worker can now be written

$$V(x) = P_y y(x) + P_g g(x). \tag{11.14}$$

The optimum work pace x is determined by maximizing $V(x)$, i.e., by satisfying the condition

$$\frac{dV(x)}{dx} = 0. \tag{11.15}$$

This condition yields

$$0 = P_y \frac{dy}{dx} + P_g \frac{dg}{dx} = P_y \frac{dy}{dx} + P_g \left(k - \frac{dy}{dx} \right) = (P_y - P_g) \frac{dy}{dx} + kP_g, \tag{11.16}$$

whence

$$\frac{dy}{dx} = \frac{-kP_g}{(P_y - P_g)}. \tag{11.17}$$

Clearly, the economic optimum output rate $(dy/dx = 0)$ only coincides with the *maximum* output (Taylor) condition in the exceptional case where $P_g = 0$. Moreover, the larger the (negative) values of P_g, the greater the deviation and the lower the optimum value of x. Much of the criticism of crude Taylorism by industrial psychologists would seem to be amply justified in view of these results. A more interesting and less obvious implication arises from the consideration that P_g is likely to be a function of the precision and performance (or complexity) of the end product. Thus, for a simple end product, such as a paper clip or a poker chip, the (negative) unit value of a defective unit is not greater, in absolute terms, than that of a "good" one. If 3 units out of a batch of 100 must be discarded, the value of the batch is essentially 97/100 of the potential maximum. Not so if the faulty part is built into a subassembly or a large machine. A faulty ball bearing installed in a large turbine or a faulty connection in a computer or radar system can cause damage and losses far in excess of the nominal value of a "good" unit. "For want of a nail the shoe was lost, for want of a shoe the horse was lost, for want of a horse the rider was lost,..." If the manufacturing task is one step in a sequence leading to a very complex product (such as a space shuttle), the value of P_g can be almost arbitrarily large and negative. In the limit as $P_g \to -\infty$ the optimum solution approaches $dy/dx = k$, which yields $y(x) = kx$. From (11.10) this condition corresponds to $g(x) = 0$. From (11.11) this requires $x = 0$, i.e., a work pace of zero.

In simple language, as errors and defects become increasingly costly, the optimum work pace becomes slower. If errors and defects are absolutely intolerable, the optimum pace (for human workers) approaches zero.

The obvious way out of this dilemma is to replace error-prone human workers by (more) reliable computer-controlled machines. To be sure, the foregoing arguments are simplified, but the underlying implication seems to be quite robust.

11.9. VALUE OF INFORMATION ADDED BY LABOR

It is particularly interesting to focus on the value of labor that adds morphological information (i.e., information regarding shape) to materials. A comprehensive analysis would entail a major empirical research effort. Some interesting clues can be extracted from engineering data, however. In machining operations, the information $H(t)$ required to achieve a tolerance T (recall Chapter 10) can be written as

$$H(t) = K\log_2\left(\frac{1}{T}\right) = -K\log_2 T, \qquad (11.18)$$

where K is a constant determined by the unit of information (e.g., bits) and T is usually defined for convenience as the maximum allowable machining error per unit (inch) of linear tool travel on the workpiece. A cost-tolerance relationship taken from a standard engineering handbook [Boltz, 1976] is given in Table 11.7. While the relative cost figures given are only approximate (taken from Figure 11.9), it is clear that the relative cost is not a simple linear function of useful information content. In fact, in the normal range of tolerances from 2^{-5} to 2^{-10}, where precision increases by a factor of $2^5 = 32$, the information content only doubles. This implies that relative cost increases something like the fifth power of relative information content or precision:

$$\frac{C_1}{C_0} \simeq \left(\frac{H_1}{H_0}\right)^5. \qquad (11.19)$$

This relationship presumably reflects a body of experience with manually operated machine tools. Therefore, it can be interpreted as an average relationship between skilled machinists' time inputs as a function of information actually embodied in the workpiece. Another way to look at it is to note that embodied information is a very small fractional power of cost:

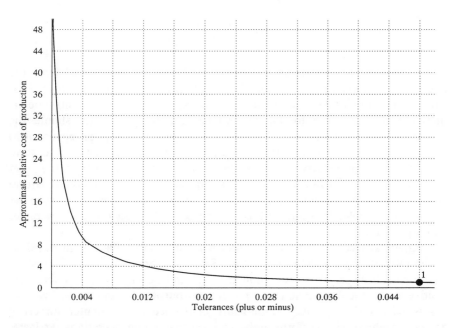

FIGURE 11.9. Cost VS. Tolerance. Source: Reproduced with Permission from Boltz, 1976.

TABLE 11.7. *Cost-Tolerance Relationship.*

Tolerance T	$H(t)$ (Bits)	Relative Cost
$.064 \approx 2^{-4}$	4	0.75
$.048 \approx .05$		1.0
$.040$		1.2
$.032 \approx 2^{-5}$	5	1.5
$.024$		2.0
$.020$		2.4
$.016$	6	3.0
$.012$		4.0
$.008$	7	6.0
$.006$		
$.004$	8	12.0
$.003$		
$.002$	9	24.0
$.0015$		
$.001 \approx 2^{-10}$	10	48.0

Source: Boltz, 1976.

$$\frac{H_1}{H_0} \sim \left(\frac{C_1}{C_0}\right)^{1/5}. \tag{11.20}$$

The time-information relationship for human labor can be inferred from the foregoing ergonomic discussion. One is roughly proportional to the other. That is, the amount of visual or tactile information required for motion control determines the time required for an elementary motion. Why, then, should the amount of information *embodied* in the product not be simply proportional to labor time? No definitive answer can be given here. However, it would seem likely that the non-linearity of the cost-precision function results from the onset of mental overload as human machinists approach the limits of their natural abilities in sensory discrimination. Each successive manual correction becomes more and more difficult as these limits are approached. Therefore, it takes longer and costs more.

A numerically controlled machine tool with internal feedback control might also be ultimately limited by the ability of its detectors or built-in "senses" to discriminate, but specialized electronic devices can be designed to be much more sensitive in a particular domain than the general-purpose human sensory system. This is why NC or CNC (numerically controlled, or computer-numerical control) machines can perform high-precision machining operations far faster than humans, although they offer little benefit in simpler operations or ones requiring lower precision. One would therefore expect the relationship between control information input and information embodied in output to be much more nearly linear for NC or CNC machines.

ENDNOTES

1. Material in this chapter has previously appeared in [Ayres, 1987].

2. Typing in a realistic environment is a far more complex activity than this statement might seem to imply. [See, for example, Salthouse, 1984].

3. For a detailed theoretical treatment see [Sheridan, 1982] and [Sheridan and Ferrell, 1974].

4. Documentation is scarce. However, anecdotal information from senior executives in several large high-tech manufacturing firms suggests that the cost of rework may account for as much as 50% of total production costs in some cases.

REFERENCES

Antis, W., J. M. Honeycutt and B. N. Koch, *The Basic Motions of MTM*, The Maynard Foundation, City, State, 1979.

Ayres, Robert U., "CIM and the Next Industrial Revolution," in: Dermer (ed.), *Competitiveness Through Technology: What Business Needs from Government*, Lexington Books, New York, 1986.

Ayres, Robert U. "Manufacturing and Labor as Information Processes" ISASA Research Report

IIASA RR-87-19. International Institute for Applied Systems Analysis, Laxenburg, Austria, July 1987.

Ayres, Robert U. *et al.*, *An Approach for Stimulating the Growth of Small and Medium-Sized Manufacturing Firms in the Pittsburgh Area*, Technical Report, Pittsburgh Countywide Corporation and Allegheny County Department of Development, Pittsburgh, Penn., August 1984.

Boltz, R. W., "Production Processes," in: Editor, *Productivity Handbook*, Industrial Press, New York, 1976. 5th edition.

Cherry, Colin, *On Human Communications*, MIT Press, Cambridge, Mass., 1978. 3rd edition.

Crossman, E. R. E. W., *The Measurement of Perceptual Load in Manual Operations, Publ., City, State, 1956.*

Crossman, E. R. E. W., "A Theory of the Acquisition of Speed-Skill," *Ergonomics*, 1961.

Cunningham, J. A., "Using the Learning Curve As a Management Tool," *IEEE Spectrum*, June 1980.

Feigenbaum, A. V., *Total Quality Control*, Technical Report, American Society for Quality Control, 1983. [First published in 1967; republished by ASQC in 1983 (3rd edition)].

Fitts, P. M., "The Informational Capacity of the Human Motor System in Controlling the Amplitude of Movements," *Journal of Experimental Psychology* 47, 1954 :381-391.

Garvin, David S., "Hard New Evidence on American Product Quality Underscores the Task Ahead For Management," *Quality on the Line*, September/October 1983 :65-75.

Grandjean, E., *Fitting the Task to the Man*, Taylor and Francis, London, 1980.

Hick, W. E., "On the Rate of Gain of Information," *Quarterly Journal of Experimental Psychology* 44, 1952 :11-26.

Jacobson, H., "The Informational Capacity of the Human Eye," *Science* 113, 1951 :292.

Kaswan, J. and S. Young, "Effects of Illumination, Exposure Duration and Task Complexity on Reaction Time," in: Haber (ed.), *Information Process Approaches to Visual Perception*, Publ., City, State, 1965.

Kelley, C. R., *Manual and Automatic Control*, John Wiley and Sons, New York, 1968.

Licklider, J. C. R. *et al.*, *Studies in Speech, Hearing and Communication*, Contract Final Report (W191-222014), Massachusetts Institute of Technology Acoustic Laboratory, Cambridge, Mass., 1954.

Marko, H., "Information Theory and Cybernetics," *IEEE Spectrum* 4, November 1967 :75.

Maynard, H. B., G. J. Stegmerten, and J. L. Schwab, *Methods-Time Measurement*, McGraw-Hill, New York, 1948.

McCormick, Ernest J., *Human Factors in Engineering*, McGraw-Hill, New York, 1970. 3rd edition.

McCormick, Ernest J. and J. R. Niven, "The Effects of Varying Intensities of Illumination upon Performance on a Motor Task," *Applied Psychology* 36(193), 1952.

McKenney, J. L. and F. W. McFarlan, "The Information Archipelago: Maps and Bridges," *Harvard Business Review*, September/October 1982.

Meister, David, *Human Factors: Theory and Practice*, John Wiley and Sons, New York, 1971.

Meister, David, *Behavioral Foundation of System Development*, John Wiley and Sons, New York, 1976.

Miller, James G., *Living Systems* 5, McGraw-Hill, New York, 1978.

Murrel, K. F. H., *Human Performance in Industry*, Reinhold, New York, 1965.

Quastler, H. and V. J. Wulff, *Human Performance in Information*, Technical Report (R-62), University of Illinois Control Systems Laboratory, Urbana, Ill., 1955.

Rook, L. W., *Reduction of Human Error in Industrial Production*, Technical Memo (SCTM 93-62(14)), Sandia Corporation, Albuquerque, N. Mex., 1962.

Salthouse, Timothy, "The Skill of Typing," *Scientific American* 250(2), February 1984 :128-135.

Salvendy, Gabriel and James L. Knight, "Psychomotor Work Capabilities," in: Gabriel Salvendy (ed.), *Handbook of Industrial Engineering*, Chapter 6.2, Wiley-Interscience, New York, 1982.

Shannon, Claude E., "Prediction and Entropy of Printed English," *Bell Systems: Technical Journal* 30, 1951 :50-64.

Sheridan, T. B., *Measuring, Modeling and Augmenting Reliability of Man-Machine Systems*, Technical Report, Massachusetts Institute of Technology, Cambridge, Mass., 1982.

Sheridan, T. B. and □. Ferrell, *Man-Machine Systems*, MIT Press, Cambridge, Mass., 1974.

Shewhart, W. A., *Economic Control of Quality in Manufactured Products*, Technical Report, American Society for Quality Control, City, State, 1980. [Originally published in 1931].

Steffy, W., "Use of Predetermined Time Standards," in: Maynard (ed.), *Handbook of Industrial Engineering*, McGraw-Hill, New York, 1971. 3rd edition.

Swain, A. D., *Design Techniques for Improving Human Performance in Production*, Publ., Albuquerque, N. Mex., 1977.

Swain, A. D. and H. E. Guttmann, *Handbook of Human Reliability Analysis with Emphasis on Nuclear Power Plant Application*, Technical Report (NUREG/CR-1278-F SAND-0200), Sandia Corporation, City, State, 1983.

Taylor, F. W., *Shop Management*, Harper and Row, New York, 1911.

Trevor, Malcolm, "Quality Control: Learning from the Japanese," *Long-Range Planning* **19**(5), 1986 :46-53.

Yaglom, A. M. and I. M. Yaglom, *Probability and Information*, D. Reidel, Boston, 1983. 3rd edition. [English translation from Russian]

Chapter 12

Evolution, Economics, and Environmental Imperatives[1]

12.1. RECAPITULATION: IDEAS OF EQUILIBRIUM IN ECONOMICS

The modern notion of *evolution* began with the geologists and was taken up by the biologists. Darwin's main concern was to discover mechanisms to account for the historical fossil record. Since the late 19th century, evolution has been, in some sense, the most important "fact of life" for much of science. Curiously, despite the fact that evolutionary ideas were well developed in the social sphere and even by early economists like Adam Smith long before they were adopted by the physical sciences, and geological and biological evolution are relatively slow by comparison with technological evolution (for instance), most theoretical economists in the late 19th century adopted a static or quasi-static world view, as exemplified by the circular-flow model of production and consumption. Economists have only recently begun to consider, in a systematic way, the economic system once again as an evolutionary system.

The reasons for this peculiar blind spot probably have their origin in the so-called marginalist revolution of the mid-19th century, and its fascination with general-equilibrium concepts and models. This fascination seems to have been based on a somewhat distorted analogy with the physics of energy. To digress briefly, physicists had discovered the applicability of extremum principles in classical mechanics (notably the "principle of least action," and the Hamilton and Lagrange formulations). A number of economists of that period, notably, Dupuit, Gossen, Cournot, Jevons, Walras, Edgeworth and Pareto, explicitly sought to develop economics along similar lines, substituting "utility" (or "pleasure") for potential energy.[2]

The problems of the energy metaphor need not concern us here. Suffice it to say that Leon Walras, in particular, succeeded in formulating economics as a static "balance of forces" between supply and demand, and focused attention on the question of whether (and under what conditions) a set of unique prices can exist that matches all demand with a source of supply and "clears" the market.

For historical reasons, it seems, economic growth has been studied primarily as a very special case (*homotheticity*) in which all structural detail remains unchanged (growth in all sectors is proportional). Essentially indistinguishable from static Walrasian equilibrium for short periods, this growth consists of a steady, smooth, monotonic "quasi-static" expansion. The homothetic growth model introduced by the famous mathematician John von Neumann, first presented in a German mathematics colloquium in the early 1930s and later published in English [von Neumann, 1945], was very influential. It led to an enormous mathematical literature. Yet such models do not at all reflect the sort of uneven, technology-driven growth—in which some sectors grow explosively, even as others decline—that actually occurs in the real world.

In the view of many economists today, the static (and quasi-static) equilibrium paradigm has largely exhausted its explanatory power. *Evolutionary* models must be *non-equilibrium* and *non-homothetic*. Clearly, Walrasian static or quasi-static equilibrium is not the condition we live in. The real world is characterized by continuous but asynchronous (and unpredictable) structural changes. Different sectors are created; they grow, mature, and decline. Moreover, the classical conditions for static equilibrium (perfect competition, perfect information, perfect rationality, etc.) do not and cannot exist.

Evolutionary growth and change, furthermore, are characterized by *non-linear* dynamic behavior. This encompasses such phenomena as quasi-stability, quasi-cyclicity, morphogenesis, and "deterministic chaos." An important phenomenon of non-linear dynamics is the so-called "butterfly effect," meaning that the slight motion of air induced by the flapping of butterfly wings may be sufficient to differentiate two long-term weather patterns from each other. It is characteristic of such systems that dynamic trajectories are much more sensitive to initial conditions than to long-range "equilibrating" forces.

Another key characteristic of evolutionary behavior is **hyperselection**, or **path dependence**, or **self-organization**. Whereas the "butterfly effect" refers to the exponential divergences of trajectories in the neighborhood of a single "strange attractor," **hyperselection** refers to the transient stage of evolution when it is possible for a system to "choose" between disjoint attractors. Still another name for the phenomenon, which is more evocative, is "lock-in/lockout." Thus, the selection of one evolutionary path or another may be largely accidental, and yet that path may be dominant for a long time (in evolutionary terms). This happens because in the real world feedback effects ("returns to adoption") favor the established choice over most competitors [see, e.g., Arthur, 1988; Arthur *et al.,* 1987].

For example—although this is not strictly an example of "lockout" — an established manufacturing technology will not automatically be displaced by a slightly better one. Only when the advantage of the new technology is quite large will a "rational" entrepreneur introduce a substitution. In the first place, there are frictional costs. The work force may need retraining, for example. In the second place, old manufacturing technology in existing plants has the advantage of variable costing with fully depreciated capital equipment, whereas new technology must compete on the basis of full cost [Salter, 1960].[3] Even in new facilities, the higher risks of unproven technologies have to be weighed against expected but uncertain gains in performance.

An example of how improvements may be too small to induce a change is provided by the familiar QWERTY typewriter keyboard [David, 1985]. This keyboard is not optimal from an ergonomic point of view. In fact, it was originally introduced to slow typists down, because early typewriter keys had a tendency to stick. But the cost of change replacement of all keyboards and retraining of all typists would not be justified by the marginal benefits of the improved keyboards.

A second example of hyperselection might be found in the history of urban transportation. In the United States, electric tramways were once as widespread and heavily traveled as they are today in Europe. But from the 1930s through the 1950s, conditions in the United States—lower fuel prices and less-restrictive land-use policies—that favored expansion of the suburbs were more favorable for diesel buses than was the case in Europe. Bus interests were supported by the powerful automotive and petroleum industries. By the 1960s trolleys and trams were obsolete in the U.S., and the tracks were being removed from most cities. This did not happen in Europe. In a larger sense, this was a case of *bifurcation*: At a critical juncture European and American cities "selected" different evolutionary branches. Today, curiously enough, the virtues of trams ("light rail systems") are being rediscovered in some U.S. cities, but once trolley tracks are gone they are, in most cases, prohibitively expensive to replace.

An example of lock-in much more relevant to current environmental concerns is the likely dependence of the world economy for many decades to come on automotive vehicles that depend, in turn, on liquid hydrocarbon fuels. Here the initial determining factor was probably the availability of low-priced gasoline. Prior to 1905, when the competition between rudimentary gasoline engines and electric cars was basically settled in favor of the former, gasoline was a cheap by-product of petroleum refining, the primary product of which was kerosene ("illuminating oil") for lighting purposes [Ayres and Ezekoye, 1991]. Later, when gasoline became the primary product of refineries, the price was kept low by subsidies to producers, notably the "depletion allowance." At the present time, fossil-fuel prices are still effectively subsidized by the lack of any charge on consumers of petroleum products to reflect the environmental damages caused by fossil-fuel use. From this perspective, it has been estimated that fossil fuels, even in Europe, are probably underpriced by a factor of 3 or more (Chizhov and Styrikovich, 1988; Hohmeyer, 1988].

In contradiction, it may be argued that the technology of electric automobiles was certainly very primitive at the beginning of the century, whence the internal combustion engine might well have prevailed on its merits even if gasoline (or diesel oil) had been much more expensive. On the other hand, electric-car technology and the supporting infrastructure would have developed much faster than it did, whereas technology for the internal combustion engine (and associated petroleum refining) would certainly have evolved more slowly. The outcome in this hypothetical case is unknowable. A much stronger argument can be made that electric propulsion would win out over gasoline engines if a fair competition between the two technologies were held today (or, better still, in the year 2000). The comparative benefits of electric drive for urban use—no air pollution, no smells, no noise—appear likely to grow even more over time. These benefits would be compounded if electricity can be produced at reasonable cost in the next century by means of photovoltaic cells.[4]

In any case, the phenomenon of hyperselection is quite relevant to the direction of long-run evolutionary change.

12.2. BROWNIAN MOTION OR GRAVITATIONAL ATTRACTION?

Economic growth is difficult to accomodate in neo-classical economics. Most of the early growth models tried to explain aggregate growth in terms of capital accumulation and an increasing labor force. This simplistic view became untenable after the empirical studies in the mid-1950s by Solomon Fabricant [1954], Moses Abramovitz [1956], Robert M. Solow [1957] and others, which established that other factors lumped under the general heading *technology* have been responsible for most of the per capita economic growth that has occurred.

Theory tried to cope with awkward facts by introducing technological change (in the guise of increasing productivity) as an *exogenous* driving factor. But this approach is also clearly unsatisfactory, inasmuch as technological change is anything but exogenous in reality. The difficulty of forecasting the rate and direction of inventive activity does not alter the fact that R&D and innovation are themselves economic activities and occur to a large extent in response to other economic conditions.

Two theoretical possibilities, representing extreme cases, could account for economic growth. The first is that evolutionary progress is stochastic: It occurs entirely as a result of short-range, localized events and decisions or "collisions." This notion can be characterized as a "drunkard's walk," or "deterministic chaos." The analog in physics is Brownian motion. The opposite extreme case is that evolution is guided by a long-range force, analogous to gravity. The analogy might be to an intergalactic spaceship drifting through a stellar cluster.

Or (a subtler idea), evolution may be subject to both short-range random collisions and a very general long-range directional constraint. Such a directional constraint is not quite equivalent to a long-range attractive force like gravitation. For one thing, it is not sufficient by itself to determine the trajectory of motion. It merely excludes some possibilities. It implies that evolutionary progress is *irreversible,* without specifying its actual direction or rate. The second law of thermodynamics (entropy law) is an example of just such a directional constraint in physics: It states that all physical processes must *approach* thermal equilibrium. (They cannot move away from equilibrium.) It is also true that in some generalized sense the distance from equilibrium is a measure of the "force" operating. But the actual rate of approach to equilibrium is not determined by the second law. It depends on the detailed characteristics of the system in question.

I noted early in Chapter 1 that this irreversibility in physics is what guarantees the existence of a non-decreasing function, which we call **entropy**. It follows, incidentally, from the existence of such a function, that the dynamics of a system can be described by an extremum principle. In thermodynamics, for instance, systems tend toward states of (local) maximum entropy or minimum free-energy.

Evolutionary biologists have postulated both irreversibilities and extremum principles. Dollo's "law of irreversibility" (1893), for example, postulated an irreversible trend toward an increasing complexity of organisms. Dollo's law was reformulated by Julian Huxley [1956]. Ludwig Boltzmann saw biological evolution in terms of a competitive struggle to capture free energy. This notion was restated as an extremum principle by A. J. Lotka [1922].

Recently, Prigogine and his colleagues have applied ideas from non-equilibrium thermodynamics to biological evolution [Prigogine *et al.,* 1972; Prigogine and Stengers, 1984]. Some of these ideas have also been reformulated in terms of extremum principles [Ebeling and Feistel, 1982, 1984], although it is too early to judge whether these principles explain observed phenomena satisfactorily. Brooks and Wiley [1986] have attempted to apply similar ideas at the ecosystem level. In some sense, all of the above are attempts to show that biological evolution is not only compatible with the laws of thermodynamics (increasing entropy), but a direct consequence of them.

In micro-economics, too, there is a fundamental irreversibility phenomenon. To state it briefly: if a pairwise exchange transaction can proceed in one direction under bounded rationality A is willing to buy X and B is willing to sell X then it cannot proceed in the reverse direction.[5] This sort of irreversibility is implicit in Walras's law and the *tâtonnement* process for price determination. It was later articulated by Ville [1951].

It should be pointed out that the approach to Walrasian static equilibrium has nothing in common (other than irreversibility) with the approach to thermal equilibrium in physics. Nor does it reflect economic or technological "progress" in any sense whatever. It is simply a conceptual device to explain how pure exchange markets might converge to a set of unique market-clearing prices.[6]

In economics, as in biology, no satisfactory extremum principle has been formulated to explain economic (or technological) evolution at the macro-level. In the real economic system, two kinds of evolutionary progress do seem to occur. There are clear indications of stochastic or chaotic processes analogous to Brownian motion [see, e.g., Chen, 1990]. There are also hints at the existence of long-range forces or, at least, directional constraints.

Traditionally, in the sciences, one tries to explain, and then predict, macro-level phenomena in terms of micro-level behavior. Economists have not been very successful in this endeavor up to now. One of the few areas where some success can be claimed is the proof that "self-organization" of markets follows from micro-level behavioral axioms. The non-decreasing "progress function" exists as a consequence of transactional irreversibility on the micro-scale. One of the questions I am raising in this concluding chapter is whether one can identify a directional constraint on the macro-scale and work back to infer the corresponding micro-mechanisms. If the *real* economic system is subject to apparent directional constraints at the macro-level, how might those constraints be characterized? And are there plausible mechanisms at the micro-level that might give rise to the appearance of such macro-constraints? These questions reverse the usual "bottom up" (micro to macro) logical order of causation imposed by traditional scientific reductionism.

In this book I have repeatedly suggested that evolution can be conceived as a kind of "motion" around a "strange attractor" whose "mass" and other characteristics can only be inferred. There is an obvious analogy with the motion of planets or asteroids in the gravitational field of an invisible object (e.g., a black hole). The analogy of *approach to equilibrium* in classical thermodynamics is also relevant, although in that case the equilibrium is static. Modern non-equilibrium thermodynamics also allows the possibility of *self-organizing* stationary states far from equilibrium [Prigogine *et al.,* 1972; Nicolis and Prigogine, 1977]. Organic life exemplifies such self-organized states. It is evident from the biological case that, if the system is sufficiently complex and non-linear, a self-organizing stationary state far from (thermodynamic) equilibrium is capable of evolutionary change. It is a natural conclusion that a self-organized economic system should also be capable of evolutionary change [Ebeling and Feistel, 1982, 1984]. In short, I suggest that the economic system must eventually evolve in accordance with certain macro-directional constraints and that this motion can also be characterized as a kind of approach to (dynamic) equilibrium. The equilibrium in question, of course, concerns the interaction between human agricultural, industrial, and consumptive activity and the physical and biological environment. An equilibrium interaction, in these terms, is one that is indefinitely *sustainable*, albeit *not* necessarily static.[7]

12.3. THE ECONOMIC SYSTEM AND THE PHYSICAL ENVIRONMENT

The dissipative character of the real economy, and its dependence on large quantities of physical materials and fuels, has two important consequences. First, since matter is conserved, all materials extracted from the environment as crops, fossil fuels, metal ores, or other minerals must eventually return to the environment as waste residuals. (See Figure 12.1.). The material cycle is closed in a very short time (weeks or months) for most organic materials, and a somewhat

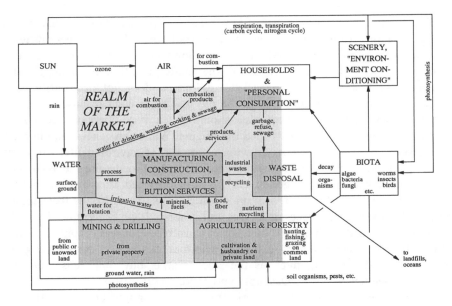

FIGURE 12.1. The Realm of the Market In Context.

longer time for some metals and minerals. But the time between extraction and return is extremely short by geologic standards.

The second consequence follows from the first, plus a massive market failure: It is the indivisibility and non-exchangeability of environmental *goods* and *bads,* and our consequent inability to attach meaningful prices to them or to the consumptive uses of raw materials and goods that are exchangeable. Because of this market failure, most of the environmental costs noted above are *externalities* in the sense that they result from economic transactions but are imposed on "third parties." Large environmental damages (hence costs, whether paid or unpaid) are imposed on the environment and, indirectly, on society. These damages and damage costs include environmental health problems like cancer and cardiobronchial problems (e.g., emphysema and asthma); climate-warming; destruction of the ozone layer; corrosion of metals and building materials due to acid rain; acidification, erosion, and salt buildup in the soil; eutrophication of lakes; accumulations of toxic metals and radioactive materials in soil and waterways; growing solid-waste mountains; and a host of other problems. To the extent that these damages are not being compensated for or repaired today, they are simply accumulating for the next generation.

In short, human economic (industrial and agricultural) activity is based on the consumptive use of non-renewable and irreplaceable resources, from fossil energy to topsoil. If the notion of long-term sustainability is a criterion of economic-environmental equilibrium, it must be said that the human economic system at present is far from being in equilibrium with the environment.

The question now arises: What, if any, relationship exists between this notion of equilibrium with the environment and the traditional Walrasian equilibrium? The former is related to materials and energy flows and accumulations; the latter is defined in terms of prices that "clear" the market, i.e., that match supply and demand. At first sight, static market equilibrium does not necessarily have anything to do with sustainability in the physical or ecological sense.

On deeper analysis, however, one should see a very close connection. A number of economists have extended the original Walrasian general-equilibrium framework. By the late 1960s, growing awareness of environmental problems underlined the need to correct for market failures by broadening the Walrasian definition of exchangeable goods (or services provided thereby) to include non-exchangeable "bads" (or disservices) such as waste emissions from industrial and consumption processes [Ayres and Kneese, 1969; Kneese et al., 1970]. Others, concerned with the problem of exhaustible resources, explicitly introduced the time dimension by extending the market to allow for exchange of future goods (or bads), much as the stock market buys and sells "futures" contracts [see Solow, 1974; Dasgupta and Heal, 1979].

Given these extensions of the static framework, it might be argued that, in principle, the market price system should be able to adjust for externalities such as environmental damages resulting from market failures, both now and in the future. In other words, a perfectly functioning market should properly balance supply and demand for both conventional "goods" and environmental "bads." In such a system, the marginal (social) cost of an incremental increase in the environmental burden of some harmful waste product should exactly balance the marginal cost of eliminating that waste residual at its source or repairing the environmental damage, whichever is cheaper.[7] This rather standard model relationship is illustrated for the static Walrasian equilibrium case by Figure 12.2.

It is a fundamental principle of efficient markets that externalities (third-party effects) of transactions should, if possible, be eliminated or compensated for by adjustment of the transaction price. By this mechanism all social costs are directly and explicitly included by sellers and absorbed by buyers. This is the basic condition for markets that achieve allocative efficiency. A way of approximating this condition is commonly termed the *polluter-pays principle* (PPP). If the polluter has to pay for all the damages, he will presumably take these downstream costs into account in his choice of technology and his pricing policy.

Of course, there is a problem with this theoretical prescription. While the Walrasian framework of analysis can be extended, in principle, the equilibration *process* still depends on the existence of explicit prices for the non-exchangeable "bads," as well as the exchangeable "goods."[9] Absent voluntary individual exchanges, we envision social choices decided by some semi-undefined (but democratic) political process, based on some undefined method of determining the willingness-to-pay of the population. The practical difficulties in determining the shadow-prices of environmental disservices and implementing a polluter pays program are so great as to possibly constitute an objection in principle.

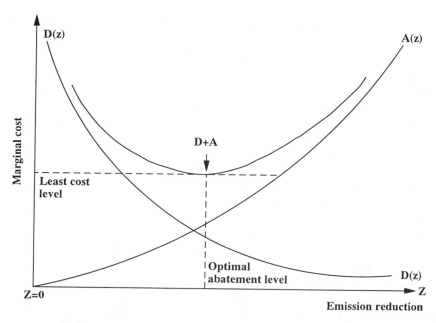

FIGURE 12.2. Optimal Abatement in Economic Equilibrium.

Skating lightly over this conceptual thin ice, however, it seems evident that most raw materials are drastically underpriced and, as a further consequence, overused. To achieve allocative efficiency, the environmental damage costs *should* be paid by the polluters (namely us), but not directly. The efficient way to accomplish this is to transfer the extra costs—not to final consumers who can only decide whether and how much to consume—but by a tax or other charge on primary producers. The latter will take these costs into account in making technological choices. They will also pass these costs along to the next tier of customers, who will also make choices between alternative materials and between raw and recycled materials. (If, for instance, fossil fuels became much more expensive, energy-conserving technologies and solar energy technologies would be selected much more often than they are now.) The chain continues until the consumer is presented with choices that better reflect all these costs.

Unfortunately, the polluter-pays principle is seldom adhered to or enforced by governments. (In a sense, this is the market failure.) From a physical perspective, the basic problem is that most pollution is due ultimately to extraction and primary processing, on the one hand, and the consumption processes, notably the burning of fossil fuels and the dissipative uses of materials, on the other hand. When the user-polluter is an industry (an intermediate), the principle can sometimes be applied. But, when the polluter is the government or the final consumer, as is increasingly the case, the polluter (so far) simply refuses to pay. If preferences are judged by behavior, consumers seem to prefer cheap energy and cheap materials to a clean environment. A loss of jobs is feared more, by most people, than a loss of environmental amenities, especially to future generations.

Taxes or emission charges, often advocated by economists, can also be faulted, on the grounds that it is difficult (or impossible) to calculate the "correct" tax rates, or to allocate the burden equitably. This reflects, in part, the conceptual problems noted earlier. On the other hand, it might be argued with some force that the present system of taxation to raise revenues for the government is so regressive and counterproductive that almost any tax on exhaustible-resource use or pollution would be an improvement, as long as it was simply used for revenue and substituted for some existing tax on labor, income, or final consumption. In short, the present tax system is far from optimal, so why worry unduly about optimizing a system of resource/environmental taxes?

In summary, it is unclear whether an economic system satisfying the polluter-pays principle can be implemented, even in principle. If such a system could be designed and implemented, a second question arises, namely, whether its application would suffice to make the economic system ecologically sustainable in the long run. Doubts arise because of the prevalence of hyperselection or "lock-in" phenomena, which drastically interfere with—or totally prevent—the convergence of the real economy to its theoretical static or quasi-static equilibrium state. A further question that arises, and that deserves deeper study, is the extent to which market failures and externalities are, themselves, major causes of hyperselection (as suggested by the case of automotive-vehicle dependence on liquid hydrocarbon fuels).

Whether convergence to a sustainable equilibrium state can be expected to occur spontaneously on the basis of market (or quasi-market) price signals alone, it is quite evident that the present economic system is far away from either equilibrium or sustainability. It is also clear that a significant reduction in environmental market failure would radically improve ecological sustainability.

12.4. TECHNOLOGICAL AND ECONOMIC EVOLUTION IN LONG-TERM PERSPECTIVE

This book seeks to examine technological and economic evolution as an example of a dynamical process subject to a directional constraint. One of the constraints I have in mind, obviously, is long-term environmental sustainability. The full range of implications of sustainability need not be spelled out in detail (even if this were possible). However, I take it that necessary, if not sufficient, conditions for sustainability include the following: Energy must be taken only from renewable sources; greenhouse gases must not be allowed to accumulate in the atmosphere; and non-biodegradable toxic materials must not be allowed to accumulate in soils or sediments. It follows that combustion of fossil fuels and dissipative uses of toxic heavy metals among other things must be phased out. In fact, all use of exhaustible stocks of materials will eventually have to cease.

Long-term sustainability also implies that, in future centuries, "spaceship Earth" materials-recycling will have to be very nearly total in.

Biological evolution was subject to similar sustainability requirements. The biosphere today is a nearly perfect materials-recycling system, but this was the result of billions of years of natural selection. The earliest organisms simply exploited (by fermentation) a stock of organic materials that had accumulated in the oceans prior to the appearance of organic life. Photosynthesis was an evolutionary "invention" that used the energy of sunlight to convert carbon dioxide into glucose to replace the exhausted resource stock. But photosynthesis yields a toxic by-product, oxygen. At first, the oxygen was used up as fast as it was produced by the oxidation of sulfides to sulfates and ferrous to ferric iron. The latter led to the formation of iron oxide (now iron ore), but when the dissolved reducing materials—essentially, the *waste assimilative capacity*—of the early ocean was exhausted, the level of molecular oxygen in the oceans began to rise. But oxygen was toxic to all existing (anaerobic) forms of life.

Another evolutionary "invention" was needed. It was a new metabolic process based on oxygen respiration (essentially, catalytic partial oxidation) in place of the older fermentation process. The newer aerobic process was more energy efficient (by a factor of 18) and, so, began to replace the old one. Thus, to simplify a complex story, the fundamental carbon cycle came into being. (Other natural cycles evolved subsequently, to recapture nitrogen, phosphorus, and other minerals.)

Evolutionary biologists tend to concentrate their attention on the micro-mechanisms of natural selection. At this level, the influence of the long-term imperative to recycle is difficult to see. It is like a faint signal in a noisy background. But, looking at the biosphere from a holistic perspective, it is clear at least in retrospect that the carbon cycle was an essential feature of organic evolution. The details might have evolved differently, but the result was inescapable. Both a photosynthetic process based on carbon dioxide and a metabolic process to utilize molecular oxygen had to evolve.

I believe a similar quasi-teleological imperative will inevitably drive future technological and economic evolution in the direction of lower energy and materials intensity, the use of solar energy, and materials recycling. Past trends provide virtually no evidence for this statement. Most of the trends, in fact, are still going in the wrong direction. Again, the signal that determines the evolutionary direction is (currently) almost drowned out by the noise of competing shorter-term imperatives.

Most economists now assume that the primary mechanism driving the evolution of technology is a combination of Schumpeterian (radical) innovation and Usherian incremental improvements (e.g., *learning*). The latter process is perhaps consistent with a quasi-static paradigm, but the radical innovations are not. Schumpeterian innovations are generally quite risky. The motivation for taking the major risks associated with innovation is assumed to be the additional profit available to an innovator by virtue of having a temporary monopoly.

But the Schumpeterian mechanism is directionally indeterminate. It implies only a search for profit opportunities, and implies absolutely nothing about where they will be found. It is the external environment that determines whether profit opportunities are to be found in any particular evolutionary *direction*. In short, the environment defines the direction of technological progress or innovation. The prevailing macro-economic theory for explaining the direction of change (the theory of *induced innovation*) is driven by factor price differentials.[10] The idea is that, if energy prices are higher relative to labor costs in country A than in country B, country A will have a relatively more favorable environment for energy-saving innovations, while country B will be relatively more favorable for labor-saving innovations. This difference in outcome would presumably result even if the *a priori* probability of basic inventions in each field were the same in each country.

Social needs are presumed (by economic theory) to express themselves through the price system. In simple terms, a rising price suggests increasing scarcity: demand increasing faster than supply (or supply falling faster than demand). A rising price suggests an inadequately filled or unfilled need and should signal a suitable response by technology adopters and/or by the technology-creation sector. It can be argued, for example, that the rising price of whale oil in the mid-19th century called forth developments in petroleum-refining technology and led to the commercialization of kerosene and the development of the petroleum industry. Similarly, it can be argued that the growing demand for automobiles after 1908 induced the petroleum-refining innovations that allowed that industry to make gasoline its primary product. (Recall, however, the counterargument that the availability of natural gasoline, a cheap by-product of kerosene—illuminating oil— enabled the gasoline engine to achieve an insurmountable lead in the competition with electric cars.)

Evidently the induced-demand hypothesis cannot be the whole story. Wrong price signals can also induce inappropriate technological responses, which in turn may be locked in by economies of scale or other increasing returns to adoption. Moreover, some needs are not clearly reflected by a rising commodity price. While prices are signals of social need, they are not the only such signals. More to the point, in some circumstances at least, prices are evidently no longer an effective signal. In the previous section, I suggested that environmental degradation itself is *ipso facto* evidence of thermodynamic disequilibrium. Other evidence of disequilibrium may be more persuasive to a neo-classical economist.

To explain this, let me revert to the neo-classical framework. Suppose, to begin with, there are no externalities and the real economy reflects a utility-maximizing, quasi-static Walrasian equilibrium, as assumed by neo-classical economics. In this case, it would automatically have selected the optimum (lowest private cost) technologies for production. Now suppose that environmental externalities (damages) are suddenly discovered. If an appropriate tax system (based on PPP) were then introduced to correct for the distortions due to environmental externalities, the choice of technology would in a Walrasian world automatically shift to minimize the sum of private and social costs. Both total costs and social costs would fall.

This shift, however, would involve some increase in private costs (or decrease in gross national income), as suggested by Figure 12.2. Under these assumptions, it would be correct to assert that reducing pollution must entail some cost in money (or a loss of income). The same logic applies to saving energy. That is, if the economy were in a Walrasian equilibrium, and if adjustments to price signals were automatic, our economy would already be using the most cost-efficient technologies. In this case, it would surely cost more money to save energy. This situation is almost universally assumed by economists (if not engineers) to exist in the real economy. Fortunately, there is good evidence that it does not [D'Errico *et al.*, 1984; Ayres, 1991].

I have argued that the economy is not in Walrasian equilibrium, due to a combination of uncompensated externalities and hyperselection (lock-in) phenomena.[10] I argue, further, that the postulated price-adjustment and convergence mechanism (*tâtonnement* and its modern variants) is relatively ineffective. In short, the standard assumptions do not hold for the real economy. In fact, the existence of major opportunities to reduce pollution or save energy, while saving money at the same time, constitutes compelling evidence of the non-equilibrium character of the economy and the distinct likelihood of "free lunches." Evidence compiled from many engineering studies suggests that such opportunities are widespread. If so, the true picture is more like Figure 12.3 (which I believe to be realistic). This situation could not exist if the economy were truly in an equilibrium state, because entrepreneurs would, presumably, have found and exploited any such opportunities.

The best evidence of disequilibrium can therefore be derived from studies indicating major discrepancies between rates of return achievable from various sorts of investments. If the economy were in or near equilibrium, rates of return would be relatively comparable, regardless of field. It is economically irrational for a profit-making, tax-paying enterprise to knowingly invest in anything yielding a rate of return less than the rate paid by safe government securities, such as T-bills. This rate currently hovers around 6 percent (U.S., 1993). Given an inflation rate of 3 percent or so, this corresponds to a real rate of return of around 3 percent and a "payback" (i.e., doubling) time of the order of 30 years, allowing for inflation.

Interestingly enough, the real rate of return for large energy-supply investments (such as electric power plants) ranges between 5 and 10 percent. This very low rate of return is possible only because various government policies have made such investments virtually risk-free.[12]

Another approach is to look for the cheapest way of supplying a given mix of energy services. If the economy were in a Walrasian equilibrium or quasi-equilibrium state, the least-cost strategy would correspond to the actual pattern of use. Such a study was carried out for the United States by the Mellon Institute in 1979 [Sant, 1979]. The resulted contradicted theory: The least-cost strategy for 1978 would have saved $43 billion in that year alone ($800 per family, or 17 percent of total expenditures for energy). In large part, those savings would have been achieved simply by using less energy, especially electricity. According to the study, the optimum mix would have reduced the electricity share from 30 to 17 percent of total energy use; the petroleum share would have fallen from 36 to 26 percent; and "conservation services" would have increased from 10 to 32 percent. In short, conservation would have cost considerably less than the country was actually paying (see Figure 12.3). Again, whatever the explanation, the situation is incompatible with the economy being in a Walrasian equilibrium.[13]

Why does even private capital in our supposedly competitive free-market economy flow into projects yielding consistently low rates of return, while *not* flowing into projects with very high returns? Whatever the explanation, the fact itself is a clear indication that the economy is not in equilibrium in any relevant sense. Either the real economy is much slower to respond to price signals than economists have ever been willing to assume, or the Walrasian paradigm is altogether inappropriate. I suspect the latter. With respect to energy conservation (and probably other cases), the consistent neglect of economically attractive opportunities seems to me to be a case of hyperselection ("lock-in") of a non-optimal trajectory.

12.5. CONCLUDING THOUGHTS AND SPECULATIONS

The mechanism driving evolution in economic systems is technological innovation. Innovations are normally induced by economic (in addition to psychological) incentives, such as the promise of a temporary monopoly and monopoly profits. These classic incentives are not always effective, however, especially where there are massive market failures. In principle, one might expect market incentives to guarantee optimum adoption of energy-conserving technologies, given that energy is an important component of cost for both the production and use of technology.

In practice, market failures have kept fossil fuels far too cheap by relieving both producers and users of fossil energy from direct responsibility for paying the costs of abating or avoiding the adverse environmental and health consequences of the excessive use of fossil fuels. It has been estimated, for instance, that the true social costs of fossil energy are several (perhaps up to ten) times their current price. Much the same statement could doubtless have been made for a number of products that have been banned (heroin, cocaine, PCBs, asbestos, chlorinated pesticides, CFC's, tetraethyl lead...), yet it is also clearly true for others — alcoholic beverages and tobacco products, for instance, that are still freely traded and even subsidized

As noted previously, the *unpaid* costs of environmental damage have been effectively deferred in most cases. Later generations will have to pay for many of them. But as these costs become larger and more visible, I hopefully conjecture

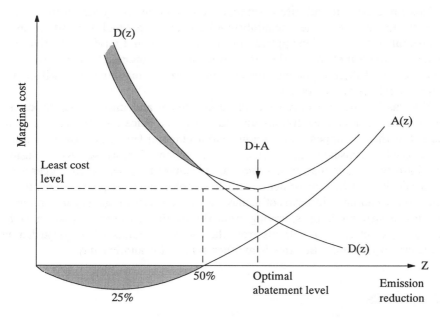

FIGURE 12.3. Optimal Abatement in Economic Disequilibruim.

that political pressure will grow to force the producers and users of fossil fuels (and such other materials as heavy metals) to pay them. Despite resistance from energy-users, it seems inevitable that in the long run these costs will have to be added—albeit gradually, indirectly, and piecemeal—to the prices of fuels and materials. This, in turn, will create significant economic opportunities for innovators in the area of "low-waste" technologies, and renewables.

To repeat: While it operates somewhat erratically, there does seem to be a long-range evolutionary imperative favoring low-energy and renewable technologies for the future industrial ecosystem, as in the biosphere. This imperative will be reflected in trends toward renewable energy sources, "decarbonization" of fuels, and "dematerialization" of products. It is entirely consistent with another trend, already well advanced, toward increasing information-intensity and knowledge-content of all economic activity. Value added means information added. In the 18th century and earlier, value was added by nature, or by crude extraction processes. It was largely attributable to physical composition: *thermodynamic information,* in the language of this book. In the 19th century and much of the present century, value was added mainly by forming and machining hard materials into ever more precise (and reproducible) shapes. Value was added in the form of *morphological information.* But the next economic transformation, which will accelerate for the foreseeable future, is toward adding value as *symbolic information.* Even in the world of information-processing technology, software is outstripping hardware.

This imperative to substitute symbolic information for cruder forms arises from the magnitude of the disequilibrium between supply and demand for environmental services (undersupplied) on the one hand, and virgin extractive raw materials (oversupplied) on the other. But, like all economic "forces," this one is subject to substantial modification amplification or attenuation by institutional and political factors in society.

Whereas biological evolution involved accidental and unconscious processes of selection, economic evolution can and must take place on a far shorter time scale. For this to happen, unconscious and accidental (myopic) processes must be replaced by conscious, far-sighted political-economic processes. A conclusion very hard to avoid is that price signals alone cannot be relied on to trigger even economically justified investments, still less ecologically necessary innovations. This is very bad news in terms of achieving long-term sustainability in economic systems. It strongly implies, among other things, that governments will have to play a more active interventionist role than most economists have regarded as necessary or desirable. But this is another subject for another day.

ENDNOTES

1. Material in this chapter appeared previously in [Ayres, 1991a].

2. For a fascinating account of this history, see Philip Mirowski's article "Physics and the Marginalist Revolution" and a subsequent debate in the literature.

3. This is a consequence of the in-built specialization or "clay-like" characteristic of most kinds of capital equipment.

4. A recently completed study suggests that power can be generated by large-scale PV "farms" on the surface of the moon (manufactured by automated factories on-site) and transmitted to earth very efficiently by phased-array microwave beams, at a total cost to the user in the neighborhood of 1 or 2% of current costs for electric power generated by conventional means. Even if this analysis is too optimistic by an order of magnitude, the option is worth serious consideration. It scarcely needs to be said that the petroleum and utility sectors would oppose this scheme vehemently. The selection of a space-based energy system would certainly be another bifurcation point.

5. It can be shown that transactional irreversibility also implies the existence of a non-decreasing function in economics. This function is analogous to the entropy function only in the sense that it is non-decreasing. However, it is not a consequence of thermodynamics. It can be thought of as a "progress function." In general terms, the progress function can be interpreted as a growing stock of useful information.

6. The *tâtonnement* process described by Walras is not implementable in real markets because no actual exchanges can occur until the market-clearing price is determined. Implementable price-adjustment processes have been described by Smale and Aubin, but in Smale's case an active "auctioneer" is postulated, while Aubin requires a passive mechanism for publishing information on prices. Both assume that there exists a unique money price, known to all parties, for the good(s) being exchanged at

every instant of time. More recently a decentralized convergence process has been described, involving only pairwise transactions in which the exchange price is not unique and is known only to the parties [Ayres and Martinàs, 1990a,b].

7. The definition of *sustainability* has attracted considerable attention in the economics literature in recent years. The neo-classical definition of sustainability is "non-decreasing utility," on the standard assumption of unlimited substitutibility between economic and environmental factors, both in production and consumption. A more radical definition (which I subscribe to) takes it for granted, both that substitutibility is quite limited—there is no technological substitute for air or water, for instance—and that some degradation is irreversible. This implies that some environmental assets, once lost, may not ever be replaceable. It follows that sustainabiity implies that irreversible processes, such as the accumulation of greenhouse gases in the atmosphere, or the buildup of toxins in soil and groundwater, must not be allowed to continue. This definition essentially amounts to a set of constraints on economic activities.

8. Of course, some kinds of damage are inherently irreversible and unrepairable. This is probably true of climate change and certainly true of loss of species diversity or of mature redwood forests. Lacking the possibility of damage repair, the market should balance marginal damages with marginal costs of source reduction.

9. Admittedly, in a decentralized pure-exchange market the actual transaction prices need not be published to ensure convergence; nor do they need to be unique.

10. Of course, inventors and investors seldom track the price indices of aggregate factors of production. They read the newspapers. There is a close connection between the intensity of public interest in a subject and the intensity of interest on the part of would-be technology creators. This link has been expressed as a societal response to social "needs."

11. In fact, the two explanations are not entirely incompatible, because the continued existence of uncompensated externalities is partly due to the enormous political power of the existing techno-structure. In simple language, the auto industry, the petroleum industry, the trucking industry, the airlines, and other industrial sectors whose continued prosperity is dependent on the continued use of petroleum as a primary fuel exert enormous political power and influence, especially in the U.S. These economic interests combine effectively to gut public policies—such as fuel taxes—that would discourage this dependence.

12. Utilities are regulated monopolies. Until the law changed in the late 1970's, utilities were allowed to set prices based on rates of return "needed" to attract capital. It was almost impossible for a utility to lose money.

13. If more evidence is needed, consider the problem of waste motor oil. Approximately 2 billion gallons are generated each year in the U.S. alone, of which ony 40% is collected for recycling or reuse. It currently costs 17 cents per gallon to collect the waste oil from service stations, of which service stations currently pay about 12 cents. Most of the waste oil collected is burned as fuel or used as road oil. Re-refining of used oil is technically and economically feasible. It can be sold at retail prices comparable to the $1 per quart ($4 per gallon) for new oil. Yet only about 10% of the

collected lube oil (4% of the total) is now recycled for this purpose. The basic problem is that the large petroleum companies who dominate the marketing system have no interest in selling re-refined oil because it competes with the sale of newly refined oil. EPA is understandably reluctant to impose more regulations on service stations for fear that unregulated do-it-yourselfers will dump even more waste oil than they do now. The obvious solution is a returnable deposit of 50 cents per quart, or so, on all lube oil purchased by DIYers. The oil industry and its allies have prevented any such rational policy.

REFERENCES

Abramovitz, Moses, "Resources and Output Trends in the United States Since 1870," *American Economic Review* **46**, May 1956.

Arthur, N. Brian, "Self-Reinforcing Mechanisms in Economics," in: Anderson, Arrow and Pines (eds.), *The Economy As an Evolving Complex System* :9-32 [Series: Santa Fe Institute Studies in the Sciences of Complexity **5**], Addison-Wesley, Redwood City, Calif., 1988.

Arthur, W. Brian *et al.*, "Path-dependent Processes and the Emergence of Macro-structure," *European Journal of Operations Research* **30**, 1987 :294-303.

Aubin, Jean-Pierre, "A Dynamical, Pure Exchange Economy with Feedback Pricing," *Journal of Economic Behavior and Organization* **2**, 1981 :95-127.

Ayres, Robert U. "Evolutionary Economics and Environmental Imperatives" *Structural Change and Economic Dynamics*, Vol 2. no. 2. 1991(a).

Ayres, Robert U., "Energy Conservation in the Industrial Sector," in: Ferrari, Tester and Woods (eds.), *Energy and the Environment in the 21st Century* :357-370, MIT Press, Cambridge, Mass., 1991(b).

Ayres, Robert U. and Ike Ezekoye, "Competition and Complementarity in Diffusion: The Case of Octane," *Journal of Technological Forecasting and Social Change* **39**, 1991 :145-158.

Ayres, Robert U. and Allen V. Kneese, "Production, Consumption and Externalities," *American Economic Review*, June 1969. [Reprinted in *Benchmark Papers in Electrical Engineering and Computer Science*, Daltz and Pentell (eds.), Dowden, Hutchison and Ross, Stroudsberg, State, Country, 1974 and Bobbs-Merrill Reprint Series, New York.]

Ayres, Robert U. and Katalin Martinàs, *Self-Organization of Markets and the Approach to Equilibrium*, Working Paper (WP-90-18), International Institute for Applied Systems Analysis, Laxenburg, Austria, 1990*b*.

Ayres, Robert U. and Katalin Martinàs, *A Computable Economic Progress Function*, Working Paper (WP-90-18), International Institute for Applied Systems Analysis, Laxenburg, Austria, April 1990*a*.

Brooks, Daniel R. and E. O. Wiley, *Evolution As Entropy: Towards a Unified Theory of Biology*, University of Chicago Press, Chicago, 1986.

Chen, Ping, *Searching for Economic Chaos: A Challenge to Mainstream Econometric Practice and Nonlinear Numerical Experiments*, Working Paper (90-08-02), Prigogine Center for Studies in Statistical Mechanics and Complex Systems + IC2 Institute, University of Texas, Austin, Tex., August 1990.

Chizhov, N. and M. Styrikovich, "Ecological Advantages of Natural Gas and Other Fuels," in: Lee *et al.* (eds.), *The Methane Age*, Chapter 11 :155-161, Kluwer Academic Publishers, Boston, 1988.

Dasgupta, Partha and G. Heal, *Economic Theory and Exhaustible Resources* [Series: Cambridge Economic Handbooks], Cambridge University Press, Cambridge, England, 1979.

David, Paul A., "Clio and the Economics of QWERTY," *American Economic Review (Papers and Proceedings)* **75**, 1985 :332-337.

D'Errico, Emilio, Pierluigi Martini and Pietro Tarquini, *Interventi di risparmio energetico nell'industria*, Report, ENEA, City, Italy, 1984.

Ebeling, Werner and Rainer Feistel, *Physik der Selbstorganization und Evolution*, Akademie-Verlag, Berlin, 1982.

Ebeling, Werner and Rainer Feistel, "Physical Models of Evolution Processes," in: Krinsky (ed.),

Self-Organization: Autowaves and Structures Far from Equilibrium, Chapter 6 :233-239, Springer-Verlag, New York, 1984.

Fabricant, Solomon, "Economic Progress and Economic Change," in: *34th Annual Report*, National Bureau of Economic Research, New York, 1954.

Hohmeyer, Olav, *Social Costs of Energy Consumption*, Springer-Verlag, Heidelberg, Germany, 1988.

Hollander, Samuel, "On P. Mirowski's Physics and the 'Marginalist Revolution,' " *Cambridge Journal of Economics*, 1989 :459-470.

Huxley, J. S., "Evolution, Cultural and Biological," in: Thomas (ed.), *Current Anthropology* :3-25, University of Chicago Press, Chicago, 1956.

Kneese, Allen V., Robert U. Ayres and Ralph d'Arge, *Aspects of Environmental Economics: A Materials Balance Approach*, Johns Hopkins University Press, Baltimore, 1970.

Lotka, Alfred J., "Contribution to the Energetics of Evolution," *Proceedings of the National Academy of Sciences* 8, 1922 :147.

Mirowski, Philip, *Physics and the Marginalist Revolution*, Cambridge Journal of Economics 8(4), 1984.

Mirowski, Philip, "On Hollander's 'Substantive Indentity' of Classical and Neo-Classical Economics: A Reply," *Cambridge Journal of Economics* 13, 1989 :471-477.

Nelson, Kenneth E., "Are There Any Energy Savings Left?" *Chemical Processing*, January 1989.

Nicolis, Gregoire and Ilya Prigogine, *Self-Organization in Non-Equilibrium Systems*, Wiley-Interscience, New York, 1977.

Ogburn, William F. and Dorothy Thomas, "Are Inventions Inevitable? A Note on Social Evolution," *Political Science Quarterly* 37, March 1922 :83-98.

Pezzey, John, *Sustainability, Intergenerational Equity and Environmental Policy*, Discussion Paper in Economics (89-7), Department of Economics, University of Colorado, Boulder, Colo., Fall 1989.

Prigogine, Ilya, Gregoire Nicolis and A. Babloyantz, "Thermodynamics of Evolution," *Physics Today* 23(11/12), November/December 1972 :23-28(Nov.) and 38-44 (Dec.).

Prigogine, Ilya and I. Stengers, *Order Out of Chaos: Man's New Dialogue with Nature*, Bantam Books, London, 1984.

Salter, W. E. G., *Productivity and Technical Change*, Cambridge University Press, New York, 1960.

Sant, R. W., *The Least-Cost Energy Strategy: Minimizing Consumer Costs Through Competition*, Report (55), Mellon Institute Energy Productivity Center, City, Va., 1979.

Smale, Stephen, "Dynamics in General Equilibrium Theory," *American Economic Review* 66, 1976 :288-294.

Solow, Robert M., "Technical Change and the Aggregate Production Function," *Review of Economics and Statistics*, August 1957.

Solow, Robert M., "The Economics of Resources or the Resources of Economics," *American Economic Review* 64, 1974.

Ville, Jean, "The Existence Conditions of a Total Utility Function," *Review of Economic Studies* 19, 1951 :123-128.

von Neuman, John, "A Model of General Economic Equilibrium," *Review of Economic Studies* 13, 1945 :1-9.

Afterword

For more than twenty years I have worked with Bob Ayres on various projects ranging from such topics as alternatives to the internal combustion engine to a general equilibrium-materials balance economic model. While he was trained as a physicist, what has always impressed me most about him is his incredible ability to grasp methods and information from a great variety of scientific disciplines. This book is the most recent display of this talent. It is a work of prodigious scholarship incorporating as it does, geophysics and geochemistry, cosmology, biochemistry, chaos theory, thermodynamics, evolutionary theory, and, of course, economics, to name a few.

As an economist I was most interested in Ayres' discussion of neoclassical economics, especially neoclassical growth models and their virtual irrelevance to the understanding of how growth, or perhaps better put, development, occurs in actual economies. He points out that economists normally do not think of economic activities and relationships in thermodynamic, or, he argues, equivalently, in information theory terms. When economists talk about equilibrium they refer to a balance between supply and demand, or (looking at it another way) between prices, wages, and profits. Neoclassical economic models consider labor, capital goods, and services to be abstractions.

Even this very abstract model of the economic system depends on resource inputs, although in most general equilibrium and neoclassical growth models, resources are assumed to be generated by labor and capital or neglected altogether. Thus, the neoclassical system is, in effect, a perpetual-motion machine.

In reality, Ayres argues, the resource inputs to the economic system are physical: they include air, water, sunlight and material substances, fuels, foods, and fiber crops, all of which embody thermodynamically available work (essergy). Outputs, on the other hand, are "final goods" whose utility is ultimately used up and thrown away or, sometimes, recycled. Available work is expended at every stage—extraction, refining, manufacturing, construction, and even final consumption. Although total energy is always conserved, essergy is not. Energy inputs like fossil fuels are rich in essergy, while energy residuals are mostly in the form of low-temperature waste heat, oxidation products, or degraded materials. Thus, the economic system in reality is absolutely dependent on a continuing flux of essergy from the environment and knowledge from structures, centrally including man. In preindustrial times, the sun provided almost all essergy in the form of wood, food crops, animals, water power, or wind power. Today, the major source, by far, is fossil fuels: petroleum, natural gas, and coal from the Earth's crust. These resources are of course exhaustible.

Thus, Ayres reasons, the real economic system resembles a "dissipative structure," in the sense described by Ilya Prigogine and Isabelle Stengers: it depends on a continuous flow of essergy (the sun or fossil fuels) as well as information in more familiar form—not instantly recognized as essergy (although the two, Ayres says, can be proven to be equivalent). And the system exhibits coherent, orderly behavior. In fact, it is self-evidently capable of growth.

Economic growth can be of two distinct kinds. First, an economic system can, in principle, expand like a balloon without technological or structural change. It simply gets bigger, as capital and labor inputs increase proportionally. This kind of quasi-static growth can lead to increased final consumption per capita while maintaining its equilibrium, but only by producing more of everything, in fixed ratios. This is possible only if there are no economies or diseconomies of scale, which is an unrealistic but common economic assumption. Also, there has to be a nonscarce input, "nature," in order for this process to continue indefinitely. (Most growth models contemplate this kind of growth.)

The second kind of economic growth adds evolutionary changes in structure. These changes are driven by innovations—new products, new processes—that result not only in quantitative increases in per capita consumption, but also in qualitative changes in the mix of goods and services generated by the economy. In general, Ayres states, this kind of growth involves increased complexity and organization.

Quasi-static growth of the first kind can be modeled theoretically as an optimal control problem with aggregate consumption (or welfare) as the objective function. The control variable is the rate of savings diverted from immediate consumption to replace depreciated capital and add new capital to support a higher level of future consumption. The rate of growth in this simple model is directly proportional to the rate of savings, which, in turn, depends on the assumed depreciation rate and an assumed temporal discount rate to compare present and future benefits.

Dynamic growth of the second, evolutionary, kind is less dependent on savings and/or capital investment. It cannot occur, however, without capital investment because new production technologies, in particular, are largely embodied in capital equipment. Technological innovation drives this kind of dynamic growth. Ayres argues that there is ample evidence that technological progress is not an autonomous (self-organizing) process, as often assumed in economic growth models (when it has been included at all). On the contrary, knowledge and inventions are purposefully created by individuals and institutions in response to incentives and signals generated within and propagated by the larger socioeconomic system.

An actual example of the importance of knowledge, as Ayres defines it, in economic development may be illuminating. This is the so-called German economic miracle following World War II. In less than a decade, the German economy recovered fully from a condition so severe that many doubted it could ever again compete in the world economy. This recovery was made possible by knowledge embodied in human skills, organizations, and infrastructure. Far greater amounts of capital became available to other countries—such as Iran—with far different results. This illustrates that financial capital and raw labor (the focus of most economic models) are feeble engines of development compared with embodied knowledge and skills.

Based on an elaboration of these ideas Ayres summarizes the main themes of the book in terms of their implications for economic development: first, because the economy is a dissipative structure, it depends on continuous essergy and material flows from (and back to) the environment. Such links are precluded by closed neoclassical general equilibrium models, whether static or quasi-static. Second, the energy and physical materials inputs to the economy have shifted, over the past two centuries, from mainly renewable sources to mainly nonrenewable sources. Third, dynamic economic growth is driven by technological change (generated, in turn, by economic forces or deliberate government policy) that also results in continuous structural change in the economic system. It follows, incidentally, that a long-term survival path must sooner or later reverse the historical shift away from renewable resources. This will be feasible only if human technological capabilities rise to levels much higher than current ones. But, since technological capability is endogenous, it will continue to increase only if the pace of deliberate investment in research and development is continued or even increased. In short, the role of knowledge-generating activity in retarding the global entropic increase seems to be growing in importance.

In Chapter 7 of the book, Ayres and a collaborator, Katalin Martinas, develop a formal model that endeavors not only to include central concepts from thermodynamics and information theory into economic growth theory but also to move them to center stage. As we try to explain long-term growth in modern economic systems, such concepts are much more deserving of attention, in my view, than are the static or quasi-static traditional concepts of capital and labor that have dominated the economic growth and natural resources literature.

While the central themes of the book deal with issues surrounding economic growth, the book should be of interest to non-economists in many fields. The exposition is very accessible because it is clearly written and makes good use of examples and analogies.

Allen V. Kneese
Washington, D.C.

Index